**Books are to be returned on or before
the last date below.**

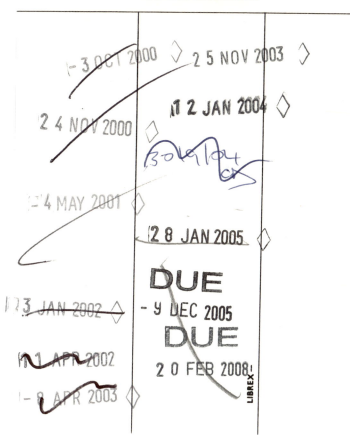

- 3 OCT 2000   2 5 NOV 2003

2 4 NOV 2000   1 2 JAN 2004

2 4 MAY 2001

2 8 JAN 2005

2 3 JAN 2002   DUE
                - 9 DEC 2005
                DUE
1 APR 2002   2 0 FEB 2008

- 8 APR 2003

LIBREX

INDUSTRIAL APPLICATIONS OF MICROEMULSIONS

SURFACTANT SCIENCE SERIES

ADDITIONAL VOLUMES IN PREPARATION

INDUSTRIAL APPLICATIONS OF MICROEMULSIONS

edited by

Conxita Solans
Consejo Superior de Investigaciones Científicas
Barcelona, Spain

Hironobu Kunieda
Yokohama National University
Yokohama, Japan

Marcel Dekker, Inc.   New York•Basel•Hong Kong

Library of Congress Cataloging-in-Publication Data

Industrial applications of microemulsions / edited by Conxita Solans.
  Hironobu Kunieda.
     p.   cm.— (Surfactant science series ; v. 66)
  Includes bibliographical references (p.   -   ) and index.
  ISBN 0-8247-9795-7 (Hardcover : alk. paper)
  1. Emulsions.   I. Solans, C.   II. Kunieda, Hironobu.
  III. Series.
  TP156.E6I5   1997
  660'.294514—dc20

                                                                96-44788
                                                                CIP

The publisher offers discounts on this book when ordered in bulk quantities.
For more information, write to Special Sales/Professional Marketing at the
address below.

This book is printed on acid-free paper.

MARCEL DEKKER, INC.
270 Madison Avenue, New York, New York 10016

Current printing (last digit):
10 9 8 7 6 5 4 3 2 1

PRINTED IN THE UNITED STATES OF AMERICA

Preface

In recent years microemulsions have attracted a great deal of attention not only because of their importance in industrial applications but also because of their intrinsic interest. They optimize the performance of a wide spectrum of industrial and consumer products and processes.

Microemulsions are isotropic and thermodynamically stable multicomponent fluids composed of water, oil, and amphiphile(s). The characteristic properties of microemulsions include spontaneous formation, optically clear appearance, large interfacial area, low interfacial tension, large solubilization capability, and low viscosity—properties that render these organized solutions unique.

Practical applications of microemulsions preceded their scientific recognition. But the related industrial development was limited for many years due to a lack of basic knowledge of these colloidal solutions. However, the rapid and continuous progress achieved over the past decade has stimulated many novel applications ranging from lubricating fluids to agricultural sprays, drug delivery systems, and reaction media. At present, industrial applications of microemulsions are under active and growing development.

The literature on microemulsions is extensive, and the field has recently experienced rapid progress. In early volumes of the Surfactant Science series, information on microemulsions was scattered throughout a chapter or parts of chapters, whereas more recently whole volumes have been devoted to the exploitation of their basic aspects. Nevertheless, there is still a shortage of information on their applications. An exhaustive review did appear in 1984, in Volume 6 of this series. But in view of the enormous advances in the field of microemulsions in the past few years, there has developed a need for a book devoted to their industrial applications. This publication satisfies that need.

The main objective of this book is to provide a comprehensive description of the most useful industrial applications of microemulsions. It is aimed at scientists, engineers, and students with either industrial or academic interests. It will also stimulate the interest of managers in industry.

The contributors are leading experts from industry and academia. They have been actively engaged in the research and development of various technological applications of microemulsions.

The two introductory chapters focus on basic concepts of microemulsions, which are essential to understanding their industrial significance. These chapters are followed by a comprehensive discussion of relevant technological and industrial applications. The scope of the coverage reflects the state of development and availability of the published material on the subject. Some of the applications (e.g., enhanced oil recovery) are well known and have been extensively reviewed in many previous publications. In such cases, only the major recent developments have been considered. Other well-established applications (e.g., pharmaceuticals, cosmetics, agrochemicals, and lubricants) for which there is not an extensive literature are explored in considerably more depth. This book also contains chapters on the relatively new areas of applications such as biotechnology, foods, textile dyeing, extraction processes, analytical determinations, and preparation of nanostructured materials.

We wish to thank all the contributors for their efforts. We also wish to thank Dr. Martin J. Schick, the editor of this series, for his encouragement in undertaking this task, and production editors, Mr. J. Stubenrauch and Mr. H. Boehm, for their help during the preparation of this volume. We are especially indebted to Mr. and Mrs. Guruswamy and to Mrs. Lidia Beltrán, who provided considerable assistance in helping to edit the book.

Conxita Solans
Hironobu Kunieda

Contents

v

Contributors

Masahiko Abe   Faculty of Science and Technology, Science University of Tokyo, Chiba, Japan

Raquel E. Antón   Laboratorio FIRP, Facultad de Ingeniería, Universidad de los Andes, Mérida, Venezuela

A. C. Auguet   Hispano Química, S.A., Barcelona, Spain

Nuria Azemar   Departamento Tecnología de Tensioactivos, Centro de Investigación y Desarrollo, Consejo Superior de Investigaciones Científicas, Barcelona, Spain

Ermanno Barni   Dipartimento di Chimica Generale ed Organica Applicata, Università degli Studi di Torino, Turin, Italy

Marc Bavière   Reservoir Engineering Department, Institut Français du Pétrole, Rueil-Malmaison, France

Karin Bonkhoff   Institut für Angewandte Physikalische Chemie, Forschungszentrum Jülich GmbH, Jülich, Germany

Jean Paul Canselier   Institut de Génie des Procédés, Ecole Nationale Supérieure d'Ingénieurs de Génie Chimique, Toulouse, France

Stephanie R. Dungan   Departments of Food Science and Technology, and Chemical Engineering and Materials Science, University of California, Davis, Davis, California

Stig E. Friberg Center for Advanced Materials Processing and Department of Chemistry, Clarkson University, Potsdam New York

F. X. Gaillard Hispano Química, S.A., Barcelona, Spain

Maria José García-Celma Departamento de Farmacia, Facultad de Farmacia, Universidad de Barcelona, Barcelona, Spain

Maria Rosa Gasco Dipartimento di Scienza e Tecnologia del Farmaco, Università degli Studi di Torino, Turin, Italy

Francesc Gusi Hispano Química, S.A., Barcelona, Spain

Xiomara Gutiérrez INTEVEP, S.A., Research and Technological Support Center of Petróleos de Venezuela, Caracas, Venezuela

Krister Holmberg Institute for Surface Chemistry, Stockholm, Sweden

Hironobu Kunieda Graduate School of Engineering, Yokohama National University, Yokohama, Japan

M. Arturo López-Quintela Department of Physical Chemistry, University of Santiago de Compostela, Santiago de Compostela, Spain

J. L. Margrave Department of Chemistry, Rice University, Houston, Texas

Hideo Nakajima Basic Research Laboratories, Shiseido Research Center, Yokohama, Japan

Vinod Pillai Center for Surface Science and Engineering, Departments of Chemical Engineering and Anesthesiology, University of Florida, Gainesville, Florida

Ramon Pons Departamento de Tecnología de Tensioactivos, Centro de Investigación y Desarrollo, Consejo Superior de Investigaciones Científicas (CSIC), Barcelona, Spain

José Quibén-Solla Laboratory of R&D, La Artística de Vigo, Vigo, Spain

Hercilio Rivas Production Department, INTEVEP, S.A., Research and Technological Support Center of Petróleos de Venezuela, Caracas, Venezuela

José Rivas Department of Applied Physics, University of Santiago de Compostela, Santiago de Compostela, Spain

Jean-Louis Salager Laboratorio FIRP, Facultad de Ingeniería, Universidad de los Andes, Mérida, Venezuela

Piero Savarino Dipartimento di Chimica Generale ed Organica Applicata, Università degli Studi di Torino, Turin, Italy

Milan J. Schwuger Institut für Angewandte Physikalische Chemie, Forschungszentrum Jülich GmbH, Jülich, Germany

Dinesh O. Shah Center for Surface Science and Engineering, Departments of Chemical Engineering and Anesthesiology, University of Florida, Gainesville, Florida

A. Shukla Department of Chemistry, Southeast College, Houston, Texas

Shyam S. Shukla Department of Chemistry, Lamar University, Beaumont, Texas

Johan Sjöblom Department of Chemistry, University of Bergen, Bergen, Norway

Conxita Solans Departamento de Tecnología de Tensioactivos, Centro de Investigación y Desarrollo, Consejo Superior de Investigaciones Científicas (CSIC), Barcelona, Spain

Günter Subklew Institut für Angewandte Physikalische Chemie, Forschungszentrum Jülich GmbH, Jülich, Germany

Th. F. Tadros Zeneca Agrochemicals, Bracknell, United Kingdom

Guido Viscardi Dipartimento di Chimica Generale ed Organica Applicata, Università degli Studi di Torino, Turin, Italy

José Luis Ziritt Reservoir Engineering Department, INTEVEP, S.A., Research and Technological Support Center of Petróleos de Venezuela, Caracas, Venezuela

INDUSTRIAL
APPLICATIONS OF
MICROEMULSIONS

1

Overview of Basic Aspects of Microemulsions

CONXITA SOLANS and RAMON PONS Departamento de Tecnología de Tensioactivos, Centro de Investigación y Desarrollo, Consejo Superior de Investigaciones Científicas (CSIC), Barcelona, Spain

HIRONOBU KUNIEDA Graduate School of Engineering, Yokohama National University, Yokohama, Japan

I. INTRODUCTION

It is now well established that large amounts of two immiscible liquids (i.e., water and oil) can be brought into a single phase, macroscopically homogeneous but microscopically heterogeneous, by addition of an appropriate surfactant or surfactant mixture. This unique class of optically clear solutions called microemulsions comprises the colloidal systems that have

1

attracted much scientific and technological interest over the past decade. This wide interest stems from their characteristic properties, namely ultra-low interfacial tension, large interfacial area, and solubilization capacity for both water- and oil-soluble compounds. These and other properties render microemulsions intriguing from a fundamental point of view and versatile for industrial applications.

Microemulsions had already been used in technological and household applications well before they were scientifically described for the first time as special colloidal dispersions by Hoar and Schulman in 1943 [1]. These authors reported the spontaneous formation of a transport or translucent solution upon mixing of oil, water, and an ionic surfactant combined with a cosurfactant (i.e., a medium chain length alcohol). At first, Hoar and Schulman [1] referred to this new type of colloidal dispersion as an oleo-phatic hydromicelle, and Bowcott and Schulman [2] referred to it with other names, such as transparent emulsions, at later stages of their studies. In 1959, about 15 years after Schulman's first publication on the subject, Schulman et al. [3] introduced the term microemulsion, the term that has prevailed for these systems.

Microemulsions form under a wide range of surfactant concentrations, water-to-oil ratios, temperature, etc.; this is an indication of the occurrence of diverse structural organizations. The picture that emerged from the earlier work on microemulsions [1–3] was that of spherical water or oil droplets dispersed in either oil (W/O) or water (O/W) with radii of the order of 100 to 1000 Å. In addition to droplet-type structures, the existence of microemulsions with bicontinuous structures in which the surfactant forms interfaces of rapidly fluctuating curvature and both the water and oil domains are continuous was later established [4].

A great deal of debate about the definition of microemulsions originated from the different concepts of the nature of these systems. Whereas Schulman et al. [1–3] viewed microemulsions as two-phase kinetically stable emulsions, Shinoda and Kunieda [5] pointed out that microemulsions could not be considered true emulsions but are one-phase systems with solubilized water or oil, identical to micellar solutions. Phase behavior studies by Friberg et al. [6–9] and Shinoda et al. [10–13] confirmed that most of Schulman's so-called microemulsions fell in the one liquid phase regions of the phase diagrams of the corresponding systems; that is, they were solubilized solutions. Adamson [14] suggested calling the microemulsions "micellar emulsions." The debate concerning thermodynamic stability of microemulsions continued in the 1980s. The definition of microemulsions suggested by Danielsson and Lindman [15] as systems of water, oil, and an amphiphile(s), which are single-phase and thermodynamically stable isotropic solutions, is quite widely accepted. However, other authors consider that the condition of thermodynamic stability is an unnecessary limitation and advocate a definition including, instead, the concept of spontaneous formation as more appropriate [16].

II. HISTORICAL BACKGROUND

A. Applications

The history of the early growth and development of microemulsions of industrial interest was extensively described by Prince [17], a pioneer in the study of theoretical and practical aspects of microemulsions. This has been the source of most of the information given in this section. The industrial development of microemulsions started in the 1930s, about 30 years before the term microemulsions was proposed by Schulman et al. [3]. However, applications of microemulsions at a domestic level were already known earlier. Indeed, it has been reported [18,19] that a very efficient recipe consisting of an oil-in-water microemulsion was widely used for washing wool more than a century ago in Australia. The formulation was made of water, soap flakes, methylated spirits, and eucalyptus oil.

The first marketed microemulsions were dispersions of carnauba wax in water. They were prepared by adding a soap (i.e., potassium oleate) to melted wax followed by incorporation of boiling water in small aliquots. The resulting opalescent formulations were used as a floor polisher and formed a glossy surface on drying. The opalescence of the dispersion obtained was interpreted as due to the presence of very small droplets (below 140 nm). The effectiveness and stability of the liquid wax formulations stimulated the development of many other formulations consisting of either O/W or W/O microemulsions [17]. An example of a particularly successful application of microemulsions of the W/O type was the formulation of cutting oils. Mineral oil-in-water emulsions had been used as effective coolants and lubricants for machine tool operations. However, after several cycles of operation, their efficiency decreased because of emulsion instability. The development of stable cutting oil formulations represented a great improvement in this area. The first formulations consisted of mineral oil (the lubricant), soap, petroleum sulfonate (an emulsifier and corrosion inhibitor), ethylene glycol (a coupling agent), an antifoam agent, and water (the coolant). Generally, the water was added by the user and the "soluble oil," the rest of the ingredients, was the commercial product [17]. Later, other formulations to which the user added both the oil and water were developed.

Simultaneously with the development of the O/W-type microemulsion formulations, a cleaning solution that was a microemulsion of the W/O type was introduced on the market. It consisted of pine oil, wood rosin, sodium oleate, and about 6% water. These solutions can be regarded as a precursor of the modern antiredeposition agents. On addition of this W/O microemulsion formulation to the washing solution, inversion to a microemulsion of the O/W type occurred, provided that the initial concentration of soap was sufficient. Soon afterward, O/W microemulsions (based on pine oil) experienced rapid development as fluid cleaning systems for floors, walls, etc. [17].

In the next decades, the 1940s and 1950s, microemulsion formulations were introduced in several areas of applications, from foods (flavor oils) to agrochemicals (pesticides), detergents (dry textile cleaning), and paints (latex particles). The task of microemulsion formulators was greatly facilitated by the commercial availability of nonionic emulsifiers. Previously, soaps were almost the only emulsifiers used in industry. The high hydrophile-lipophile balance (HLB) of soaps rendered formulation of microemulsions difficult, requiring the presence of long-chain alcohols as cosurfactants.

The application of microemulsions that led to the greatest expectations was, without doubt, that in tertiary oil recovery [20]. A considerable amount of oil is trapped in the porous rocks of oil reservoirs after primary and secondary oil recovery; a surfactant solution is then injected. In order to remove this residual oil successfully, the interfacial tensions between oil and water should be lower than 10^{-2} mN/m. The main advantage of a microemulsion over other surfactant solutions is the ultralow interfacial tension (lower than 10^{-3} mN/m) achieved when it coexists with an aqueous and an oil phase [21,22]. The application of microemulsions in oil recovery offered a large economic potential that stimulated enormously the development of theoretical and experimental research in the field of microemulsions. Even though microemulsions were considered appropriate systems for oil recovery since the early 1940s, increased interest in this application developed in the 1960s. This has been reflected in numerous patents and publications. The early developments of the applications of microemulsions in the area of enhanced oil recovery have been extensively reviewed [20,23]. Chapters 14–16 of this volume describe recent developments.

The main developments of practical applications of microemulsions up to the mid-1980s are described in a comprehensive review by Gillberg [24]. This area has experienced continuous progress. The objective of the various chapters of this book is to provide information about the most significant advances in the field of microemulsion applications in the past decade.

B. Formation

The formation and thermodynamic stability of microemulsions were the issues that attracted most of the interest in the early research in this area. In this context, one of the important contributions by Schulman et al. [1–3] was to realize that a reduction of the interfacial tension by three to four orders of magnitude is a requirement for the stability of these systems. This view was a natural consequence of their experimental approach to microemulsion formation. A typical experiment consisted of adding a medium chain length alcohol to an emulsion consisting of water, oil, and a soap as the emulsifier. At a certain concentration of alcohol a transition takes place spontaneously from a turbid emulsion to a transparent microemulsion. The spontaneous formation and thermodynamic stability of microemulsions were attributed to a further decrease of interfacial tension between water and oil by the effect of added alcohol, up to negative values. The requirement for transient negative interfacial tensions for microemulsion formation was examined theoretically and experimentally by different groups [25–28].

The understanding of the basis for the thermodynamic stability of microemulsions was improved considerably with the development of several thermodynamic theories [26–28]. Ruckenstein and Chi [26] considered the free energy of formation of microemulsions to consist of three contributions: (1) interfacial free energy, (2) energy of interaction between droplets, and (3) entropy of dispersion. Analysis of the thermodynamic factors showed that the contribution of the interaction energy between droplets was negligible and that the free energy of formation can be zero or negative if the interfacial tension is very low (of the order of 10^{-2}–10^{-3} mN/m), although not necessarily negative.

These studies led to the conclusion that microemulsions are thermodynamically stable because the interfacial tension between oil and water is low enough to be compensated by the entropy of dispersion. Surfactants with well-balanced hydrophile-lipophile (H-L) properties have the ability to reduce the interfacial tension to the values required for microemulsion formation. Surfactants with unbalanced H-L properties are unable to reduce the oil-water interfacial tension to values lower than about 1 mN/m; this is why a cosurfactant is often required to form microemulsions.

Considering microemulsions directly related to micellar solutions rather than to emulsions [5–13] was a significant contribution to elucidation of the problem of the formation and stability of microemulsions. It was clearly shown that formation of microemulsions may take place on increasing the amount of oil added to a micellar solution without a phase transition. Furthermore, phase behavior studies of nonionic surfactant systems as a function of temperature showed that the hydrophile-lipophile properties of ethoxylated nonionic surfactants are highly temperature dependent [5,10–12]. Shinoda and Saito [12] introduced the concept of HLB temperature or phase inversion temperature (PIT) as the temperature at which the hydrophile-lipophile properties of the surfactants are balanced. At this temperature, maximum solubilization of oil in water and ultralow interfacial tensions are achieved. Further studies showed that the effects produced by temperature in nonionic surfactant systems were produced by salinity in ionic surfactant systems [22,29]. The study of the phase behavior of surfactant systems has made it possible to rationalize the formation of microemulsions and to predict their properties. Because of the interest for practical applications, Chapter 2 is devoted entirely to the general trends of microemulsion formation.

III. PROPERTIES RELEVANT TO APPLICATIONS

Spontaneous formation, clear appearance, thermodynamic stability, and low viscosity are some characteristics of microemulsions that render these systems attractive and suitable for many industrial applications. The widespread use of and interest in microemulsions are based mainly on the high solubilization capacity for both hydrophilic and lipophilic compounds, on their large interfacial areas, and on the ultralow interfacial tensions achieved when they coexist with excess aqueous and oil phases. The properties of microemulsions have been extensively reviewed [30–36] and are described throughout this book. Therefore, only a brief summary is given in this section.

In some applications, microemulsions and emulsions could be used. However, microemulsions have important advantages. Low energy input is required for their preparation (spontaneous formation) and stability [24]. Their isotropic or clear appearance not only is an aesthetic property of interest for consumer products but also allows applications such as photochemical reactions, for which emulsions are unsuitable. For other applications—for instance, when the surfactant system plays the role of a reservoir of surfactant molecules—microemulsions and micellar solutions can be equally suitable. However, for applications requiring high solubilization power, microemulsions are with any doubt superior [19]. Chapters 5–10 show the importance of solubilization for the use of microemulsions in pharmacy, food, cosmetics, agrochemicals, and textile dyeing.

The ultralow interfacial tension achieved in microemulsion systems has application in several phenomena involved in oil recovery as well as in other extraction processes. Enhanced oil recovery, soil decontamination, and detergency are processes that benefit from ultralow interfacial tensions. These aspects are treated in detail in Chapters 14–18.

The compartmentalized structure of microemulsions with hydrophobic and hydrophilic domains offers a great potential for applications as microreactors. The possibility that water- and oil-soluble reactants are in contact at large interfaces may lead to a remarkable increase in rates of heterogeneous reactions [37]. The application of this property in biotechnology is considered in Chapter 4. The fluidity of the surfactant layers may be also important in diffusion-controlled reactions. The characteristic size of microemulsions (i.e., droplet radius) can be controlled by changing composition parameters, temperature, salinity, etc. This has application in the preparation of nanoparticles of a desired size. This aspect is covered in Chapters 11–13. Other applications of microemulsions that benefit from their various properties are those in the fields of analytical determinations (Chapter 3) and lubrication (Chapter 19).

IV. STRUCTURE

Two main general structures have been proposed and are accepted: discrete microemulsions and bicontinuous microemulsions. A schematic picture of a ternary phase diagram at constant temperature corresponding to a typical water/nonionic surfactant/oil ternary system is shown in Fig. 1. Microemulsions poor in either water or oil have a globular structure. Microemulsions containing similar amounts of oil and water and relatively high amounts of surfactant present bicontinous structures. Frequently, liquid crystalline phases are also present in the phase diagrams.

Discrete microemulsions consist of domains of one of the pseudophases (water or oil) dispersed in the other pseudophase. These structures are generally found when the main component of one of the pseudophases (water or oil) is present in a higher proportion than the main component of the other pseudophase and little surfactant is present. The structure of this type of microemulsion resembles that of emulsions in that one phase is dispersed in another phase. However, as already stated, they are essentially different in many aspects, in particular, for concerning their stability. The structure of emulsions depends on their history and they evolve with time, whereas microemulsions are thermodynamically stable and their structure is independent of their preparation history. Moreover, other differences arise from other aspects. Emulsion droplets are spherical or nearly spherical; this form minimizes the interface, which gives a highly energetic term because of the interfacial tension. In microemulsions, because of the very low interfacial tension, the energetic term related to the interfacial

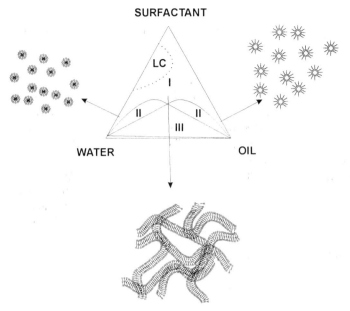

FIG. 1 Schematic ternary phase diagram for a typical water/nonionic surfactant/ oil system at the HLB temperature. Microemulsion structure is shown in the normal regions of occurrence; left to right: O/W globular microemulsions, bicontinuous microemulsions, and W/O globular microemulsion.

tension and total surface is of less importance and therefore nonspherical droplets can be present without a large energy contribution. Because of the small size of the droplets and the low contribution of total surface to the total energy, the geometry of the surfactant molecules at the interface plays an important role.

For microemulsions it is useful to consider the so-called critical packing parameter. This concept, put forward by Israelachvilli et al. [38], considers that the amphiphile molecules can be regarded as a two-piece structure: polar head and hydrophobic tail. The possible geometry of a film formed by the amphiphile molecules depends on their intrinsic geometry. The critical packing parameter (CPP) is calculated as CPP = $V/a_o l$, where a_o is the optimal area of the polar head, l the length of the hydrophobic tail, and V its volume. The area per polar head is usually measured at an air-water or oil-water interface using the Gibbs isotherm [39]. The length of the hydrophobic tail can be calculated from the values obtained by Tanford [40], and the volume of the hydrocarbon tail can be calculated from the density of bulk hydrocarbon. Critical packing parameters lower than 1/3 give a tendency to form globular structures, values around $1/2$ favor cylindrical structures, and values close to 1 favor planar layers. Attention should be paid to the indiscriminate use of this parameter. This parameter evaluates the natural geometry of the amphiphile by itself. In microemulsions, hydrocarbon penetration and cosurfactant presence may completely change the structure from the natural tendency. Oil penetration in the hydrocarbon tail produces an increase in the apparent hydrophobic volume and thus an increase in the critical packing parameter. Cosurfactants, such as medium-chain alcohols, coadsorb at the interface, producing an overall reduction of the critical packing parameter. The concentration of surfactant and the ratio of the pseudophases play important roles in the structure as well. High amounts of ionic surfactant produce a high ionic strength with a subsequent reduction of the polar head area and reduction of the critical packing parameter. High amounts of the internal pseudophase may produce phase separation if the total concentration of surfactant is low. Other variables that influence the natural curvature of the amphiphile are electrolyte concentration (mainly for ionic surfactants [41], although they influence nonionics as well) and temperature (mainly affecting nonionic surfactants [42,43]).

In contrast to the discrete microemulsion structure, which is relatively easy to treat theoretically, the structure of bicontinuous microemulsions is more difficult to visualize and therefore its theoretical treatment is complicated. In a bicontinuous microemulsion both the aqueous and oil phases are continuous. This continuity means that it is possible to go from one extreme of the system to the other by either an oil path or an aqueous path. This structure has an extremely large interfacial area, which is possible because of an extremely low interfacial tension, close to zero. Contrary to what some authors have maintained, in a microemulsion there is not a negative interfacial tension because this would mean production of energy as the interface increases. This could be the case in microemulsion formation but not at equilibrium. Near-zero interfacial tension imply, at the same time, that the interfaces are unstable and can form and disappear without an energy increase. Interfacial energy of the order of kT has been considered to be a condition for the formation of bicontinuous structures [44]. Conditions for the formation of bicontinuous structures are a ratio of oil and water pseudophases close to one, large amounts of surfactant (enough to cover the interface), and zero natural curvature of the interface.

The theoretical treatment of a bicontinuous structure is complicated. Since this structure was proposed by Scriven [4], several models have been proposed. The "random lattice" theory of Talmon and Prager [45,46] is based on a tessellation of the space by a Voronoi structure; the cells of this tessellation are occupied by either oil or water in a random way. This model was improved by De Gennes and Taupin [47], whose model is based on a cubic lattice. The cubes are occupied by either water or oil in a random way. There is a critical water/oil ratio for which percolation occurs; that is, an infinite path is possible in both phases.

The bicontinuous structure is consistent with most of the experimental observations on these systems. For instance, the self-diffusion coefficients of oil, water, and surfactant are well explained [42]. In these systems the self-diffusion coefficients of oil and water are close to the self-diffusion coefficients of these molecules in the pure liquids and the self-diffusion coefficient of the surfactant is about one order of magnitude lower. This indicates "free" diffusion for water and oil, i.e., infinite domains, and the lower diffusion coefficient of the surfactant is related to the positioning of the molecules at the interface (see more information below).

A useful picture of the structural transitions can be obtained by considering a surfactant solution in water with an increasing amount of oil. Surfactants above the critical micelle concentration form micelles in which some oil can be solubilized. The limit of solubility in the micelles depends on the nature of the surfactant and the number of micelles. Ionic surfactants usually have large head groups and have a strong tendency to form spherical aggregates in water. The incorporation of oil increases the size of the aggregates and therefore reduces the curvature. To reach a large amount of solubilizate the oil must penetrate the surfactant tail effectively, a co-surfactant should be added, electrolyte should be added, the temperature should be changed, or a combination of the four. On increasing the amount of oil pseudophase the percolation point will be reached and a bicontinuous structure formed. As said before, a large amount of surfactant is needed to prevent phase separation before this point is reached. Further increase of the oil pseudophase will make the system reach the percolation threshold for the water domains and discrete water domains will be formed.

V. METHODS OF CHARACTERIZATION

Microemulsions have been studied using a great variety of techniques. This suggests that characterization of microemulsions is a rather difficult task. This is due to their complexity, namely the variety of structures and components involved in these systems, as well as the limitations associated with each technique. Therefore, complementary studies using a combination of techniques are usually required to obtain a comprehensive view of the physicochemical properties and structure of microemulsions. In this section, some of the most common methods and techniques used to identify and characterize microemulsions are briefly described.

A. Phase Behavior

Phase behavior studies, with phase diagram determinations, are essential in the study of surfactant systems. They provide information on the boundaries of the different phases as a function of composition variables and temperature, and, more important, structural organization can be also inferred [48]. In addition, phase behavior studies allow comparison of the efficiency of different surfactants for a given application. It is important to note that simple measurements and equipment are required in this type of study. The boundaries of one-phase regions can be assessed easily by visual observation of samples of known composition. However, long equilibration times in multiphase regions, especially if liquid crystalline phase are involved, can make these determinations long and difficult.

The phase behavior of interest for microemulsion studies involves at least three components: water, surfactant, and oil. Although most of the formulations of practical interest consist of more than three components, study of simple systems with the basic three, four, etc. components from which they are formulated is a prerequisite to understanding the behavior of complex systems. The phase behavior of three-component systems at fixed temperature and pressure is best represented by a ternary diagram (Fig. 1) and by a triangular prism if temperature is considered as a variable (see Fig. 2 in Chapter 2). Other useful ways of representing the phase behavior are to keep constant the concentration of one component (Fig. 1 in Chapter 2) or the ratio of two components (Fig. 3 in Chapter 2). As the number of components increases, the number of experiments needed to define the complete phase behavior becomes extraordinary large and the representation of phase behavior is extremely complex. One approach to characterizing these multicomponent systems is by means of pseudoternary diagrams that combine more than one component in the vertices of the ternary diagram.

Most of the phase studies concerning microemulsions have been limited to the determination of one-liquid-isotropic phase boundaries. However, information about the number and compositions of the coexisting phases in equilibrium is of the utmost interest in characterizing these systems [48,49].

B. Scattering Techniques

Scattering methods have been widely applied in the study of microemulsions. These include small-angle X-ray scattering (SAXS), small-angle neutron scattering (SANS), and static as well as dynamic light scattering. In the static scattering techniques, the intensity of scattered radiation $I(q)$

is measured as a function of the scattering vector q, $q = (4\pi/\lambda) \sin\theta/2$, where θ is the scattering angle and λ the wavelength of the radiation. The general expression for the scattering intensity of monodispersed spheres interacting through hard sphere repulsion is $I(q) = nP(q)S(q)$, where n is the number density of the spheres, $P(q)$ is the form factor, which expresses the scattering cross section of the particle, and $S(q)$ is the structural factor, which takes into account the particle-particle interaction. $P(q)$ and $S(q)$ can be estimated by using appropriate analytical expressions. The lower limit of size that can be measured with these techniques is about 2 nm. The upper limit is about 100 nm for SANS and SAXS and a few micrometers for light scattering. These methods are very valuable for obtaining quantitative information on the size, shape, and dynamics of the components. There is a major difficulty in the study of microemulsions with the use of scattering techniques: dilution of the sample, to reduce interparticle interaction, is not appropriate because it can modify the structure and the composition of the pseudophases. Nevertheless, successful determinations have been achieved by using a dilution technique that maintains the identity of droplets [31] and extrapolating the results obtained at infinite dilution to obtain the size, shape, etc., or by measurements at very low concentrations.

Small-angle X-ray scattering techniques have long been used to obtain information on droplet size and shape [50,51]. Using synchrotron radiation sources, with which sample-to-detector distances are bigger (4 m instead of 30–50 cm as with laboratory-based X-ray sources), significant improvements have been achieved. With synchrotron radiation more defined spectra are obtained and a wide range of systems can be studied, including those in which the surfactant molecules are poor X-ray scatterers [42,52].

Small-angle neutron scattering (SANS) allows selective enhancement of the scattering power of the different microemulsion pseudophases by using protonated or deuterated molecules (contrast variation technique). Therefore, this technique allows determination of the size and shape of the droplets as well as the characteristics of the amphiphilic layer without great perturbation of the system [42,53–55].

Static light scattering techniques have also been widely used to determine microemulsion droplet size and shape. In these experiments the intensity of scattered light is generally measured at various angles and for different concentrations of microemulsion droplets. At sufficiently low concentrations, provided that the particles are small enough, the Rayleigh approximation can be applied. Droplet size can be estimated by plotting the intensity as a function of droplet volume fraction [42,54,56,57].

Dynamic light scattering, also referred to as photon correlation spectroscopy (PCS), is used to analyze the fluctuations in the intensity of scattering by the droplets due to Brownian motion. The self-correlation function is measured and gives information on the dynamics of the system. This technique allows the determination of z-average diffusion coefficients, D. In the absence of interparticle interactions, the hydrodynamic radius of the particles, R_H, can be determined from the diffusion coefficient using the Stokes-Einstein equation: $D = kT/6\pi\eta R_H$, where k is the Boltzmann constant, T is the absolute temperature, and η is the viscosity of the medium. Although dynamic light scattering measurements are relatively easy and fast, extrapolation of results to infinite dilution is not possible in most microemulsion systems and R_H values obtained should be corrected because of interparticle interactions [42,54,58,59].

C. Nuclear Magnetic Resonance

Nuclear magnetic resonance techniques have been used to study the structure and dynamics of microemulsions. Self-diffusion measurements using different tracer techniques, generally radioactive labeling, supply information on the mobility of the components (self-diffusion coefficient). A limitation of this technique is that experiments are time-consuming and the use of labeled molecules in multicomponent systems such as microemulsions is not practical [60]. However, the Fourier transform pulsed-gradient spin-echo (FT-PGSE) technique, in which magnetic field gradients are applied to the sample, allows simultaneous and rapid determination of the self-diffusion coefficients (in the range of 10^{-9} to 10^{-12} m^2 s^{-1}), of many components [61]. In water-in-oil microemulsions, water diffusion is slow and corresponds to that of the droplets (of the order of 10^{-11} m^2 s^{-1}), oil diffusion is high (of the order of 10^{-9} m^2 s^{-1}), and the diffusion of surfactant molecules, located at the interface, is of the same order as that of the droplets. In contrast, in oil-in-water microemulsions the diffusion coefficients of water are higher than that of oil. In bicontinuous microemulsions the diffusion coefficients of water and oil are both high (of the order of 10^{-9} m^2 s^{-1}) and the diffusion coefficient of the surfactant has been found to be intermediate between the value of nonassociated surfactant molecules and the value for a droplet-type structure (of the order of 10^{-10} m^2 s^{-1}) [62–65]. These changes can be observed in Fig. 2.

D. Electron Microscopy

Several electron microscopic techniques have been attempted for the characterization of microemulsions. Because of the high lability of the samples and the danger of artifacts, electron microscopy used to be considered a misleading technique in microemulsion studies. However, images showing clear evidence of the microstructure have been obtained [42,44]. Freeze fracture electron microscopy, a well-established method in the biological

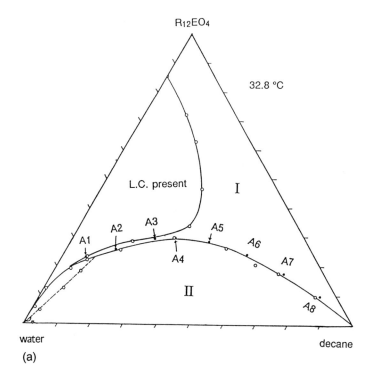

(a)

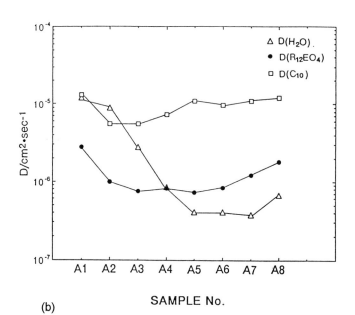

(b)

field, has been successfully applied to microemulsions. Careful control of the temperature of the sample before freezing and ultrarapid cooling followed by fracture and replication of the fracture face yield images of the microstructure of these systems.

E. Other Methods

Interfacial tension measurements are useful in the study of the formation and properties of microemulsions. Ultralow values of interfacial tensions are correlated with phase behavior, particularly the existence of surfactant phase or middle-phase microemulsions in equilibrium with aqueous and oil phases [29,66]. Ultralow interfacial tensions can be measured with the spinning-drop apparatus. Interfacial tensions are derived from the measurement of the shape of a drop of the low-density phase, rotating in a cylindrical capillary filled with the high-density phase [67]. Figure 3 shows an example of the changes of interfacial tension in a ternary system as a function of temperature.

Electrical conductivity has been widely used to determine the nature of the continuous phase and to detect phase inversion phenomena. The distinction between O/W (high-conductivity) and W/O (low-conductivity) emulsions is quite straightforward. However, in microemulsions the behavior is more complex. A sharp increase in conductivity in certain W/O microemulsion systems was observed at low volume fractions [68]. This behavior was interpreted as an indication of a "percolative behavior" or exchange of ions between droplets before the formation of bicontinuous structures. When the conductivity of nonionic surfactant systems is measured, water is generally replaced by an electrolyte solution. If the electrolyte concentration is kept low (10^{-2}–10^{-3} M), no effect on the structure is produced [42]. Dielectric measurements have also been used in the determination of structural aspects of microemulsions and results similar to those obtained from conductivity measurements have been obtained.

Viscosity measurements as a function of volume fraction have been used to determine the hydrodynamic radius of droplets, as well as interactions between droplets and deviations from spherical shape by fitting the results to appropriate models [32]. Some microemulsions show Newtonian behavior, and their viscosities are similar to that of water. For these microemul-

---

FIG. 2 (a) Phase diagram of a water/$R_{12}EO_4$ (tetraethyleneglycol dodecyl ether)/ decane system at 32.8°C. (From Ref. 65.) (b) Self-diffusion coefficients for water, $R_{12}EO_4$, and decane in a single-phase region surrounding the main miscibility gap in (a). The sample number is indicated in (a). (From Ref. 65.)

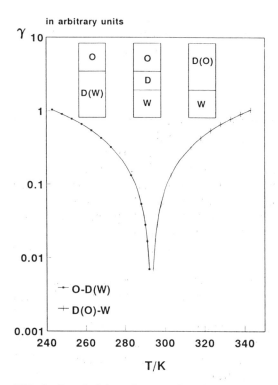

FIG. 3 Interfacial tension as a function of temperature for a model ternary water/ nonionic surfactant/oil system.

sions, the hydrodynamic volume of the particles can be calculated from Einstein's equation for the relative viscosity η_r ($\eta_r = 1 + 2.5\phi$, where ϕ is the particle volume fraction) if ϕ is lower than ~0.1 or from modifications of this equation if it is higher.

VI. SUMMARY

Microemulsions have experienced continuous scientific and industrial development since their introduction. The knowledge gained on the fundamental aspects of these systems has made it possible to improve some established applications and to develop new ones. Judging from the growth of the research activity in this area, it can be predicted that this trend will continue in the future.

REFERENCES

1. T. P. Hoar and J. H. Schulman, Nature *329*:309 (1943).
2. J. E. Bowcott and J. H. Schulman, Z. Electrochem. *59*:283 (1955).
3. J. H. Schulman, W. Stoekenius, and L. M. Prince, J. Phys. Chem. *63*:1677 (1959).
4. L. E. Scriven, Nature *263*:123 (1976).
5. K. Shinoda and H. Kunieda, J. Colloid Interface Sci. *42*:381 (1973).
6. G. Gillberg, H. Lehtinen, and S. E. Friberg, J. Colloid Interface Sci. *33*:215 (1970).
7. S. I. Ahmad and K. Shinoda, J. Colloid Interface Sci. *47*:32 (1974).
8. S. Friberg, in *Microemulsions. Theory and Practice* (L. M. Prince, ed.), Academic Press, New York, 1970, pp. 133–145.
9. S. Friberg and I. Buraczewska, in *Micellization, Solubilization and Microemulsions* (K. L. Mittal, ed.), Plenum, New York, 1977, vol. 2, p. 791.
10. H. Saito and K. Shinoda, J. Colloid Interface Sci. *24*:10 (1967).
11. K. Shinoda and T. Ogawa, J. Colloid Interface Sci. *24*:56 (1967).
12. K. Shinoda and H. Saito, J. Colloid Interface Sci. *26*:70 (1968).
13. K. Shinoda and S. Friberg, Adv. Colloid Interface Sci. *4*:281 (1975).
14. A. W. Adamson, J. Colloid Interface Sci. *29*:261 (1969).
15. I. Danielsson and B. Lindman, Colloids Surfaces *3*:391 (1981).
16. S. E. Friberg, Colloids Surfaces *4*:201 (1982).
17. L. M. Prince, in *Microemulsions. Theory and Practice* (L. M. Prince, ed.), Academic Press, New York, 1977, pp. 21–32.
18. D. F. Evans, D. J. Mitchell, and B. W. Ninham, J. Phys. Chem. *90*:2817 (1986).
19. D. Langevin, in *Reverse Micelles* (P. L. Luisi and B. E. Straub, eds.), Plenum, New York, 1984, pp. 287–303.
20. D. O. Shah and R. S. Schechter, eds., *Improved Oil Recovery by Surfactant and Polymer Flooding,* Academic Press, New York, 1977.
21. H. Kunieda and K. Shinoda, J. Colloid Interface Sci. *75*:601 (1980); H. Kunieda and K. Shinoda, Bull. Chem. Soc. Jpn. *55*:1777 (1982).
22. M. Bourel and R. S. Schechter, *Microemulsions and Related Systems,* Marcel Dekker, New York, 1988.
23. D. O. Shah, ed., *Surface Phenomena in Enhanced Oil Recovery,* Plenum, New York, 1979.
24. G. Gillberg, in *Emulsions and Emulsion Technology* (K. J. Lissant, ed.), Marcel Dekker, New York, 1984, pp. 1–43.
25. C. A. Miller and L. E. Scriven, J. Colloid Interface Sci. *33*:360 (1970).
26. E. Ruckenstein and J. C. Chi, J. Chem. Soc. Faraday Trans. 2 *71*:1690 (1975).
27. J. Th. G. Overbek, Faraday Disc. Chem. Soc. *65*:7 (1978).
28. E. Ruckenstein and R. Krishnan, J. Colloid Interface Sci. *76*:201 (1980).
29. C. A. Miller and P. Neogi, *Interfacial Phenomena, Equilibrium and Dynamic Effects,* Marcel Dekker, New York, 1985, pp. 140–179.
30. I. D. Robb, ed., *Microemulsions,* Plenum, New York, 1977.

31. J. T. G. Overbeek, P. L. de Bruy, and F. Verhoeckx, in *Surfactants* (Th. F. Tadros, ed.), Academic Press, New York, 1984, pp. 111–131.
32. Th. F. Tadros, in *Surfactants in Solution,* (K. L. Mittal and B. Lindman, eds.), Plenum, New York, 1984, pp. 1501–1532.
33. D. O. Shah, ed. *Macro- and Microemulsions: Theory and Applications,* ACS Symposium Series 272, American Chemical Society, Washington, DC 1985.
34. B. H. Robinson, Nature *320*:309 (1986).
35. S. E. Friberg and P. Bothorel, eds., *Microemulsions: Structure and Dynamics,* CRC Press, Boca Raton, FL, 1987.
36. R. Leung, M. Jeng Hou, and D. O. Shah, in *Surfactants in Chemical/Process Engineering* (D. T. Wasan, M. E. Ginn, and D. O. Shah, eds.), Marcel Dekker, New York, 1988, pp. 315–367.
37. B. H. Robinson, in *Reverse Micelle* (P. L. Luisi and B. E. Straub, eds.), Plenum, New York, 1984, p. 73.
38. J. N. Israelachvilli, D. J. Mitchell, and B. W. Niham, J. Chem. Soc. Faraday Trans. I *72*:1525 (1976).
39. S. Ross and I. D. Morrison, *Colloidal Systems and Interfaces*, Wiley, New York, 1988, p. 176.
40. C. Tanford, *The Hydrophobic Effect*, Wiley, New York, 1980, p. 52.
41. L. Auvray, J. P. Cotton, R. Ober, and C. Taupin, J. Phys. *45*:913 (1984).
42. M. Kahlweit, R. Strey, D. Haase, H. Kunieda, T. Schmeling, B. Faulhaber, M., Borkovec, H. F. Eicke, G. Busse, F. Eggers, Th. Funck, H. Richmann, L. Magid, O. Söderman, P. Stilbs, J. Winkler, A. Dittrich, and W. Jahn, J. Colloid Interface Sci. *118*:436 (1987).
43. J. C. Ravey and M. Buzier, in *Surfactants in Solution* (K. L. Mittal and B. Lidman, eds.), Plenum, New York, 1984, vol. 3, p. 1759.
44. R. Strey, Colloid Polym. Sci. *272*:1005 (1994).
45. Y. Talmon and S. Prager, J. Chem. Phys. *69*:2984 (1978).
46. Y. Talmon and S. Prager, J. Chem. Phys. *76*:1535 (1982).
47. P. G. De Gennes and C. Taupin, J. Phys. Chem. *86*:2294 (1982).
48. R. G. Laughlin, *The Aqueous Phase Behavior of Surfactants* Academic Press, New York, 1994, pp. 102–176.
49. H. Kunieda and K. Shinoda, J. Dispers. Sci. Technol. *3*:233 (1982).
50. D. S. Bohlm, Diss. Abs. Int. Ser. B *51*:1946 (1990).
51. T. N. Zemb, S. T. Hyde, P. J. Derian, I. S. Barnes, and B. W. Ninham, J. Phys. Chem. 91:3814 (1987).
52. R. Hilfiker, H. F. Eicke, W. Sager, U. Hofmeier, and R. Gehrke, Ber. Bunsenges Phys. Chem. *94*:677 (1990).
53. A. M. Howe, C. Toprakcioglu, J. C. Dore, and B. H. Robinson, J. Chem. Soc. Farad. Trans I *82*:2411 (1986).
54. D. J. Cebula, D. Y. Myers, and R. H. Ottewill, Colloid Polym. Sci. *260*:96 (1982).
55. E. Caponetti, W. L. Griffith, J. H. Johnson, R. Triolo, and A. L. Compere, Langmuir *4*:606 (1988).

56. R. C. Baker, A. T. Florence, R. H. Ottewill, and Th. F. Tadros, J. Colloid Interface Sci. *100*:332 (1984).
57. Th. F. Tadros, P. F. Luckman, and C. Yanaranop, in *Structures, Microemulsions and Liquid Crystals,* ACS Symposium Series 384, American Chemical Society, Washington, DC, 1989, p. 22.
58. A. M. Cazabat, D. Langevin, and A. J. Puchelon, J. Colloid Interface Sci. *73*: 1 (1980).
59. N. J. Chang and E. W. Kaler, Langmuir *2*:184 (1986).
60. B. Lidman, N. Kamenka, T. M. Kathopoulis, B. Brun, and P. G. Nilsson, J. Phys. Chem. *84*:2485 (1980).
61. P. Stilbs, Prog. NMR Spectrosc. *19*:1 (1987).
62. P. G. Nilsson and B. Lidman, J. Phys. Chem. *86*:271 (1982).
63. K. Fontell, A. Ceglie, B. Lidman, and B. Ninham, Acta Chem. Scand. Ser. A *40*:247 (1986).
64. T. Warnheim, U. Henriksson, and O. Soderman, in *Organized Solutions* (S. E. Friberg and B. Lindman, eds.), Marcel Dekker, New York, 1992, pp. 221–235.
65. C. Solans, R. Pons, S. Zhu, H. T. Davis, D F. Evans, K. Nakamura, and H. Kunieda, Langmuir *9*:1479 (1993).
66. H. Kunieda and S. E. Friberg, Bull. Chem. Soc. Jpn. *54*:1010 (1981).
67. J. L. Cayias, R. S. Schechter, and W. H. Wade, in *Adsorption at Interfaces,* ACS Symposium Series, American Chemical Society, Washington, DC, 1975, 8, p. 234.
68. M. Clausse, J. Heil, J. Peyrelasse, and C. Boned, J. Colloid Interface Sci. *87*: 584 (1982).

2

How to Prepare Microemulsions: Temperature-Insensitive Microemulsions

HIRONOBU KUNIEDA Graduate School of Engineering, Yokohama National University, Yokohama, Japan

CONXITA SOLANS Departamento de Tecnología de Tensioactivos, Centro de Investigación y Desarrollo, Consejo Superior de Investigaciones Científicas (CSIC), Barcelona, Spain

I. INTRODUCTION

Microemulsions are isotropic surfactant solutions that solubilize considerable amounts of water and/or oil [1,2]. In order to apply microemulsions for practical purposes, it is important to form temperature-insensitive microemulsions of large solubilization. When hydrophilic surfactants are used, oil-swollen-type or oil-in-water (O/W) microemulsions form, whereas lipophilic surfactants produce water-swollen-type or water-in-oil (W/O) microemulsions. When the hydrophile-lipophile property of the surfactant is just balanced in a given water-oil system, a microemulsion, which is also called a surfactant phase or middle-phase microemulsion, coexists with excess water and oil phases [1]. In the three-phase body, the solubilization capacity of the surfactant reaches its maximum in both nonionic and ionic surfactant systems [1–5] and ultralow interfacial tensions are also attained [6,7]. In general, the phase behavior of microemulsions is very sensitive to the temperature change at the maximum solubilization. The more the solubilization increases, the more temperature sensitive the phase behavior becomes.

It is also known that the type of an ordinary macroemulsion is changed from O/W to W/O around the three-phase region. The O/W type is stable on one side of the three-phase body and the W/O type becomes stable on the opposite side. Therefore, it is also important to obtain the temperature-insensitive three-phase body to produce macroemulsions stable against temperature change for practical applications.

There have been attempts to form temperature-insensitive microemulsions of large solubilization [8–10]. If the composition of a single-phase microemulsion is not changed much and phase separation does not occur over a wide range of temperature, the microemulsion is called a temperature-insensitive microemulsion. Middle-phase microemulsions have a so-called bicontinuous structure in which both micro-oil and water domains are continuous and surfactant molecules are adsorbed at the water-oil interface [11,12]. The composition of the three-phase body or the single-phase microemulsion is generally temperature sensitive in both single nonionic and ionic surfactant systems, as mentioned later in the next section. To obtain temperature-insensitive microemulsions of large solubilization, it is necessary to mix surfactants. In this case, phase behavior is more complicated because of the difference in distribution of surfactants among the micro-water and micro-oil domains and the water-oil interface inside the microemulsion [13]. One has to take the distribution into account in order to construct temperature-insensitive microemulsion systems.

In this chapter, we systematically describe the analysis of the three-phase behavior in order to form a temperature-insensitive microemulsions of large solubilization. First, the basic phase behavior of single surfactant is described in both nonionic and ionic surfactant microemulsion systems. The phase behavior of a water/ionic surfactant/cosurfactant/oil system is also described. Second, we analyze the three-phase behavior in a mixed surfactant system. We apply the theory to commercial polyoxyethylene-type nonionic surfactant, mixed polyoxyethylene-type and sucrose-type nonionic surfactant, and mixed nonionic-ionic surfactant systems. We also describe the effect of mixing surfactant on the maximum solubilization in microemulsion systems.

II. SINGLE SURFACTANT SYSTEMS

A. Single Nonionic Surfactant Systems

Polyoxyethylene-type nonionic surfactants are the most widely distributed nonionic surfactants in industrial fields. If the hydrophile-lipophile property of the nonionic surfactant is just balanced in a given water-oil system, the middle-phase microemulsion forms without addition of cosurfactant. The phase diagram of a water/pure homogeneous nonionic surfactant (trioxyethylene octyl ether)/decane system as a function of temperature is shown in Fig. 1, where the weight fraction of oil in water + oil is plotted [6].

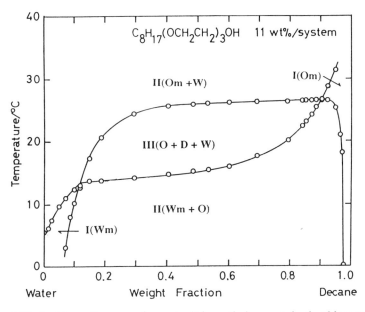

FIG. 1   Phase diagram of a water/trioxyethylene octyl ether/decane system as a function of temperature. The concentration of surfactant in the system is 11 wt%. I, II, and III indicate one-phase, two-phase, and three-phase regions, respectively. Wm and Om are oil-swollen micellar solution and water-swollen reverse micellar solution phases. W and O are excess water and oil phases. D is the middle-phase microemulsion or surfactant phase.

Polyoxyethylene-type nonionic surfactants become relatively lipophilic with increasing temperature because dehydration takes place as a result of the conformational change in hydrophilic polyoxyethylene chains [14]. At lower temperatures, surfactant forms aqueous micelles and the oil-swollen aqueous micellar solution phase (Wm) coexists with excess oil phase (O). On the other hand, at higher temperatures the surfactant forms a water-swollen reverse micellar solution (Om) that coexists with excess water phase (W). At the transition temperature, the solubilization reaches its maximum and the microemulsion (D) coexists with excess water and oil phases. The degree of freedom for the three-phase body is 1 in the ternary system at constant pressure, and the compositions of the individual phases in the three-phase tie triangle are invariant at constant temperature.

The upper and lower boundaries of the three-phase body correspond to the critical end temperatures of microemulsion-water and microemulsion-oil, respectively, and they are fixed in a ternary system because the degree of freedom for the critical end points is zero at constant pressure [15]. In other words, one cannot change the width of the three-phase temperature in the ternary system.

The three-phase equilibrium of a water/single nonionic surfactant/oil system is shown schematically in Fig. 2 [16]. The curves represent the loci

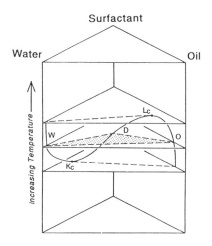

FIG. 2 Schematic representation of a three-phase body in the space of composition and temperature in a ternary water/nonionic surfactant/oil system. K_c and L_c are critical end points for D-W and D-O, respectively. Curves W, D, and O represent the loci of the phases forming the three-phase triangle. The three-phase body exists at temperatures between two critical end points.

of water (W), microemulsion (D), and oil (O) phases forming a three-phase body. The composition of the middle-phase microemulsion (D) changes from water-rich to oil-rich regions with changing temperature as shown in Fig. 2. Two of the three phases are merged at the critical end point (K_c or L_c) and the three-phase body is terminated. This is why ultralow interfacial tensions are attained in the three-phase body that consists of the stack of three-phase tie triangles [7]. If the two critical end points approach and eventually coincide with each other, three coexisting phases are simultaneously identical at the tricritical point [7,17,18]. There are, in general, three ways to terminate the three-phase regions: at critical end points, at the tricritical point, and through four-phase regions [2,19–21].

On the other hand, if two critical end temperatures are considerably separated, the three-phase region exists over a wide range of temperature. In this case, the slope of the locus of microemulsion (the curve K_c-L_c in Fig. 2) is high against temperature and the composition of the microemulsion phase does not change much. Figure 3 shows the phase diagrams for water/triethylene glycol octyl ether/oil in which the water/oil ratio is kept constant. Point α is the maximum solubilization point, at which the microemulsion solubilizes equal weights of water and oil. Points a and b in the tetradecane system show the two critical end temperatures for D-W and D-O, respectively. When the separation between the two critical end temperatures is large, the solubilization capability is decreased as shown in Fig. 3. Therefore, it is difficult to form a temperature-insensitive three-phase microemulsion of large solubilization in a single polyoxyethylene-type nonionic surfactant system.

An ordinary O/W macroemulsion is stable at temperatures below the three-phase temperature, whereas a W/O macroemulsion is stable at higher temperatures [22,23]. The stability and the type of emulsion are also highly dependent on temperature change. Hence, it is also difficult to form temperature-insensitive stable macroemulsions in a single polyoxyethylene-type nonionic surfactant system. The three-phase temperature in a nonionic surfactant system is called the phase inversion temperature (PIT) in emulsion as well as the hydrophile-lipophile balance (HLB) temperature.

B. Single Ionic Surfactant Systems

It has been accepted that microemulsions are temperature insensitive in ionic surfactant systems, although it is not always correct. Three-phase microemulsions are formed in a brine/double-chain ionic surfactant/oil [4] or brine/ionic surfactant/cosurfactant/oil [8,24,25] systems. Inorganic salts are usually used for the formation of three-phase microemulsions in ionic

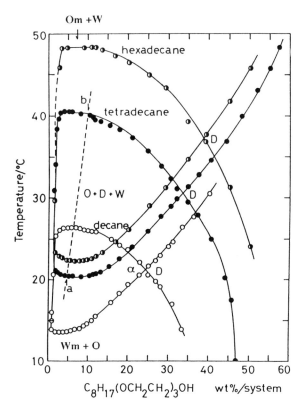

FIG. 3 Phase diagrams of water/trioxyethylene octyl ether/oil systems as a function of temperature. The water/oil ratio is 50/50 (w/w). The oils are decane, tetradecane, and hexadecane.

surfactant systems because salt functions to suppress the hydrophilicity of ionic surfactant and to destroy the formation of liquid crystals [26–28].

Figure 4 shows the typical three-phase behavior of a brine/Aerosol OT/isooctane system at constant salinity and surfactant concentration [4]. Although Fig. 4 resembles Fig. 1, the effect of temperature is opposite to that in polyoxyethylene-type nonionic surfactant systems. At higher temperatures, surfactant tends to dissolve in water, whereas it forms reverse micelles in oil at lower temperatures. With rising temperature the dissociation of ionic groups increases, and ionic surfactants tend to be more hydrophilic at higher temperatures at a fixed salinity. Compared with the former single nonionic surfactant systems, the degree of freedom increases

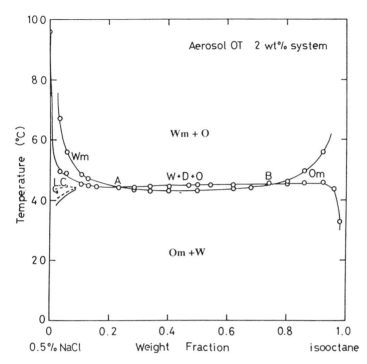

FIG. 4 Phase diagram of a 0.5 wt% NaCl aq./Aerosol OT/isooctane system as a function of temperature. The concentration of Aerosol OT is 2 wt%/system. L.C. is the region in which liquid crystal is present. Points A and B are maximum solubilization points in this phase diagram. Note that the effect of temperature on the phase behavior is opposite to that in nonionic surfactant systems.

and the three-phase temperature is a function of salinity. At constant temperature, a phase diagram similar to that of a polyoxyethylene-type surfactant is obtained by increasing salinity [29]. The effect of NaCl concentration on the three-phase temperature is shown in Figure 5 [4]. The three-phase temperature range tends to be wider when it appears at high temperature. This means that the maximum solubilization is reduced. Hence, as with single nonionic surfactant systems, when the three-phase behavior is less temperature sensitive, the solubilization is decreased in a single ionic surfactant system.

In ternary or pseudoternary systems including a single surfactant component, the three-phase body becomes temperature insensitive because of the separation of two critical end temperatures. However, the solubilization

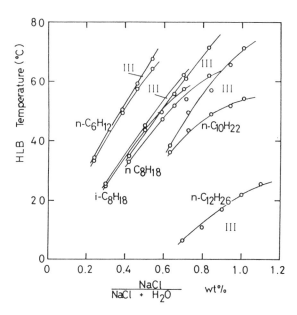

FIG. 5 Effect of salinity and type of oil on the three-phase temperature in brine/ Aerosol OT/oil systems. III means a three-phase region.

capability is decreased. It is difficult to form a temperature-insensitive three-phase microemulsion of large solubilization by using a single surfactant.

However, if an ionic surfactant of moderate HLB (hydrophile-lipophile balance) or a nonionic surfactant whose hydrophilic groups are different from the polyoxyethylene chain is used, temperature-insensitive microemulsions can be formed, although the solubilization capability is relatively low compared with three-phase microemulsions. For example, Fig. 6 shows the phase diagram of a water/didodecyldimethylammonium chloride/oil system as a function of temperature [30]. The solubilization limit of the water-swollen microemulsion (Om) is not changed much with the change in temperature. Because a three-phase microemulsion does not appear in this system, the maximum solubilization is not very large. Although polyoxyethylene-type nonionic surfactants are influenced by temperature, other type of nonionic surfactants whose hydrophilic groups are not polyoxyethylene chains are less affected. For example, sucrose ester–type nonionic surfactant does not show the cloud point phenomenon in water [31]. The microemulsion can be also temperature insensitive, as mentioned later.

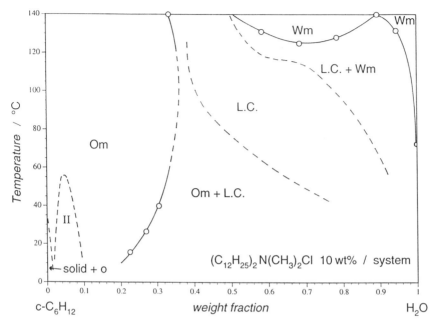

FIG. 6 Phase diagram of a water/didodecyldimethylammonium chloride/cyclohexane system as a function of temperature. L.C., liquid crystal.

III. BRINE/IONIC SURFACTANT/COSURFACTANT/ OIL SYSTEMS

Microemulsions are also produced by combining ionic surfactants and cosurfactants such as middle-chain alcohols, and lipophilic nonionic surfactants. Inorganic salts are also used for the formation of three-phase microemulsions in these systems. Lipophilic cosurfactants are usually required to form microemulsions of large solubilization because most ionic surfactants are very hydrophilic. Inorganic salts also suppress the dissociation of ionic hydrophilic groups of surfactants and function to make ionic surfactants less hydrophilic. As a result, inorganic salts promote cosurfactant penetration into the palisade layer of the surfactants in a water-oil system and the monomeric solubility of the cosurfactant in oil tends to decrease [8]. The salts also destroy lamellar liquid crystals, which are often produced in a mixed ionic surfactant–middle-chain alcohol system.

Average curvatures of surfactant layers for a middle-phase microemulsion and a lamellar liquid crystal are considered to be the same: zero. In fact, a lamellar liquid crystal swollen with water and oil often appears in the vicinity of the middle-phase microemulsion at a higher surfactant concentration in the phase diagrams [21,32,33]. In general, the solubilization capability of a longer chain ionic surfactant is much larger than that of a shorter chain one. However, when one uses a longer chain surfactant, the lamellar liquid crystal extends to the dilute region and isotropic microemulsions tend to disappear.

The stability of lamellar liquid crystals is highly related to the hydrocarbon chain length of amphiphiles [34,35]. It is known that the lamellar liquid crystal extends to the dilute region and coexists with the water phase in a water/ionic surfactant/cosurfactant (alcohol) system [36]. The liquid crystal melts and changes to an isotropic solution at higher temperatures. This melting temperature depends on the chain lengths of the ionic surfactant and cosurfactant. The correlation between the melting temperature of the lamellar liquid crystal and the hydrocarbon chain length of the surfactant and cosurfactant is shown in Table 1 [26]. Even if the sums of the hydrocarbon chain lengths of the ionic surfactant and cosurfactant are the same, the melting temperature of the lamellar liquid crystal is very low when the hydrocarbon chains are very different. Hence, the combination of a longer chain ionic surfactant and a shorter chain cosurfactant is good for forming isotropic microemulsions of large solubilization.

Even in this pseudoquaternary system, the three-phase behavior is in general temperature sensitive. Figure 7 shows the phase diagram of a brine/ionic surfactant/hexanol/isooctane system as a function of temperature [6]. At fixed salinity, the phase behavior looks similar to that in Fig. 3. However, the temperature effect is opposite and the surfactant becomes

TABLE 1   The Melting Temperature of Lamellar Liquid Crystal in Water/Ionic Surfactant/Alcohol Systems (°C)

	C_5OH	C_6OH	C_7OH	C_8OH	C_9OH	$C_{10}OH$	$C_{12}OH$
C_5SO_3Na	<0	<0	40.2	57.4	68.3	82.1	112.0
C_6SO_3Na	<0	47.9	70.9	84.5	94.2		
C_7SO_3Na	13.0	66.0	87.9	101.6	110.2		
C_8SO_3Na	27.0	78.0	100.8	114.8	123.4		
C_9SO_3Na	32.9	85.9	110.1	124.7	133.8		
$C_{10}SO_3Na$	34.6					148.1	
$C_{12}SO_3Na$	28.7	94.8					
$C_{14}SO_3Na$	20.4	87.1					
Sodium cholate						<0	
DTAB[a]	43.7	90.0					

[a]DTAB, dodecyltrimethylammonium bromide.

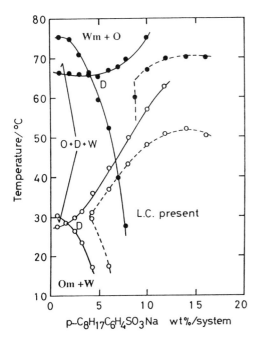

FIG. 7 Effect of concentration of p-$C_8H_{17}C_6H_4SO_3Na$ on the phase behavior of an equiweight mixture of brine [(○) 1 wt% NaCl; (●) 3 wt% NaCl] and 2,2,4-trimethylpentane (isooctane) containing 30 wt% hexanol as a function of temperature.

hydrophilic at higher temperatures. As with the polyoxyethylene-type nonionic surfactant system, the microemulsion in this system is temperature sensitive.

IV. TEMPERATURE-INSENSITIVE MICROEMULSIONS

A. Analysis of the Three-Phase Behavior in a Mixed Surfactant System

There have been attempts to combine two surfactants whose HLBs are change in opposite directions with temperature. However, when surfactants are mixed, their distributions in aggregates or a continuous medium are different, as shown schematically in Fig. 8. Figure 8 shows three coexisting phases in a mixed surfactant system. Usually, the monomeric solubilities of surfactants in water are negligibly small as far as three-phase behavior

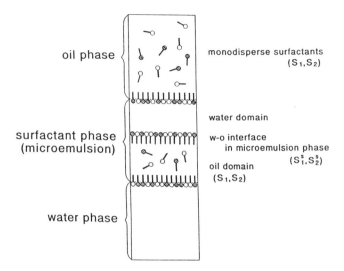

oil phase

surfactant phase
(microemulsion)

water phase

monodisperse surfactants
(S_1, S_2)

water domain

w-o interface
in microemulsion phase
(S_1^s, S_2^s)
oil domain
(S_1, S_2)

FIG. 8 Surfactant distribution in three coexisting phases.

is concerned. On the other hand, the monomeric solubility of a nonionic surfactant or cosurfactant in oil is often quite large. In the surfactant phase or middle-phase microemulsion, surfactants are distributed among the micro-water and micro-oil domains and the oil-water interface (surfactant layer). One can assume that the composition of the micro-water domain is the same as that in the excess water phase of the three-phase body: pure water. It can also be assumed that the composition of the micro-oil domain is the same as that of the excess oil phase.

If lipophilic (1) and hydrophilic (2) surfactants are mixed, the condition for forming the particular three-phase triangle in the midst of the three-phase body is represented by [37–39]

$$W_1 = S_1^S + \frac{S_1 S_2^S - S_2 S_1^S}{1 - S_1 - S_2} R_{\text{ow}} \left(\frac{1}{X} - 1 \right) \qquad (1)$$

where W_1 is the weight fraction of lipophilic surfactant in total surfactant, R_{ow} is the weight fraction of oil in water-oil, X is the weight fraction of total surfactant in the system, and S_1 and S_2 are the solubilities of lipophilic and hydrophilic surfactants in the excess oil phase, respectively. S_1 and S_2 can also be regarded as the surfactant concentrations in the micro-oil domain inside the microemulsion phase. S_1^S and S_2^S are the weight fractions of lipophilic and hydrophilic surfactants at the water-oil interface inside the microemulsion phase. In the case of an ionic surfactant–cosurfactant mixture, S_2 can be regarded as zero because ionic surfactant is practically insoluble in oil.

If the system contains n surfactants, the plane on which the particular three-phase tie triangle is situated is represented by [38,39]

$$W_i = S_i^S + \frac{S_i - S_i^S \Sigma_i S_i}{1 - \Sigma_i S_i}\left(\frac{1}{X} - 1\right)R_{ow} \qquad (i = 1, 2, \ldots, n) \qquad (2)$$

where W_i is the weight fraction of the ith surfactant in the total surfactant, S_i is the weight fraction of the ith surfactant in the excess oil phase, and S_i^S is the weight fraction of the ith surfactant at the water-oil interface inside the microemulsion. Equations (1) and (2) are applicable to both nonionic and ionic surfactant systems.

B. Mixture of Polyoxyethylene-type Nonionic Surfactants

Usually, a commercial polyoxyethylene-type nonionic surfactant is a mixture of homologues with different lengths of polyoxyethylene chains. In the case of a pure homogeneous nonionic surfactant system, the HLB temperature is constant even if the water/oil ratio and surfactant concentration are changed. However, the HLB temperature is not constant or the three-phase body is distorted in a mixed surfactant system as shown in Fig. 9 [13,40]. The difference in the distribution of each surfactant between aggregates and oil causes this distortion because the three-phase temperature is directly related to the mixing ratio of surfactant in aggregates or the water-oil interface inside the microemulsion phase in a polyoxyethylene-type nonionic surfactant system [38,40]. With increasing temperature, the monomeric solubilities in oil of the three-phase body for lipophilic surfactants tend to decrease, whereas those for hydrophilic surfactants tend to increase [40]. The stability of the macroemulsion becomes temperature insensitive in a commercial surfactant system compared with pure surfactant as shown in Fig. 9 because the HLB temperature of pure C_{12} EO$_5$ is fixed: 30.5°C. At constant temperature, S_i and S_i^S are constant because the surfactant mixing ratio, W_i, is fixed. Therefore, the compositions of the three-phase body and the single-phase microemulsion are changed considerably with increasing temperatures.

According to Eq. (2), the following relation holds for a commercial polyoxyethylene-type nonionic surfactant [40]:

$$T_{HLB} = B\left(\frac{1}{X} - 1\right)R_{ow} + \sum_i T_i W_i \qquad (i = 1, 2, \ldots, n) \qquad (3)$$

where T_{HLB} is the HLB temperature or the three-phase temperature in the

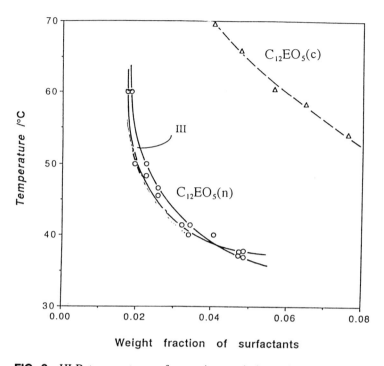

FIG. 9 HLB temperatures of water/pentaethylene glycol dodecyl ether/heptane systems as a function of surfactant concentration. The water/oil ratio is 50/50 (w/w). $C_{12}EO_5(c)$ is commercial surfactant. $C_{12}EO_5(n)$ is also commercial surfactant but unreacted alcohol was extracted. The HLB temperature for $C_{12}EO_5(c)$ was determined by electroconductivity.

mixed surfactant system and T_i is the HLB temperature of the ith surfactant in the given water-oil system. The B is originally a function of temperature but is almost constant for many commercial polyoxyethylene-type nonionic surfactant systems [38,41]. If a large amount of lipophilic surfactant, whose monomeric solubility in oil is large, is present, B tends to be large.

Figure 10 shows the midtemperature curve of the three-phase body (T_{HLB}) plotted against $1/X - 1$. We can obtain straight lines and the extrapolated value at the left-hand axis ($X = 1$) corresponds to $\Sigma T_i W_i$. This value should coincide with the values for the $C_{12}EO_5$ system regardless of whether the surfactant is homogeneous or a mixture. As shown in Fig. 10, the values approach each other. In contrast to the case of a pure homogeneous polyoxyethylene-type nonionic surfactant, the three-phase tem-

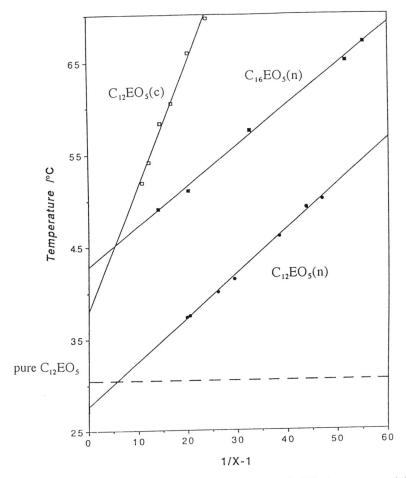

FIG. 10 HLB temperature plotted against $1/X - 1$. $C_{16}EO_5$ is a commercial pentaoxyethylene hexadecyl ether. The HLB temperature for the pure $C_{12}EO_5$ system is indicated by the dashed line.

perature of commercial surfactant is shifted toward higher values according to Eq. (3).

C. Mixture of Ionic-Nonionic or Ionic-Ionic Surfactants

On the other hand, S_1^s and S_2^s values do not change with temperature in a mixed system of nonionic and ionic surfactants [8], because, as mentioned before, temperature has opposite effects on the HLBs of ionic and polyoxyethylene-type nonionic surfactants. However, this does not mean that temperature-insensitive microemulsions can be obtained in mixed surfactant systems, because nonionic surfactants (cosurfactants) are soluble in the micro-oil domains in the microemulsion. According to Eq. (1), if both S_1, for nonionic surfactant, and S_1^S, the surfactant mixing ratio at the water-oil interface inside the microemulsion, do not change greatly with temperature, a temperature-insensitive microemulsion can be obtained. In ordinary brine/ionic surfactant/cosurfactant/oil systems, S_1 for the cosurfactant increases monotonically with increasing temperature although S_1^S does not change much. Hence, if a nonionic surfactant, whose solubility in oil is

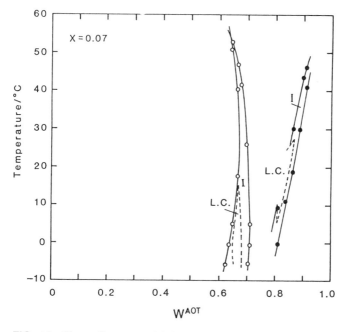

FIG. 11   Phase diagrams of brine [(1 wt% NaCl)/$C_{12}EO_6$/Aerosol OT/isooctane (O) and (1 wt% NaCl)/SDS/Aerosol OT/isooctane (●)] as a function of temperature. The weight fraction of total surfactant is 0.07 in the system. The water/oil weight ratio is 50/50. I, single-phase microemulsion; L.C., multiphase region including liquid crystal.

low, is used, temperature-insensitive microemulsions can be obtained as shown in Fig. 11. In Fig. 11, the combinations of $C_{12}EO_6$–Aerosol OT and sodium dodecylsulfate (SDS)–Aerosol OT are used. In the former system, $C_{12}EO_6$ acts as a hydrophilic surfactant and the monomeric solubility in oil is very low.

D. Mixture of Sucrose Monoalkanoate and Polyoxyethylene-type Nonionic Surfactants

As mentioned before, the hydrophile-lipophile property of polyoxyethylene-type nonionic surfactants is greatly influenced by temperature change, whereas the other type of nonionic surfactant is expected not to be highly influenced by temperature. Sucrose monoalkanoate is a strongly hydrophilic nonionic surfactant. The phase behavior of mixed sucrose monoalkanoate and tetraethyleneglycol dodecyl ether ($C_{12}EO_4$) is shown in Figure 12 [42]. The three-phase body is skewed upward and then shifted to higher concentrations at higher temperatures. Therefore, the three-phase body is not very temperature sensitive. Applying Eq. (1) to the system, we determined the monomeric solubility of $C_{12}EO_4$ in the excess oil phase and the surfactant mixing ratio S_1^S at the oil-water interface in the microemulsion. The results are shown in Fig. 13. The monomeric solubilities of sucrose monododecanoate in both water and oil are very small [42] and are negligible when the three-phase behavior is discussed. S_1 is increases monotonically with increasing temperature. This result is quite different from that for a polyoxyethylene-type nonionic surfactant mixture. In the latter system, the monomeric solubility of hydrophilic surfactant increases whereas that of lipophilic surfactant decreases with rising temperature. This change in the sucrose monoalkanoate system resembles that in an ionic surfactant–cosurfactant system [8].

Because the HLB temperature for $C_{12}EO_4$ alone is around 10°C, S_1^S approaches unity at lower temperatures. However, at higher temperatures, S_1^S becomes almost constant as shown in Fig. 13b. This means that the mixing ratio of surfactant at the water-oil interface is not changed with temperature. Hence, if more lipophilic surfactant, whose HLB temperature is below zero, is used, the mixing ratio may remain unchanged over a wide range of temperature. In addition, if the monomeric solubility of lipophilic surfactant is not changed much, temperature-insensitive microemulsions can be formed. In fact, a temperature-insensitive microemulsion of large solubilization is formed in a water/sucrose monoalkanoate/hexanol/oil system [43].

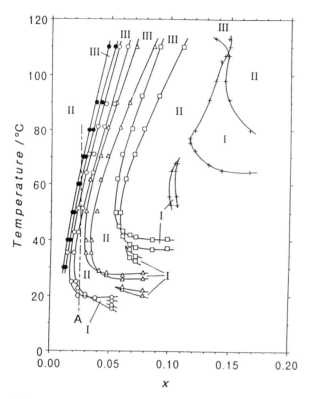

FIG. 12 Effect of temperature on the phase behavior of a water/sucrose monododecanoate/$C_{12}EO_4$/heptane system. The water/heptane ratio is 50/50 (w/w). The surfactant mixing ratios are fixed in each system. I, II, and III indicate one-, two-, and three-phase regions, respectively. W_1 = 0.5 (+), 0.6 (□), 0.7 (△), 0.8 (○), and 0.9 (●). W_1 is the weight fraction of $C_{12}EO_4$ in total surfactant.

V. SOLUBILIZATION CAPACITY OF MICROEMULSION IN A MIXED SURFACTANT SYSTEM

As mentioned before, the solubilization reaches its maximum when the three-phase body appears in a given water-oil system. It is important to know the effect of mixing the surfactants on the maximum solubilization. We compared the effect of mixing of surfactants on the maximum solubilization by using pentaoxyethylene dodecyl ether ($C_{12}EO_5$). Figure 14 shows the phase diagrams of homogeneous $C_{12}EO_5$ systems with different

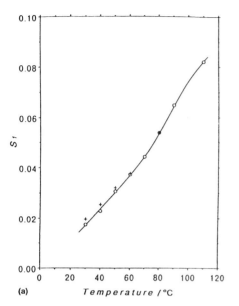
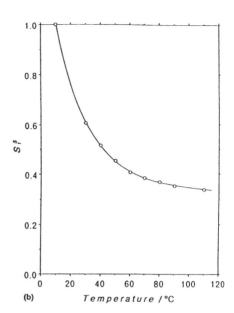

FIG. 13 (a) Monodisperse solubility of $C_{12}EO_4$, S_1, in the heptane phase in the three-phase region as a function of temperature. (○) Values calculated from Eq. (1); (+) experimental data. (b) S_1^s in a water/sucrose monododecanoate/$C_{12}EO_4$/ heptane system as a function of temperature.

oils. The three-phase temperatures are dependent on the type of oil. Point a shows the maximum solubilization point at which the microemulsion solubilizes equal weights of water and oil with the least surfactant, and its surfactant concentration is denoted by X_a. The HLB (point a) temperature is 52.9°C in a homogeneous water/$C_{12}EO_5$/hexadecane system [39,44]. To evaluate the solubilization capability, one has to compare the solubilization at the same temperature.

Figure 15 shows phase diagrams of mixed surfactant systems in which the temperature is adjusted to the HLB temperature for the corresponding homogeneous $C_{12}EO_5$ system [39]. In mixed surfactant systems, the phase behavior is changed by changing the mixing ratio of surfactants as a result of the increase in the degree of freedom. Point b in Fig. 15 also indicates the maximum solubilization point, and its surfactant concentration is denoted by X_b.

The solubilization capability can be evaluated from the following equations:

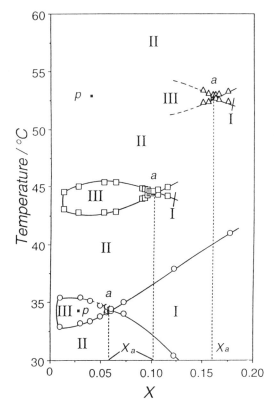

FIG. 14 Phase diagrams of water/pentaethylene glycol dodecyl ether ($C_{12}EO_5$)/ isooctane (○), water/$C_{12}EO_5$/dodecane (□), and water/hexadecane (△) systems as a function of temperature. The water/oil weight ratio is 50/50. X is the weight fraction of surfactant in the system. X_a is the composition at maximum solubilization at point a. I, II, and III indicate one-, two-, and three-phase regions, respectively. The whole three-phase body is not shown in a hexadecne system.

$$\frac{1 - X_a}{2X_a} \quad \text{for the homogeneous surfactant system} \qquad (4)$$

$$\frac{1 - X_b}{2X_b} \quad \text{for the mixed surfactant system}$$

or

$$\frac{1 - X_a}{2C} \quad \text{for the homogeneous surfactant system} \qquad (5)$$

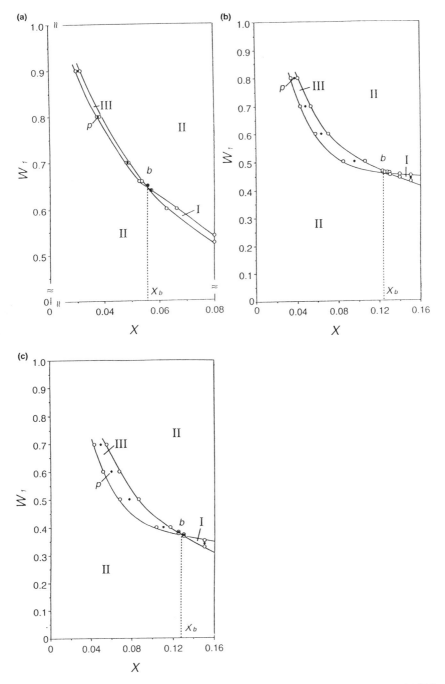

FIG. 15 Phase diagram of a water/$C_{12}EO_8$/$C_{12}EO_2$/isooctane system at 34.3°C (a), a water/$C_{12}EO_8$/$C_{12}EO_2$/hexadecane system at 52.9°C (b), and a water/$C_{12}EO_8$/$C_{12}OH$/hexadecane system at 52.9°C (c). Point b is the maximum solubilization point at each temperature.

TABLE 2 Apparent and Net Solubilization Capability of Polyoxyethylen-type Nonionic Surfactant[a]

Surfactant	Oil[a]	Temp. (°C)	$(1 - X_a)/2X_a$	$(1 - X_a)/2C$	
$C_{12}EO_5$	I	34.3	8.0	8.9	
$C_{12}EO_5$	H	52.9	2.6	2.8	
			$(1 - X_b)/2X_b$	$(1 - X_b)/2(C_1 + C_2)$	m, n[b]
$C_{12}EO_2/C_{12}EO_8$	I	34.3	8.4	15.8	12.0, 4.7
$C_{12}EO_2/C_{12}EO_8$	H	52.9	3.5	4.5	12.0, 5.0
$C_{12}OH/C_{12}EO_8$	H	52.9	3.4	4.8	12.0, 4.9
$C_{14}EO_5/C_{10}EO_5$	H	52.9	2.7	2.9	12.2, 5.0

[a] I, isooctane; H, hexadecane.
[b] m, n, the average hydrocarbon and polyoxyethylene chains of mixed surfactant at water-oil interface inside the microemulsion.

$$\frac{1 - X_b}{2(C_1 + C_2)} \quad \text{for the mixed surfactant system}$$

where C, C_1, and C_2 are the weight fraction of the surfactants in the system at the water-oil interface inside the microemulsion phase. Equation (4) indicates the apparent solubilization capability: the weight of water or oil solubilized in the microemulsion per unit weight of total surfactant. In this case, the monomeric surfactant in the micro-oil domains is not excluded. On the other hand, Eq. (5) gives the net solubilization capability: the weight of solubilized water or oil per unit weight of surfactant at the water-oil interface.

The results are shown in Table 2. The solubilization capability increases with mixing of surfactants as shown in Table 2. In particular, the net solubilization capability increases when surfactants with very different HLBs ($C_{12}EO_8$ and $C_{12}OH$) are mixed. However, in this case, because of the increase in monomeric solubility of dodecanol in oil, the apparent solubilization capability decreases compared with the mixed $C_{12}EO_8$ and $C_{12}EO_2$ system. Consequently, in order to obtain large solubilization, mixing of surfactants with very different HLBs and selection of a lipophilic surfactant whose monomeric solubility in oil is low are important.

The average hydrocarbon chain and polyoxyethylene chain lengths of mixed surfactant at the water-oil interface inside a microemulsion are shown in the last column of Table 2. The values are in good agreement with homogeneous surfactant, $C_{12}EO_5$. That is, the surfactant mixing ratio at the water-oil interface is directly related to the HLB temperature.

REFERENCES

1. K. Shinoda and H. Kunieda, J. Colloid Interface Sci. *42*:381 (1973).
2. M. Bourrel and R. S. Schechter, *Microemulsions and Related Systems*, Marcel Dekker, New York, 1988, chapter 1.
3. K. Shinoda and H. Kunieda, in *Microemulsions* (L. M. Prince, ed.), Academic Press, New York, 1977, chapter 4.
4. H. Kunieda and K. Shinoda, J. Colloid Interface Sci. *75*:601 (1980).
5. H. Kunieda and K. Shinoda, Yukagaku (J. Jpn. Oil Chem. Soc.) *29*:676 (1980).
6. H. Kunieda and S. E. Friberg, Bull. Chem. Soc. Jpn. *54*:1010 (1981).
7. H. Kunieda and K. Shinoda, Bull. Chem. Soc. Jpn. *55*:1777 (1982).
8. H. Kunieda, K. Hanno, S. Yamaguchi, and K. Shinoda, J. Colloid Interface Sci. *107*:129 (1985).
9. R. E. Anton, A. Graciaa, J. Lachaise, and J. L. Salager, J. Dispers. Sci. Technol. *13*:565 (1992).

10. K.-H. Oh, J. R. Baran, Jr., W. H. Wade, and V. Weerasooriya, J. Dispers. Sci. Technol. *16*:165 (1995).
11. U. Olsson, K. Shinoda, and B. Lindman, J. Phys. Chem. *90*:4083 (1986).
12. R. Strey and W. Jahn, J. Phys. Chem. *92*:2294 (1988).
13. H. Kunieda and K. Shinoda, J. Colloid Interface Sci. *107*:107 (1985).
14. G. Karlström, J. Phys. Chem. *89*:4962 (1985).
15. H. Kunieda, Bull. Chem. Soc. Jpn. *56*:625 (1983).
16. H. Kunieda and M. Yamagata, Colloid Polym. Sci. *271*:997 (1993).
17. H. Kunieda and K. Shinoda, Bull. Chem. Soc. Jpn. *56*:980 (1983).
18. H. Kunieda, J. Colloid Interface Sci. *116*:224 (1988).
19. K. E. Bennett, H. T. Davis, and L. E. Scriven, J. Phys. Chem. *86*:3917 (1982).
20. H. Kunieda, H. Asaoka, and K. Shinoda, J. Phys. Chem. *92*:185 (1988).
21. H. Kunieda, K. Nakamura, and A. Uemoto, J. Colloid Interface Sci. *163*:245 (1994).
22. K. Shinoda and H. Saito, J. Colloid Interface Sci. *26*:70 (1968).
23. H. Saito and K. Shinoda, J. Colloid Interface Sci. *32*:647 (1970).
24. P. A. Winsor, *Solvent Properties of Amphiphilic Compounds*, Butterworths, London, 1954, p. 68.
25. R. L. Reed and R. N. Healy, in *Improved Oil Recovery by Surfactant and Polymer Flooding* (D. O. Shah and R. S. Schechter, eds.), Academic Press, New York, 1977, p. 383.
26. H. Kunieda and K. Nakamura, J. Phys. Chem. *95*:1425 (1991).
27. H. Kunieda and K. Nakamura, J. Phys. Chem. *95*:8861 (1991).
28. H. Kunieda, H. Ito, S. Takebayashi, and M. Kodama, Colloid Polym. Sci. *271*: 952 (1993).
29. K. Shinoda and H. Kunieda, J. Colloid Interface Sci. *118*:586 (1987).
30. H. Kunieda and K. Shinoda, J. Colloid Interface Sci. *70*:577 (1979).
31. M. J. Rosen, *Surfactants and Interfacial Phenomena*, Wiley, New York, 1989, chapter 1.
32. H. Kunieda and K. Shinoda, J. Dispers. Sci. Technol. *3*:233 (1982).
33. F. Lichterfeld, T. Schmeling, and R. Strey, J. Phys. Chem. *90*:5762 (1986).
34. H. Kunieda and F. Harigai, J. Colloid Interface Sci. *134*:585 (1990).
35. H. Kunieda, in *Mixed Surfactant Systems* (K. Ogino and M. Abe, eds.), Marcel Dekker, New York, 1993, chapter 8.
36. P. Ekwall, L. Mandell, and K. Fontell, Mol. Liquid Cryst. *8*:157 (1969).
37. H. Kunieda and Y. Sato, in *Organized Solutions* (S. E. Friberg and B. Lindman, eds.), Marcel Dekker, New York, 1992, pp. 67–88.
38. H. Kunieda and M. Yamagata, Langmuir *9*:3345 (1993).
39. H. Kunieda, A. Nakano, and M. Akimaru, J. Colloid Interface Sci. *170*:78 (1995).
40. M. Muto, N. Naito, and H. Kunieda, Yukagaku (Jpn. Oil Chem. Soc.) *43*:502 (1994).

41. H. Kunieda and N. Ishikawa, J. Colloid Interface Sci. *107*:122 (1985).
42. H. Kunieda, N. Ushio, A. Nakano, and M. Miura, J. Colloid Interface Sci. *159*:37 (1993).
43. M. A. Pes, K. Aramaki, and N. Nakamura, J. Colloid Interface Sci., submitted.
44. H. Kunieda, M. A. Pes, and A. Nakano, Langmuir, in press.

3

The Applications of Microemulsions for Analytical Determinations

A. SHUKLA   Department of Chemistry, Southeast College, Houston, Texas

SHYAM S. SHUKLA   Department of Chemistry, Lamar University, Beaumont, Texas

J. L. MARGRAVE   Department of Chemistry, Rice University, Houston, Texas

I. INTRODUCTION

Titrimetry has been the most extensively used procedure in chemical analysis since 1800s. The inherent advantages of titrimetry over the direct instrumental evaluation of a physiochemical property are accuracy, reliability, ease and rapidity of performance, possible specificity through adjustment of conditions, adaptability to a large variety and range of sizes of samples, and simplicity of interpreting the results. For example, titrimetric methods are preferred over gas chromatographic methods for determination of purity of pharmaceutical compounds [1]. Underwood [2] and Pellizetti and Pramauro [3] titrated a number of nearly water-insoluble organic compounds in micellar media. These authors found that potentiometric titrations are sluggish owing to drift in the pH value measured, and rapid equilibrium was seen in both visual and photometric titrations. In 1985, Shukla and Meites [4] used thermometric titrimetry to titrate such kinds of compounds and got rapid and accurate results.

There are three major reasons why nonaqueous solvents are of analytical interest. The first is that they can dissolve sparingly water-soluble compounds. The second arises from the leveling effect of water [5]. Leveling effect can be decreased or eliminated by use of aprotic solvents. If the proton in an acid transfers to the solvent completely, the acid is said to be leveled. For example, in acetonitrile, the transfer is not complete for hydrochloric acid and it appears to be a weak acid. The third is that an acid that is too weak for an aqueous titration can be titrated in nonaqueous media. Acidity of such compounds can be drastically modified, and their titratability would be increased by choosing a solvent more basic than water. A similar strategy could be adopted for weak bases as well.

Even with all these advantages, there are some serious drawbacks to these titrations, most of which can be ascribed to the solvents themselves. Ideally, an amphiprotic solvent should possess a high dielectric constant and also a low value of the autoprotolysis constant, since the magnitude of the latter determines the range of potentials that will be available in that solvent. Unfortunately, these two ideals are not mutually compatible [6]. Aprotic solvents, apart from their weak solvating effects, do not interact with either acids or bases.

It is worthwhile to note that the majority of nonaqueous solvents are toxic, flammable, or otherwise hazardous. In today's environmentally conscious world, there is a problem of disposing of nonaqueous solvents after their use. The cost of these solvents is generally high and so is the cost of disposal. Micellar systems and microemulsions could be used for the same purposes as nonaqueous solvents but without encountering the toxicity, high cost, and disposal and technical problems associated with the latter.

II. HISTORICAL BACKGROUND

A. Solubilization and pK_a Shifts in Micelles and Microemulsions

Surfactant molecules, such as cetyltrimethylammonium bromide (CTAB) and sodium dodecyl sulfate (SDS), have a long hydrophobic residue and a polar head group. In aqueous media, the hydrophobic residues aggregate to form a micellar interior or core, while the head groups and some of the counterions [2] remain at the surface.

Nonpolar compounds, such as aliphatic hydrocarbons, are dissolved in the core of the micelle. Semipolar and polar compounds, e.g., fatty acids and alkanols, are taken up in what is often termed the palisade layer of the micelle, oriented with their hydrophobic moieties toward the center of the micelle and their polar groups in its surface [7,8]. If the sparingly water-soluble compound has a polar functional group, the functional group tends to be located in the vicinity of the head group of the surfactant molecules and its dissociation is, consequently, affected by the electrostatic field generated by the surfactant head groups. Thus, the pK_a value of the functional group is shifted in a direction commensurate with the electrostatic charge on the surfactant group. For acidic functional groups in a cationic micellar system, the pK_a values are expected to be lower, and the reverse occurs in an anionic micelle. For bases the opposite is true.

Microemulsions [10] were introduced by Hoar and Schulman [11] in 1943, and the systems were also called swollen micellar solutions and transparent emulsions. They have properties similar to those of micellar solutions but their solubilizing capacity is far superior.

The microemulsions are clear, thermodynamically stable dispersions of two immiscible liquids containing appropriate amounts of surfactants or surfactants and cosurfactants, and these systems can be classified as water in oil (W/O) and oil in water (O/W). Several structural models of microemulsions have been proposed, and a useful compilation of different studies on this subject may be found in the books by Robb [12] and Shah and Schechter [13]. For W/O microemulsions, the hydrocarbon tails of the surfactant and cosurfactant dissolve in the oil phase while the polar heads are buried in the water phase. The opposite is true for the O/W microemulsions, in which tails of the surfactant (and cosurfactant) are embedded in the oil microdroplets while the polar groups are oriented toward the bulk water. Thus, O/W microemulsions are very similar to aqueous micellar structures containing a hydrophobic interior and polar head groups at the surface oriented toward bulk water. Generally speaking, the dispersed phase consists of small droplets with a diameter in the range of 10–100 nm as opposed to a typical diameter of micelles in the range of 2–4 nm. It is one of the reasons why the microemulsion can solubilize larger solubilizates at relatively high concentrations. A number of factors influence solubilization. The more important ones are the types of surfactants, nature of the solubilizate, temperature, and added electrolytes and nonelectrolytes. It is important that measurement of the maximum solubility be subject to adequate temperature control.

B. Techniques for Detection of End Points

Many techniques are available for detecting end points. Watters [14] has reviewed more than 20. The techniques most commonly used for detection of end points are visual, potentiometric, and spectrophotometric. The potentiomatric technique is widely applicable and provides results that are inherently more accurate than those obtained with corresponding methods employing indicators. It is particularly useful for titration of colored or turbid solutions and for detecting the presence of unsuspected species in a solution. However, potentiometric titrations, in which the instrument response depends on the logarithm of concentration via the Nernst equation, show a sigmoidal type of titration curve in which the end point coincides with the point of inflection, if the pK_a values are in the range 4–10. The linear portions of a photometric titration curve are actually positioned more accurately far from the equivalence point where reaction between A and B may be considered complete. A number of authors [15,16] have estimated the limitations of the potentiometric method and their findings have been summarized by Kolthoff and Furman [17]. The precision with which a titration may be performed is governed by the product $C_a K_t$, where C_a is the concentration and K_t is the overall dissociation constant. Limits of feasibility for visual, potentiometric, and photometric titrations are shown in Table 1 [18]. These values were obtained through practical experience.

TABLE 1 Minimum Values of $C_a K_t$ Required for a Feasible Titration

Technique	$C_a K_t$
Visual	10^6
Potentiometric	10^4
Photometric	10^2
Thermometric	10^2

The fundamental law of monochromatic light absorption is Beer's law. The most important consequence of Beer's law for photometric titration is that the absorbance, the quantity directly measured on a spectrophotometer, is proportional to the concentration of the absorbing species linearly rather than logarithmically as in potentiometric methods. It was reasoned that a sharp break might be observed in systems in which the change in hydrogen ion concentration would ordinarily be too small. The high sensitivity of the photometric method makes it attractive for the titration of very dilute solutions.

The shape of a photometric titration curve depends on the absorption properties of the reactant, titrant, and products of the reaction at the wavelength used. The shapes of such kinds of self-indicating photometric absorbance titration curves are described in undergraduate analytical chemistry textbooks.

The equation of a type "a" curve has been deduced by L'Her [19]. For the reaction A + B = C, assume that C is the only absorbing species at the wavelength used and that the absorbance is corrected for volume changes using the factor $(V + v)/V$ (where v is the volume of titrant B added and V is the initial volume of A). The absorbance increases linearly up to the equivalence point and then remains constant. The end point is the intersection of two straight lines. However, a finite value of K_t tends to round off the end-point breaks; but, because no particular importance is attached to points taken in the near vicinity of the equivalence point, the straight-line portions of the curve may be extended to locate the end point accurately.

If the color change is not sharp or if other substances present interfere with the detection of the end point, indicator titration may be advantageous. Goddu and Hume [20], in their 1954 paper on self-indicating acid-base titration, described the proper use of indicators in photometric titrations. They suggested that the indicators should not begin the color change until very near the equivalence point, and usually a relatively large quantity of indicator is added, so that the whole amount is not converted. Thus, for strong acid (titrand)–strong base titrations they advocated the use of thymol blue ($pK_a = 9.0$) rather than bromthymol blue ($pK_a = 7.1$).

From 1956 the theory of acid-base photometric titration has been dominated by Higuchi [21] and his school. His highly successful approach has been to express the ratio of the acid form to the base form of an indicator as a linear function of the volume of titrant added and the equilibrium constant for the indicator-titrand interaction using data obtained some distance away from the equivalence point.

Consider the titration of a weak acid HA with a strong base; the indicator-titrand reaction may be written as the proton exchange process

$$A^- + IH^+ = HA + I \tag{1}$$

$$K_{ex} = \frac{[HA][I]^-}{[A][IH^+]} = \frac{K_i}{K_a} = K \tag{2}$$

where K_i and K_a are the indicator and titrand dissociation constants, respectively. It is advantageous to select an indicator value 2 or 3 units greater than the acid pK value. For $K = 0.001$, $[IH^+]/[I]$ may easily be determined to an accuracy of 1 part per 100. It is very easy to locate the end point.

For maximum sensitivity in a photometric titration, the wavelength of light used should be that which produces maximum absorption by the most strongly absorbing species. Frequently, however, such sensitivity is not required and other wavelength giving lower values may be used to avoid interfering absorbers or provide a wider concentration range. This is important because serious photometric errors may result in high-absorbance regions [19]. Any real spectrophotometer, because of an imperfect monochromator, will give low readings for high-absorbance samples. Any light leakage will increase this effect, which appears as an apparent decrease in absorbance index with concentration (failure of Beer's law).

Thermometric titrations are conceptually simple. The course of the titration is followed by monitoring the change in temperature associated with the reaction. Since almost all reactions are accompanied by changes of enthalpy, the method is of universal applicability. The temperature is measured with a thermistor. The solutions do not need to be transparent. The titration curves are segmented like the curves in spectrophotometric titrations and have the same basic features.

C. Mathematical Modeling of Potentiometric Titration Data

An excellent critical discussion of linear and nonlinear regression analysis in analytical chemistry has been published by Meites [22]. To obtain ionization constants from potentiometric acid-base titration data, a mathematical model can be designed and used to fit the experimental data. This can be done by minimizing the sum of the squared deviations between experimental and calculated data. The general theory behind the derivation of the equation for acid-base titration can be found in Daniels and Alberty [23].

The mathematical model for the titration of a weak monobasic acid with a strong base is

$$f(X) = X^3 + AX^2 - BX - C \tag{3}$$

where

$$A = C_s V_s / (V_s + V_r) \cdot f + K_a$$
$$B = C_s \cdot (1 - f) + V_s K_a / (V_s + V_r) + K_w$$
$$C = K_a \cdot K_w$$

C_s = concentration of titrand
V_s = volume of titrand
C_r = concentration of titrant
V_r = volume of titrant
K_a = dissociation constant of titrand (acid)
K_w = ionization constant of water
$f = C_r V_r / C_s V_s$

In this model, the assumed activity coefficients are 1 and the parameters are K_a, K_w, C_s, and C_r. Based on these equations, a subroutine can be written and used with the multiparametric curve-fitting program CFT4A written by Meites [22].

The subroutine is used to calculate the hydrogen ion concentration, and this value is sent to the main body of the program to minimize the sum of the squared deviations. After convergence, the program provides the correct values of parameters that fit the mathematical model. This method is a precise way to get good results; however, it is important to design a proper model and set appropriate initial values.

D. Research Goals

SDS and CTAB are inexpensive and nontoxic chemicals. They could be used to prepare the aqueous micellar solutions and O/W microemulsions as solvents to solubilize sparingly water-soluble compounds. The amount of solubilizate that can be solubilized is usually directly proportional to the concentration of the surfactant in the micellar form. For larger solubilizates, the solubilization process may be inefficient. In this respect, microemulsions are especially useful. The first part of the present work is, therefore, directed toward the study of the solubilizing capacity of both micellar systems and microemulsions. The major goal is to examine the utility of microemulsions for analytical determinations. Titrimetric determinations were chosen to be performed. A series of acids and bases of varying strength in these media were titrated. Also, this research examined the range of concentrations that can be determined by titrimetric analysis.

A mathematical model was designed and used with nonlinear regression analysis to obtain dissociation constants of acids and bases from potentiometric acid-base titration data. From these values of dissociation constants, the pK_a shift properties of micellar systems and microemulsions could be studied quantitatively. This approach with an appropriate technique should allow the titration of very weak acids and bases that are sparingly water soluble.

Surfactant micelles and microemulsions have been used as solvents to perform titrimetric analyses by our groups for several years. Development of analytical procedures that parallel those of nonaqueous titrimetry but have several advantages over them seems promising. If useful, these procedures could be adopted for routine analyses and replace most of the analyses that require nonaqueous solvents.

III. EXPERIMENTAL PROCEDURES

A. Materials

CTAB and SDS were obtained from the Aldrich Chemical Co. Microemulsions used for the titration were O/W type and were prepared using oil, water, pentanol, and either CTAB [CTAB/hexadecane/water/butanol in a 17.8:4.0:60.0:18.2 (w/w/w/w) ratio] or SDS [SDS/paraffin oil (Saybolt viscosity 335/365)/water/pentanol in a 9.22:4.21:60.34:26.23 (w/w/w/w) ratio]. These formulations were arrived at by trial and error. The correction for the blanks was sometimes significant, especially for lower concentration of titrand.

Other chemicals were obtained from Aldrich Chemical Co., Sigma Chemical Co., and Matheson Coleman & Bell. They were ordinary reagent grade and were used without further purification. Standard solutions of hydrochloric acid and sodium hydroxide were prepared in conventional ways, with due care exercised to avoid contamination by carbon dioxide, and were standardized by using potassium hydrogen phthalate as a primary standard. The indicators were dissolved in a mixture of 9:1 (v/v) undenatured ethanol (95% pure)/water. The concentration of phenolphthalein was 0.1% (w/v) and of bromcresol green 0.2% (w/v).

B. Titration Procedure

The solubilities of acetophenone and benzophenone in water and surfactant solutions were determined by measuring the absorption of saturated solutions spectrophotometrically at the absorption maxima for the π-π* transition at 26°C. Solutions were prepared by shaking an excess of substrate with solution for 2.5 h. The solutions were then allowed to equilibrate overnight and then centrifuged at low speed for 10 min. Aliquots were removed from each sample and diluted for absorption measurement. Absorption spectra of acetophenone and benzophenone in different systems were obtained on a Cary 17 spectrophotometer using 1.00-cm matched quartz cells. Accuracy of the absorption maxima is considered to be ± 0.5 nm.

The spectrophotometer used for photometric titrations was a Bausch & Lomb Spectronic 20. It can be used only in the visible range. The spectrophotometer was set at the desired wavelength and allowed to warm up, and the dark current was adjusted. The instrument can be set to real zero absorbance or some other starting value. As only changes in absorbance are involved in locating the end point, the absolute value of the absorbance need not be determined as long as it is within the range of linearity of the instrument. For convenience, the absorbance is set at zero for the starting solution. In our titration system, the instrumental cell was too small to hold the whole volume of titrand and titrant. The same volume of titrand (10 mL) was put into every test tube and added to each with the same amount of indicator (if necessary). The absorption of each solution was recorded after each addition of titrant. In some cases, 40 mL of titrand was put into a beaker and large amounts of titrant were added successively. After each addition, the solution was mixed thoroughly, and an aliquot of this solution was transferred to the cell. Once the absorbance was recorded, the solution was poured back into the beaker. Great care was taken to avoid loss of the solution during transfer.

The pH measurements were performed with an Orion 91-05 pH meter and a combination electrode. A sluggish response was observed; readability in the expanded scale mode was about ± 0.002 pH. Potentiometric titrations were done at room temperature ($26 \pm 2°C$). Each sample was titrated at least three times. Some of the chemicals were difficult to dissolve in surfactant solutions; therefore, stirring or breaking with supersonic waves became necessary.

Mathematical analysis of potentiometric titration data was performed using a nonlinear regression analysis program called CFT4A written in BASIC. The program is written in such a way that it can be adopted for different situations by simply writing a subroutine to describe the system. Like linear regression analysis, CFT4A minimizes the sum of the squared deviations. The output of the program contains the values of the parameters that best describe the data.

For thermometric titrations, the solutions were placed in a thermostatted jacket. Titrant was delivered by a syringe pump. The titrant was brought into the titrand vessel through coiled tubing surrounding the vessel. The tip of the thermistor was placed in the solution and constituted one arm of a Wheatstone bridge. The three arms of the bridge were simple resistors.

After the solution was equilibrated to the temperature of the jacket, delivery of titrant began. The unbalanced voltage of the bridge was fed to a recorder or a computer to obtain the titration curve.

IV. RESULTS AND DISCUSSION

The solubilities [24] of acetophenone and benzophenone are given in Table 2. The ratio of the solubility in surfactant solution to that in water, S_s/S_w, is also shown. S_s is the solubility in organized media and S_w is that in water. It is clear from Table 2 that the solubilities increase in the order CTAB microemulsion > 0.1 M CTAB > water. The microemulsion in particular exhibits very high solubilizing capacity. It has also been observed that the extent of solubility increases with increasing length of hydrophobic residues [25,26], and 0.1 M CTAB produced high solubilities compared with 0.1 M SDS. Acetophenone is smaller than benzophenone; perhaps for this reason, the former has higher solubility than the latter in the same solvent.

For titrations, two techniques for determining end points have been chosen. These techniques are potentiometric and spectrophotometric. Owing to slight acidity (CTAB) and basicity (SDS) of the surfactant solutions, their blank values could be negligible in the titration at high concentrations of titrand. However, larger errors would be obtained when the concentration is lower than 0.01 M. Thus, the blank values are an important factor in titrations at lower concentrations. For convenience, the blank values were measured by potentiometric titrations and the results were corrected for the contribution from the blanks.

TABLE 2 Solubilities of Acetophenone and Benzophenone in Micellar Systems and Microemulsions

Solvent	$\lambda \pm 0.5$ nm (maximum)	Solubility $\pm$ 10% M (at 26 $\pm$ 0.5°C)	S_s/S_w
		Acetophenone	
Water	246	0.0535	
0.1 M CTAB	244	0.2967	5.55
CTAB microemulsion	242.5	13.53	253
		Benzophenone	
Water	257	0.001068	
0.1 M CTAB	255	0.1127	105
CTAB microemulsion	253.5	0.798	747
0.1 M SDS	257	0.0333	31.2

5757575757575757

ganic acids and phenols in a CTAB microemulsion and amines in an SDS
microemulsion are given in Tables 3 and 4. The concentrations of acids
and phenols were approximately 0.1–0.05 M. The accuracy obtained in
each titration is generally good. Results of similar titrations at lower concentrations are shown in Tables 5 and 6. The pKa values of the compounds
are listed in Table 7. According to these pK_a values, the phenols are the
weak acids. However, their titration in a surfactant system did not pose
any undue problem. Generally, it is not difficult to locate the end points.
However, the end points for the titrations of 0.001 M analyte in CTAB
microemulsions are not sharp, and we may encounter larger errors in locating them. The blank appears to be the source of the problem.

Results of these potentiometric titrations showed us that the surfactant
systems could be used as an alternative to nonaqueous solvents to solubilize the sparingly water-soluble organic compounds and to perform analytical determinations. Generally, detection of end points was slightly better

TABLE 3 Potentiometric Titrations of Organic Acids and Phenols (~0.1 M) in
O/W Microemulsions Containing CTAB and Butanol

| Compound titrated | Amount (mmol) | | Error (%) |
	Taken	Found	
Capric acid	2.5000	2.5338	1.35
	2.5000	2.4750	−1.72
	2.5000	2.4623	−1.51
		Average result = −0.63 ± 1.72	
1-Naphthylacetic acid	2.5000	2.4899	−0.40
	2.5000	2.4675	−1.30
	2.5000	2.4780	−0.88
		Average result = −0.86 ± 0.45	
p-Toluic acid	2.5000	2.4675	−1.30
	2.5000	2.4612	−1.55
		Average result = −1.43 ± 0.12	
Palmitic acid	2.5000	2.4824	−0.71
	2.5000	2.4713	−1.15
		Average result = −0.93 ± 0.23	
p-Nitrophenol	2.5000	2.4913	−0.35
	2.5000	2.4733	−1.07
	2.5000	2.4824	−0.71
		Average result = −0.71 ± 0.36	
		Average of all results = − 0.89 ± 0.81	

TABLE 4 Potentiometric Titration of Amines (~0.05 M) in O/W
Microemulsions Containing SDS and Pentanol

Compound titrated	Amount (mmol)		Error (%)
	Taken	Found	
n-Octylamine	1.2575	1.2375	−1.59
	1.2575	1.2500	−1.60
	2.2575	1.2425	−1.19
		Average result = −1.13 ± 0.50%	
n-Decylamine	1.2500	1.2100	−3.20
	1.2500	1.2325	−1.40
	1.2500	1.2325	−1.40
		Average result = −2.00 ± 1.04%	
Triethylamine	1.4863	1.5163	+2.02
	1.4863	1.5163	+2.02
	1.4863	1.5215	+2.37
		Average result = +1.90 ± 0.54	
		Average of all results = −0.41 ± 1.88%	

in micellar solutions than in microemulsions. However, the solubility in
the former is less than in the latter (e.g., a 0.01 M solution of palmitic acid
cannot be prepared in 0.1 M CTAB).

Photometric titrations are defined as those in which the absorbance of
the solution, varying with the changing concentrations of absorbing solutes,

TABLE 5 Potentiometric Titration of Amines in O/W Microemulsions
Containing SDS and Pentanol

Compound titrated	Amount (mmol)		Error (%)	Average result (%)
	Taken	Found		
Triethylamine	0.4080	0.4102	+0.49	
	0.4080	0.4082	+0.05	
	0.4080	0.4071	+0.24	+0.26
n-Octylamine	0.4000	0.3999	−0.25	
	0.4000	0.3958	−1.05	
	0.4000	0.3979	−0.53	−0.61
n-Decylamine	0.4000	0.3979	−0.53	
	0.4000	0.3959	−1.00	
	0.4000	0.3959	−1.00	−0.84

TABLE 6 Potentiometric Titration of Organic Acids and Phenols in O/W
Microemulsions Containing CTAB and Butanol

Compound titrated	Amount (mmol)		Error (%)	Average result (%)
	Taken	Found		
p-Toluic acid	0.4000	0.3942	−1.45	
	0.4000	0.3962	−0.95	
	0.4000	0.3952	−1.20	−1.20
	0.04000	0.03860	−3.50	
	0.04000	0.03860	−2.45	
	0.04000	0.03947	−1.28	−2.41
1-Naphthylacetic acid	0.4000	0.3962	−0.95	
	0.4000	0.3982	−0.45	
	0.4000	0.3992	−0.70	−0.70
	0.04000	0.04018	+0.45	
	0.04000	0.04058	+1.45	
	0.04000	0.03977	−0.58	+1.32
Capric acid	0.4000	0.3942	−1.45	
	0.4000	0.3982	−0.45	
	0.4000	0.3982	−0.45	−0.78
	0.04000	0.04099	+2.48	
	0.04000	0.03979	−0.52	
	0.04000	0.04139	+3.48	+1.81
Palmitic acid	0.3960	0.3904	−1.40	
	0.3960	0.3949	−0.27	
	0.3960	0.3927	−0.83	−0.83
	0.03960	0.03954	−0.15	
	0.03960	0.03994	+0.86	
	0.03960	0.04014	+1.36	+0.69
p-Nitrophenol	0.4000	0.4022	+0.55	
	0.4000	0.3982	−0.45	
	0.4000	0.4002	−0.05	+0.00
	0.04000	0.04018	+0.45	
	0.04000	0.03977	−0.58	
	0.04000	0.04058	+1.45	−0.44

is plotted against volume of titrant added. This method commonly presents
a sharper break than potentiometric titration. Thus this method can be used
for systems in which the change in hydrogen ion concentration would be
small.

All spectrophotometric titrations were performed in the visible region
of the electromagnetic spectrum. Of all of the compounds used for titration,

TABLE 7 pK_a Values of Some of the Organic Compounds in Aqueous Solutions

Compounds	pK_a [27]	Temperature (°C)
Triethylamine	11.01	18
Octylamine	10.65	25
n-Decylamine	10.64	25
o-Nitrophenol	7.17	25
p-Nitrophenol	7.15	25
p-Toluic acid	4.36	25
1-Naphthylacetic acid	4.24	25
n-Capric acid	~5	
Palmitic acid	~5	

only *para*- and *ortho*-nitrophenols can absorb visible light. For the others, we added indicators to detect the end points. For weak acids, the choice of indicators is limited. Phenolphthalein is generally suitable in this case because of its high pK_a value, 9.7 [5]. In 1970, Jagner [28] used bromcresol green indicator to titrate sulfate in seawater and obtained good results, which showed how to detect the end point in a photometric titration of a weak base with an acid. The blank of the CTAB surfactant solution showed trace amounts of acidity. Therefore, the blank was measured by photometric titration and the results were appropriately corrected. Tables 8 and 9 summarize the results obtained in photometric titrations of several amines and organic acids.

TABLE 8 Photometric Titration of Amines in O/W Microemulsions Containing SDS and Pentanol (Bromcresol Green Indicator)

Compound titrated	Amount (mmol)		Error (%)	Average result (%)	Wavelength (nm)
	Taken	Found			
Triethylamine	0.3060	0.3106	+1.5		
	0.3060	0.3045	+0.49		
	0.3060	0.3045	+0.49	+0.83	531
n-Octylamine	0.3000	0.2984	−0.53		
	0.3000	0.2954	−1.53		
	0.3000	0.2923	−2.57	−1.54	531
n-Decylamine	0.3000	0.2984	−0.53		
	0.3000	0.2996	−0.13		
	0.3000	0.2984	−0.53	−0.40	531

TABLE 9 Photometric Titration of Organic Acids and Phenols in O/W Microemulsions Containing CTAB and Butanol

Compound titrated	Amount (mmol)		Error (%)	Average result (%)	Wavelength (nm)
	Taken	Found			
p-Toluic acid	0.3000	0.2986	−0.47		
	0.3000	0.2975	−0.83		
	0.3000	0.2975	−0.83	−0.71	562
	0.3000	0.3012	+0.40		
	0.3000	0.3024	+0.80		
	0.3000	0.3068	+1.07	+0.76	562
1-Naphthylacetic acid	0.3000	0.2986	−0.47		
	0.3000	0.2986	−0.47		
	0.3000	0.2975	−0.83	−0.59	562
	0.03000	0.03086	+2.87		
	0.03000	0.03042	+1.40		
	0.03000	0.02974	−0.87	+1.13	562
Capric acid	0.3000	0.2969	−1.03		
	0.3000	0.2975	−0.86		
	0.3000	0.2986	−0.47	−0.79	562
	0.01000	0.0982	−0.80		
	0.01000	0.0990	−1.40		
	0.01000	0.0986	−1.80	−0.13	563
Palmitic acid	0.1000	0.0998	−0.20		
	0.1000	0.0998	−0.20		
	0.1000	0.1002	+0.20	−0.06	563
	0.01000	0.01021	+2.10		
	0.01000	0.01014	+1.40		563
	0.01000	0.01014	+1.40	+1.63	
o-Nitrophenol	0.1000	0.0984	−1.60		
	0.1000	0.0986	−1.40		
	0.1000	0.0982	−1.80	−1.60	531
p-Nitrophenol	0.1000	0.0985	−1.50		
	0.1000	0.0981	−1.90		
	0.1000	0.0985	−1.50	−1.63	484

Photometric titrations yielded good accuracy and precision. They exhibited sharp breaks in curves even for titrations that gave some trouble in the potentiometric method. In addition, there is no problem such as drift in the pH response in potentiometric titrations. The method works for low concentrations and small quantities, and it is rapid.

It is worthwhile to note that many titrations would not even be feasible if the pK shift of the surfactant medium were absent. For example, it is well established [4] that surfactant systems shift the pK of acids and bases [2–4]. The pK values of amines and acids are respectively increased and decreased by as much as 1.5 units, causing their strength to increase. In Table 6, the lower amount of p-nitrophenol (0.04 mmol) corresponds to a concentration of 0.001 M and the product $C_aK_t \sim 10^4$ in the aqueous media, making the titration barely feasible (Table 1). However, in surfactant media this value is about 25 times greater, corresponding to pK shift of approximately 1.5. Thus the value of C_aK_t is enhanced and titration becomes feasible.

The results of thermometric titrations are presented in Tables 10–14. Generally, the accuracy and precision of the analyses are comparable to those of any other titrimetric procedure.

TABLE 10 Thermometric Titrations of Organic Acids and Phenols (~0.10 M) in O/W Microemulsions Containing CTAB and Butanol

Compound titrated	Amount (M)		Error (%)	Average (%)
	Taken	Found		
o-Chlorophenol	0.1001	0.0994	−0.7	−0.55
	0.1001	0.0997	+0.99	
Palmitic acid	0.1066	0.1051	−1.41	0.45
	0.1066	0.1061	−0.50	
	0.1066	0.1072	+0.56	
p-Toluic acid	0.0999	0.1002	+0.30	+0.30
	0.0999	0.1002	+0.30	
	0.0999	0.1002	+0.30	
p-Nitrophenol	0.1000	0.1002	+0.20	+0.3
	0.1000	0.1002	+0.20	
	0.1000	0.1005	+0.50	
o-Toluic acid	0.1004	0.0999	−0.49	+0.40
	0.1004	0.1019	+1.50	
	0.1004	0.0102	+0.19	
o-Nitrophenol	0.0999	0.1002	+0.30	+0.56
	0.0999	0.1017	+1.80	
	0.0999	0.0995	−0.40	
Decanoic acid	0.0999	0.1020	+2.10	+0.57
	0.0999	0.0991	−0.80	
	0.0999	0.1003	0.40	

TABLE 11 Thermometric Titrations of Organic Acids and Phenols (~0.05 M) in O/W Microemulsions Containing CTAB and Butanol

Compound titrated	Amount (M)		Error (%)	Average (%)
	Taken	Found		
o-Chlorophenol	0.0495	0.04951	+0.02	+0.69
	0.0495	0.0505	+2.02	
	0.0495	0.04952	+0.04	
Palmitic acid	0.0499	0.0507	+1.6	+0.47
	0.0499	0.0490	−1.8	
	0.0499	0.0507	+1.6	
p-Toluic acid	0.0499	0.04991	+0.02	−0.65
	0.0499	0.0493	−1.07	
	0.0499	0.0495	−0.9	
p-Nitrophenol	0.0499	0.0493	−1.2	−0.48
	0.0499	0.0501	+0.24	
o-Toluic acid	0.0499	0.0498	−0.2	−0.2
	0.0499	0.0498	−0.2	
o-Nitrophenol	0.0499	0.0489	−1.8	−0.87
	0.0499	0.0497	−0.4	
	0.0499	0.0497	−0.4	
Decanoic acid	0.0499	0.0498	−0.2	+0.40
	0.0499	0.0498	−0.2	
	0.0499	0.0507	+1.6	

V. NONLINEAR REGRESSION ANALYSIS OF POTENTIOMETRIC TITRATION OF ACIDS AND BASES IN CTAB SURFACTANT SYSTEMS

Mathematical modeling of the reactions and the analysis of data based on those models offer some unique and exceptional advantages. With wide availability of computers, such methods of data analysis are becoming almost routine. The mathematical equation describing the potentiometric titration data [Eq. (3)] is a polynomial equation that was used as part of a nonlinear regression analysis program called CFT4A written by Prof. L. Meites. The independent variable was the volume of titrant and the dependent variable was pH. The concentration of the titrand (C_s), its dissociation constant (K_a), and the ion product of water (K_w) were defined as parameters. The output of the program contains the numerical values of these parameters that best describe the data. The results are shown in Table 15.

TABLE 12 Thermometric Titrations of Organic Acids and Phenols (~0.01 M) in O/W Microemulsions Containing CTAB and Butanol

Compound titrated	Amount (M)		Error (%)	Average (%)
	Taken	Found		
o-Chlorophenol	0.0101	0.0100	−0.99	+0.33
	0.0101	0.0102	+0.99	
	0.0101	0.0102	+0.99	
Palmitic acid	0.0113	0.0116	−0.85	−1.13
	0.0113	0.0116	−0.85	
	0.0113	0.0116	−1.30	
p-Toluic acid	0.0101	0.0102	+0.99	+0.34
	0.0101	0.0102	−0.99	
	0.0101	0.0100	−0.93	
p-Nitrophenol	0.0101	0.0102	+0.99	+0.33
	0.0101	0.0100	−0.99	
	0.0101	0.0102	+0.99	
o-Toluic acid	0.0100	0.01003	+0.16	+0.03
	0.0100	0.0099	−0.96	
	0.0100	0.0101	+1.00	
o-Nitrophenol	0.0100	0.0101	+1.00	+0.33
	0.0100	0.0101	+1.00	
	0.0100	0.0099	−1.00	
Decanoic acid	0.0101	0.0100	−1.00	−0.70
	0.0101	0.00998	−1.20	
	0.0101	0.0101	0.00	

It is clear that the nonlinear regression analysis program is able to calculate the concentration of titrand accurately. Generally, the pK_a values for the acids in CTAB media also decreased; i.e., acids became stronger.

VI. SUMMARY

Potentiometric, spectrophotometric, and thermometric titrations provide good accuracy and precision in analyses. The results clearly prove that microemulsions can be alternatives to many nonaqueous solvents and procedures. With appropriate selection of the system, microemulsions are environmentally friendly, safe, and inexpensive, and their disposal does not present a serious problem. The method of analyses in microemulsions is therefore very attractive.

TABLE 13 Thermometric Titrations of Amines (~0.05 M) in O/W Microemulsions Containing SDS and Pentanol

Compound titrated	Amount (M)		Error (%)	Average (%)
	Taken	Found		
Triethylamine	0.0439	0.0435	+1.5	+0.94
	0.0439	0.0488	+1.2	
	0.0439	0.0483	+0.11	
p-Anisidine	0.0499	0.0494	−0.93	−0.1
	0.0499	0.0493	−0.43	
	0.0499	0.0504	+1.11	
n-Octylamine	0.0504	0.0505	+0.19	+0.03
	0.0504	0.0503	+0.59	
	0.0504	0.0499	−0.99	

ACKNOWLEDGMENTS

Research support from Lamar Research Council and Gulf Coast Hazardous Substance Research Center is greatly appreciated. S. J. Lie and B. Patel greatly contributed to the experimental work. A.S. appreciates greatly the constant support, encouragement, and opportunity provided by Dr. Paul Pai, Mr. Steven Grega, Dr. James Patterson and Dr. Sylvia Ramos.

TABLE 14 Thermometric Titrations of Amines (~0.01 M) in O/W Microemulsions Containing SDS and Pentanol

Compound titrated	Amount (M)		Error (%)	Average (%)
	Taken	Found		
Triethylamine	0.0099	0.0098	−1.0	−2.33
	0.0099	0.0095	−4.0	
	0.0099	0.0093	−2.0	
p-Anisidine	0.0101	0.0102	+0.99	+0.99
	0.0101	0.0102	+0.99	
	0.0101	0.0102	+0.99	
n-Octylamine	0.0103	0.0104	+1.08	+1.08
	0.0103	0.0104	+1.08	

TABLE 15 Values of Parameters in Acid-Base Titration Equation after CFT4A Nonlinear Regression Analysis

Compounds	Para. no.	Values	Theoretical values	Errors or change in pK
p-Toluic acid	3	C_s = 1.01E-2	0.01	+1.00%
(a) 0.01 M in 0.1 M		K_a = 7.88E-5	4.33E-5	−0.26
CTAB		K_w = 3.13E-14	1.00E-14	
(b) 0.01 M in CTAB	3	C_s = 1.023E-2	0.01	+2.30%
microemulsion		K_a = 1.94E-5	4.33E-5	−0.36
		K_w = 8.23E-14	1.00E-14	
1-Naphthylacetic acid	3	C_s = 9.95E-3	0.0099	+0.51%
(a) 0.0099 M in 0.1 M		K_a = 1.31E-4	5.75E-5	−0.36
CTAB		K_w = 3.79E-14	1.00E-14	
	2	C_s = 1.41E-4	5.75E-5	−0.39
		K_a = 5.72E-14	1.00E-14	
(b) 0.01 M in CTAB	3	K_w = 1.02E-2	0.01	+2.00%
microemulsion		C_s = 2.50E-5	5.75E-5	+0.36
		K_a = 4.85E-14	1.00E-14	
Capric acid	3	K_w = 1.01E-2	0.01	+1.00%
(a) 0.01 M in 0.1 M		C_s = 1.66E-5	—	
CTAB		K_a = 3.14E-14	1.00E-14	
(b) 0.01 M in CTAB	3	K_w = 1.03E-2	0.01	+3.00%
microemulsion		C_s = 3.96E-6	—	
		K_a = 2.77E-14	1.00E-14	
p-Nitrophenol	3	K_w = 1.01E-2	0.01	+2.00%
(a) 0.01 M in 0.1 M		C_s = 1.02E-2	0.01	
CTAB		K_a = 5.33E-7	7.08E-8	−0.88
(b) 0.01 M in CTAB	3	K_w = 3.44E-14	1.00E-14	
microemulsion		C_s = 0.01	0.01	0.00%
		K_a = 1.10E-7	7.08E-8	−0.19
		K_w = 2.07E-14	1.00E-14	

REFERENCES

1. P. J. Kline and W. H. Soine, in *Pharmaceutical Analysis: Modern Method* (J. W. Munson, ed.), Marcel Dekker, New York, 1981, part A, p. 73.
2. A. L. Underwood, Anal. Chim. Acta *93*:267 (1977).
3. E. Pellizetti and E. Pramauro, Anal. Chim Acta *117*:430 (1980).
4. S. S. Shukla and L. Meites, Anal. Chim Acta *174*:225 (1985).
5. C. A. Streuli, in *Treatise on Analytical Chemistry* (I. M. Kolthoff and P. J. Elving, eds.), Wiley, New York, 1975, part 1, vol. 11, pp. 7035–7116.
6. E. P. Serjeant, *Potentiometry and Potentiometric Titrations*, Wiley, New York, 1984, pp. 363–370.

7. M. E. L. McBain and E. Hutchinson, *Solubilization and Related Phenomena*, Academic Press, New York, 1955.

8. P. H. Elworthy, A. T. Florence, and C. P. MacFarlane, *Solubilization by Surface-Active Agents and Its Application in Chemistry and Biological Science*, Chapman & Hall, London, 1968.

9. M. K. Sharma and D. O. Shah, in *Macro- and Microemulsion: Theory and Application* (D. O. Shah, ed.), American Chemical Society, Washington, DC, 1985, pp. 10–16.

10. L. H. Prince, in *Microemulsion, Theory and Practice* (L. M. Prince, ed.), Academic Press, New York, 1977, pp. 1–19, 33–49.

11. T. P. Hoar and J. H. Shulman, Nature *152*:102 (1943).

12. I. D. Robb, *Microemulsion*, Plenum, New York, 1982.

13. D. O. Shah and R. S. Schechter, eds., *Improved Oil Recovery by Surfactant and Polymer Flooding* Academic Press, New York, 1977.

14. J. I. Watters, in *Treatise on Analytical Chemistry* (I. M. Kolthoff and P. J. Elving, eds.), Wiley, New York, 1975, part 1, vol. 11, pp. 6845–6974.

15. E. D. Eastman, J. Am. Chem. Soc. *47*:332 (1925); *56*:2646, (1934).

16. S. Kilpi, Z. Phys. Chem. *172A*:277 (1935); *173A*:427 (1935).

17. I. M. Kolthoff and N. H. Furman, *Potentionietric Titrations*, 2nd ed., Wiley, New York, 1931.

18. M. A. Leonard, in *Comprehensive Analytical Chemistry* (G. Svehla, ed.), Elsevier, New York, 1977, vol. 8, pp. 207–270.

19. M. L'Her, Les Titrages Spectrophotométriques, Dept. de Chim., Centre d'Études Nucl., Fontenay-aux-Roses, France, Rapp. CEA R No. 3140 (1967).

20. R. F. Goddu and D. N. Hume, Anal. Chem. *28*:1506 (1954); *26*:1740 (1954).

21. T. Higuchi, C. Rehm, and C. Barnstein, Anal. Chem. *28*:1506 (1956).

22. L. Meites, The General Non-Linear Regression Program CFT4A, The George Mason Institute, Fairfax, VA, 1984.

23. F. Daniels and R. A. Alberty, *Physical Chemistry*, 4th ed., Wiley, New York, 1975, pp. 223–226.

24. J. H. Fendler, E. J. Fender, G. A. Infante, Pong-Su Shih, and Larry K. Patterson, J. Am. Chem. Soc. *97*:89 (1975).

25. J. W. McBain and A. A. Green, J. Am. Chem. Soc. *68*:1731 (1946).

26. H. P. Klevens, Chem. Rev. *47*:1 (1950).

27. *Handbook of Chemistry and Physics*, 66th ed., CRC Press, Boca Raton, FL, 1985–1986, pp. D161–D163.

28. D. Jagner, Anal. Chim. Acta *52*:483 (1970).

4
Microemulsions in Biotechnology

KRISTER HOLMBERG   Institute for Surface Chemistry, Stockholm, Sweden

I. INTRODUCTION

During the past decade there has been considerable interest in the use of microemulsions, particularly of the L2 type (water droplets in oil), for various applications related to biotechnology. To a considerable degree this interest stems from the observation that many proteins can be solubilized in microemulsions based on apolar solvents such as aliphatic hydrocarbons without denaturation or loss of function. This is remarkable in view of the fact that most proteins are sparingly soluble in apolar solvents and that transfer of proteins into these solvents frequently results in irreversible denaturation and loss of biological activity.

Wells and his coworkers [1–3] seem to have been the first to work systematically with proteins in microemulsions, although the concept of solubilizing a protein in a hydrocarbon solvent had been described earlier [4,5]. The majority of papers in this field relate to the use of microemulsions as a medium for enzyme-catalyzed reactions but several other uses of microemulsions in biotechnology have also been investigated. This review places an emphasis on enzymatic reactions in microemulsions but it also covers two newer areas of scientific as well as practical interest: use of microemulsion for immobilization of proteins and in bioseparations.

In the biologically oriented literature L2 microemulsions are often referred to as reverse micelles. It has been suggested that the borderline between reverse micelles and microemulsion droplets should be defined by the water-to-surfactant ratio; above molar ratio 15 the system should be referred to as a microemulsion [6]. In this review no such distinction is made. All systems containing oil and water together with surfactant are termed microemulsions, regardless of the relative component proportions.

II. ENZYMATIC REACTIONS IN MICROEMULSIONS

Microemulsions were introduced into the field of biocatalysis with papers from Martinek et al. [7], Luisi et al. [8], and other groups around 1980 and the research interest has grown steadily since then. The topic has been the subject of several reviews [9–11].

The major potential advantages of employing enzymes in media of low water content are

1. Increased solubility of nonpolar reactants
2. Possibility of shifting thermodynamic equilibria in favor of condensation
3. Improvement of thermal stability of the enzymes, enabling reactions to be carried out at higher temperatures

In fact, the use of enzymes in water-poor media is not unnatural. Many enzymes, including lipases, esterases, dehydrogenases, and oxidoreductive enzymes, often function in the cells in microenvironments that are hydrophobic in nature. Also, the use of enzymes in microemulsions is not an artificial approach per se. In biological systems many enzymes operate at the interface between hydrophobic and hydrophilic domains, and these interfaces are often stabilized by polar lipids and other natural amphiphiles.

Enzymatic catalysis in microemulsions has been used for a variety of reactions, such as synthesis of esters, peptides, and sugar acetals; transesterifications; various hydrolysis reactions; and steroid transformations. The enzymes employed include lipases, phospholipases, alkaline phosphatase, pyrophosphatase, trypsin, lysozyme, α-chymotrypsin, peptidases, glucosidases, and oxidases.

By far the most widely used class of enzymes in microemulsion-based reactions is the lipases; of microbial as well as of animal origin. This review focuses on lipase-catalyzed reactions with emphasis on work published since 1988. Prior to discussing specific reactions some general aspects of enzymatic catalysis in microemulsions will be illuminated.

A. Enzymatic Activity in W/O Microemulsions

Many investigations have been carried out regarding the effect of water content and nature of organic solvent on the catalytic behavior of enzymes in water-poor media. The role of the enzyme-bound water in the biocatalysis has still not been fully clarified; however, the water dependence clearly differs from one enzyme to another. For instance, some lipases exhibit high activity and good stability in organic solvents containing only traces of water, whereas other lipases reach optimal activity at a relatively high ratio of water to organic solvent [12]. A related enzyme, α-chymotrypsin, has been studied in detail with regard to the effect of hydration on activity [13]. Circular dichroism measurements indicate that at a very low water content the protein is essentially frozen. The first molecules of water added interact predominantly with ionizable groups. This leads to even higher rigidity of the protein molecule. A minimum in activity was displayed at a water-to-surfactant molar ratio of 5. (The molar ratio of water to surfactant is often referred to as W_o). Further addition of water first leads to hydration of hydrogen-bonding sites, and subsequently nonpolar regions are covered by a water monolayer. The catalytic activity increases with increasing W_o up to a value of around 10.

It has been demonstrated by several groups that the catalytic activity of enzymes confined in microemulsion water droplets varies with overall water content. For a wide variety of enzymes it has been found that there is a bell-shaped dependence of activity on W_o [14–16]. Figure 1 shows a representative example.

In general, it seems that maximum activity occurs around a value of W_o at which the size of the droplet is somewhat larger than that of the en-

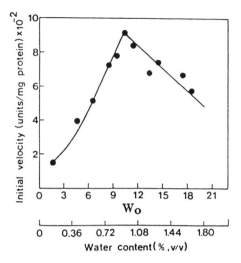

FIG. 1 Initial velocity of lipase in W/O microemulsions (50 mM AOT in isooctane) of varying W_o. The overall enzyme concentration is the same at all W_o values. (From Ref. 16.)

trapped enzyme. In many cases the enzyme exhibits enhanced activity (superactivity) compared with that in bulk water.

The anomalous activity characteristics have been attributed to conformational changes of the solubilized enzymes [17], but more recent spectroscopic studies seem to indicate that this is not the main cause. Solubilization of an enzyme into reverse micelles does not normally lead to major conformational alterations, as indicated, e.g., by fluorescence and phosphorescence spectral investigations [13,18]. The situation is complex however, and it has been shown by circular dichroism (CD) measurements that the influence of the oil-water interface on enzyme conformation may vary between enzymes belonging to the same class [19]. In the case of human pancreatic lipase the conformation of the polypeptide chain is hardly altered after the enzyme is transferred from a bulk aqueous solution to the microenvironment of reverse micelles. Conversely, the CD spectra of the lipases from *Candida rugosa* and *Pseudomonas* sp. are considerably different in reverse micelles compared with those in aqueous solution, indicating that both enzymes lose their native structure in the microemulsion environment [19].

It has been proposed that repulsive interactions between charges at the droplet surface and charged substrate ions are the main reason for the anomalous behavior of solubilized enzymes. (The majority of studies of enzyme activity in water-in-oil (W/O) microemulsions have been performed with the anionic surfactant AOT.) Both the superactivity and the bell-shaped curve of activity versus W_o have been theoretically explained in this way [20]. However, later investigations with α-chymotrypsin in AOT-based microemulsions showed that superactivity could be obtained with uncharged as well as with negatively charged substrates [21]. Hence, the superactivity cannot be due solely to a high local substrate concentration arising from electrostatic repulsion between the negatively charged substrate and the negatively charged surfactant palisade layer.

B. Choice of Solvent

The organic solvent used in the microemulsion formulation should be non-polar. The hydrophobicity of the solvent seems to be a key factor in the catalytic activity of the enzyme. A good correlation between hydrophobicity of the organic solvent and biocatalytic activity is obtained by use of log P values of solvents [22,23]. (P is the partition coefficient of the solvent in the water/octanol system.) In general, enzyme stability and activity in microemulsions are poor with relatively hydrophilic solvents, in which log $P < 2$; moderate in solvents in which log P is between 2 and 4; and high in hydrophobic solvents with log $P > 4$. Even relatively small changes in solvents, such as going from cyclohexane to nonane, can give rise to large improvements in enzyme stability, as has been demonstrated for lipase from *Chromobacterium viscosum* [24]. The rationale behind this division is that very hydrophobic solvents do not distort the essential water layer around the enzyme, thereby leaving the catalyst in an active state. This limits the choice of organic solvent to aliphatic hydrocarbons with seven or more carbon atoms. On the other hand, lower hydrocarbons are preferred from a workup point of view because they can be readily removed by evaporation after reaction. As a compromise, heptane, octane (in particular isooctane), and nonane are the solvents of choice and these hydrocarbons have been used in almost all enzymatic reactions in microemulsion media.

C. Influence of the Surfactant

The choice of surfactant is of importance for the rate of many enzymatic reactions in microemulsions. For instance, it has been found that whereas lipase-catalyzed hydrolysis of triglycerides is rapid in microemulsions based on AOT, it is extremely sluggish when normal nonionic surfactants are used [25]. The difference in behavior between microemulsions based on anionic and nonionic surfactants was attributed to differences in accessibility of the triglyceride to the enzyme. In the AOT system the lipase was anticipated to have good access to the interface between the oil and

water domains. On the other hand, in the systems based on nonionic surfactants the poly(ethylene glycol) (PEG) chains stretching out from the
interface into the droplets would effectively prevent the enzyme from en-
tering into the interfacial region. The steric repulsion between the protein
and the PEG layer paralleled the well-known protein-repelling effect of
PEG chains grafted on solid surfaces. Later work, using $C_{12}E_5$ as the surfactant, confirmed the view that straight-chain alcohol ethoxylates are unsuitable for use in lipase-catalyzed hydrolysis reactions in microemulsions.
At all compositions tested only partial hydrolysis of triglyceride was obtained [26]. Contrary to what would be expected, an increase in the overall
water content of the microemulsion formulation led to a decreased rate of
lipase-catalyzed triglyceride hydrolysis, as can be seen from Fig. 2.

Nuclear magnetic resonance (NMR) self-diffusion measurements were
performed on microemulsions containing from 1% up to 40% water while
keeping the surfactant concentration constant. As shown in Table 1, all
self-diffusion coefficients, including that of water, D_w, decrease with increasing water content although $D_{palm\ oil}$ seems to have reached a plateau

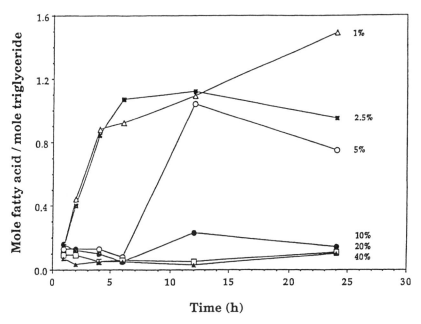

FIG. 2 Effect of water content on the formation of fatty acids from palm oil.
Reactions were run at 37°C and the surfactant concentration was kept constant at
17%. (From Ref. 26.)

TABLE 1 Self-Diffusion Coefficients D (10^{-11} m^2 s^{-1}) at 37°C for the Components of a Microemulsion Containing 17% C$_{12}$E$_5$, 5% Palm Oil, 1–40% 0.1 M Phosphate Buffer, pH 8, Isooctane Constituting the Balance[a]

Water (%)	R	D_{water}	D_w/D_w°	$D_{\text{hydrocarbon}}$	D_H/D_H°	$D_{\text{surfactant}}$	$D_{\text{palm oil}}$
1.05	1.4	—	—	168	0.71	36	37
2.09	2.8	21	0.085	169	0.72	26	28
5.01	6.7	13	0.048	—	—	18	21
10.0	13.4	16	0.059	—	—	14	17
19.7	26.1	16	0.059	142	0.60	12	20

[a]R is the molar ratio of water to surfactant. Self-diffusion values for neat coponents are $D_{\text{water}}(D_w^\circ)$ 271, disooctane (D_H°) 235, and $D_{\text{palm oil}}$ (as 5% palm oil in isoctane) 60.3. Confidence interval 60%.

at about 5% water. From the D_w/D_w° and D_H/D_H° values it is obvious that whereas the diffusion of water is highly restricted throughout the series, that of isooctane is of the same magnitude as that of neat solvent. Furthermore, diffusion constants of the surfactant closely follow that of water. Evidently, the system is a W/O type of microemulsion and structure changes in the direction of more closed water domains upon water addition. The conclusion that may be drawn from this study is that in the systems with more well-defined water-in-oil spheres, i.e., those of high water content, hydrolysis of the triglyceride was inhibited compared with reaction in the more bicontinuous structures. This inhibition may be related to incorporation of surface-active reaction products in the palisade layer between the water and oil microdomains. It may also be due to the fact that C$_{12}$E$_5$ seems to be a competitive inhibitor for lipases [27].

In recent work, the effect of the surfactant on lipase-catalyzed hydrolysis of palm oil in microemulsion was further investigated [28]. Three surfactants were used, one anionic, one nonionic, and one cationic. As shown in Fig. 3, all three compounds were double tailed with similar hydrophilic-lipophilic balances, giving large regions of L2 microemulsions with isooctane and water at 37°C.

NMR self-diffusion measurements indicated that all microemulsions consisted of closed water droplets and that the structure did not change much during the course of reaction. Hydrolysis was fast in microemulsions based on branched-chain anionic and nonionic surfactants but very slow when a branched cationic or a linear nonionic surfactant was employed (Fig. 4). The cationic surfactant was found to form aggregates with the enzyme. No such interactions were detected with the other surfactants. The

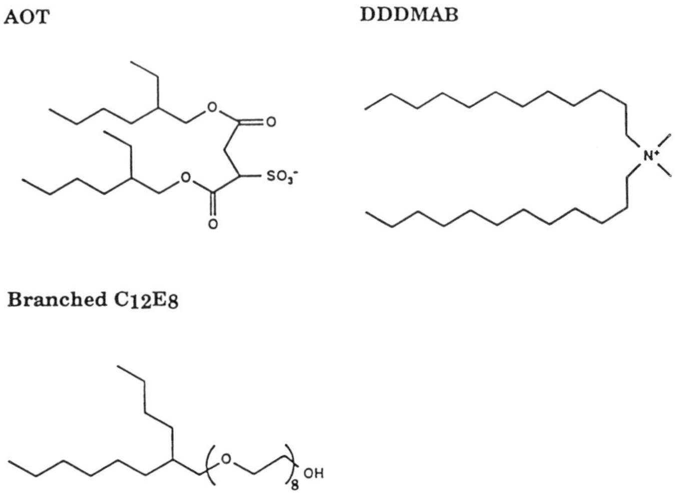

AOT   DDDMAB

Branched C12E8

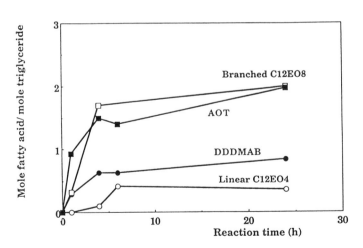

FIG. 3 Structures of a branched anionic (AOT), cationic (didodecyldimethylam-monium bromide, DDDMAB), and nonionic (branched $C_{12}E_8$) surfactant.

FIG. 4 Degree of palm oil hydrolysis at 37°C as a function of time in lipase-containing microemulsions based on isooctane, palm oil, distilled water, and surfactant in the weight proportion 50:5:8:37. The double-tailed surfactants of Fig. 3, as well as straight-chain C_2E_4, were used. (From Ref. 28.)

straight-chain, but not the branched, alcohol ethoxylate was a substrate for the enzyme. Evidently, this type of enzyme-catalyzed reaction should preferably be performed in a microemulsion based on an anionic or branched nonionic surfactant.

D. Structure of Protein-Containing Reverse Micelles

Most enzymatic reactions in microemulsions have been carried out in W/O-type systems with the enzyme confined in the water pools. Luisi and co-workers [29] have studied how the size and structure of reverse micelles change upon uptake of a large guest molecule, such as a protein. Generally, one is then dealing with three different water pools with different radii R: initial, R_0; protein-containing micelles, R_f; and unfilled micelles, R_e (f stands for full and e for empty). It has been demonstrated that R_0 splits after addition of enzyme into the larger R_f and the smaller R_e. This means that the micelles accommodating the protein grow in size at the expense of the unfilled micelles. The size difference between filled and unfilled micelles varies with the protein but is of the order of 2:1.

An alternative model, suggested by Levashov et al. [30], assumes that such an increase in size occurs only when the inner cavity of the initial empty micelle is smaller than the protein molecule. If the size of the initial water cavity is equal to or exceeds that of the protein molecule, protein entrapment may not lead to any substantial increase in the size of the water droplet.

This "water shell model," picturing the enzyme molecules residing in the interior of the water droplet surrounded on all sides by water, will apply only to hydrophilic proteins that do not contain large hydrophobic domains. Many enzymes, including lipases, are surface active and interact strongly with oil-water interfaces. Such proteins are, of course, not confined to the interior of the water pools. In fact, lipases need a hydrophobic surface in order to open the lid covering the active site [31]. No water layer is therefore likely to separate the lipase active site from the continuous hydrocarbon domain.

E. Lipase-Catalyzed Ester Synthesis

Synthesis of triglycerides from glycerol and fatty acids has been attempted in microemulsions of very low water content [32–36]. The main reaction product is monoglyceride, with smaller amounts of diglyceride formed as well. The yield of triglyceride is negligible. The same reaction proceeds well in monolayers, giving triglyceride in fair yield [35]. It is believed that triglycerides are not readily formed in microemulsions because the intermediate diglyceride is too lipophilic and has too low surface activity to stay at the hydrocarbon-water interface. Once formed, it rapidly partitions into the continuous hydrocarbon domain leaving behind the enzyme, which is located in or at the surface of the water pools. The monolayer situation is very different because the hydrophobic diglyceride has no other choice than to stay at the surface. The static air-water interface should constitute an ideal environment for the lipase-catalyzed reaction as illustrated schematically in Fig. 5.

Lipase-catalyzed synthesis of esters of monofunctional alcohols proceeds in good yield in microemulsions [37–39]. In a systematic study of esterification of different alcohols (primary, secondary, tertiary) with fatty acids of various chain lengths, a pronounced selectivity was obtained [40]. The three different lipases used—*P. simplicissimum, R. delemar,* and *R. arrhizus*—exhibited different preferences with regard to chain length of the acid and type of alcohol. Studies using spectroscopic techniques indicated that the selectivity is related to localization of the enzyme molecule within the micellar microstructure. Hence, the hydrophilic-lipophilic character of the protein, not its specificity as expressed in aqueous solution, is responsible for the selectivity. This illustrates the important point that regioselectivity of bioorganic reactions may differ between homogeneous and

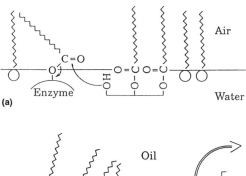

(a)

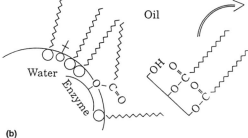

(b)

FIG. 5 Arrangement of substrates/surfactants at the interfaces of monolayers and microemulsions.

microheterogeneous media. Surprisingly, esterification of fatty acids with simple sugars, such as glucose and mannitol, in AOT-based microemulsions failed completely [37]. No reaction at all was seen using two different lipases. This is probably due to poor phase contact between the very hydrophilic sugar molecule in the water pool and the fatty acid that resides in the hydrocarbon domain. Sugar monoesters can be produced in high yields by lipase-catalyzed esterification in a water-free medium [41].

The main practical problem in large-scale use of biocatalysis for synthesis of hydrophobic esters is that of workup. Separating surfactant from product is not a trivial issue because normal purification procedures, such as extraction and distillation, tend to be troublesome due to the well-known problems of emulsion forming and foaming caused by the surfactant.

An interesting extension of the use of microemulsions for synthesis is to use microemulsion-based gels (MBGs) as reaction media [42–44]. The MBGs are made by mixing normal W/O microemulsions with aqueous gelatin solutions above the gelling temperature of the gelatin and then allowing the mixture to cool, for instance in a column, so that a stiff gel is formed. Various spectroscopic techniques have revealed that the microemulsion structure is retained in the gel. The gel can be seen as an immobilized enzyme-containing microemulsion. The gel is resistant to hydrocarbon solvents. By charging a hydrocarbon solution of an acid and an alcohol at the top of the column, the corresponding ester can be recovered from the eluent. This is an elegant way to avoid the problem of separation of product and surfactant from the reaction mixture. Lipase-containing MBGs have been used to synthesize, on a preparative scale, a variety of different esters under mild conditions, and both regio- and stereoselectivity have been demonstrated. The MBGs have not yet been applied to triglyceride synthesis.

F. Lipase-Catalyzed Ester Hydrolysis

Monoglycerides can be obtained in high yields by lipase-catalyzed hydrolysis of the corresponding triglyceride oil in microemulsions of W/O type. By using a 1,3-specific enzyme the hydrolysis takes place in a fairly regioselective manner, conversion into 2-monoglyceride being completed in 2–3 h at a temperature of 37°C [45]. Prolonged reaction time results in a decrease in monoglyceride yield, a process that is due not to lack of regioselectivity of the enzyme but to an acyl group migration in the 2-monoglyceride yielding 1-monoglyceride. The latter compound is a good substrate for the regiospecific enzyme and complete hydrolysis to fatty acids and glycerol will eventually take place (Fig. 6).

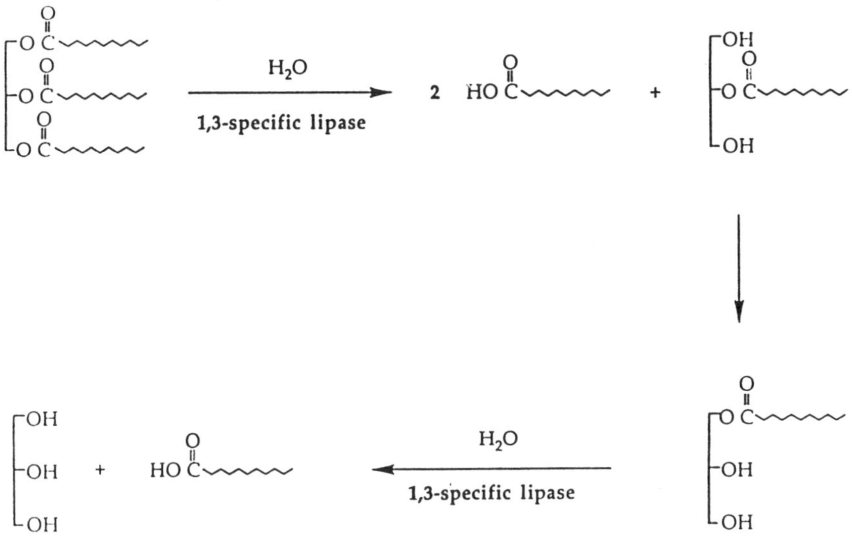

FIG. 6 Enzymatic hydrolysis of a triglyceride in combination with migration of the acyl group from the 2- to the 1-position.

Kinetic studies indicate that effects of temperature and pH, as well as reaction constants, resemble those in aqueous systems. The activity versus W_0 curve shows a typical maximum at W_0 between 10 and 15, the value varying slightly with the type of lipase used [45,46]. In the AOT/water/isooctane system, which is commonly used for hydrolysis of triglycerides, $R = 11$ was found to correspond to a ratio of about 2 nm [47].

Lipase activity was generally higher in AOT-based systems than in microemulsions based on nonionics or cationics with both triglycerides and nitrophenyl alkanoate esters as substrates. AOT also gave excellent enzyme stability [27,47,48].

A hydrophilic substrate, acetylsalicylic acid, has been subjected to lipase-catalyzed hydrolysis in W/O microemulsion [49]. For comparison, the reaction was also carried out in aqueous buffer. Because hydrolysis of acetylsalicylic acid proceeds spontaneously without added catalyst (intramolecular catalysis), reactions without lipase were performed as controls. It was found that addition of lipase did not affect the rate of reaction in aqueous buffer. However, the reaction in microemulsion was catalyzed by the lipase and the rate was linearly dependent on lipase concentration. This is a further illustration of the fact that microemulsions, with their large oil-water interfaces, are suitable media for lipase-catalyzed reactions. The same reactions were also performed using α-chymotrypsin as the catalyst. This enzyme, which also catalyzes ester hydrolysis but, unlike lipase, functions independently of a hydrophobic surface, was not more active in the microemulsion than in the buffer solution.

G. Lipase-Catalyzed Glycerolysis

Water-free microemulsions can be designed with glycerol as a polar component. Such systems are of potential interest for lipid conversions, since they open the possibility of performing glycerolysis instead of hydrolysis, provided that glycerol can replace water as an activator and stabilizer of the enzyme.

Lipase-catalyzed glycerolysis of triglycerides would yield a mixture of 1- and 2-monoglycerides. As monoglycerides are important emulsifiers in the food industry, this synthesis is of potential industrial interest. Attempts to perform the reaction failed, however; very little enzyme activity was obtained in the completely nonaqueous system [50]. Use of a combination of water and glycerol as the polar component of an AOT-based microemulsion led to simultaneous hydrolysis and glycerolysis. Using ^{3}H-labeled material, monoglyceride originating from added glycerol could be distinguished from that originating from the starting triglyceride as shown in Fig. 7. It was found that glycerolysis and hydrolysis occur at approximately the same rate. The two reactions probably go through the same intermediate. The work shows that, although water can be replaced by glycerol as

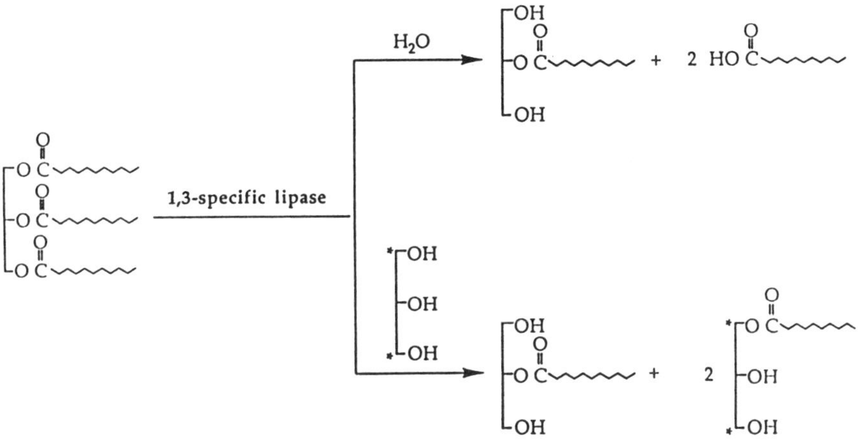

FIG. 7 Enzymatic glycerolysis using ^{3}H-labeled glycerol. Hydrolysis occurs in parallel. (From Ref. 50.)

both the polar component of the microemulsion and the solvolytic agent in the reaction, it cannot be fully replaced in its role as activator of the enzyme.

H. Lipase-Catalyzed Transesterification

Lipase-catalyzed transesterification, i.e., replacement of one acyl group in a triglyceride by another acyl group, can be performed in microemulsions of low water content [32,51,52]. Special attention has been directed to production from an inexpensive starting material (such as palm oil) of a triglyceride mixture that corresponds to natural cocoa butter. This reaction requires a partial replacement of palmitoyl groups by stearoyl groups in 1(3)-position, while leaving the 2-position essentially unaffected. A high degree of conversion was obtained in lipase-containing microemulsions based on either an anionic or a nonionic surfactant [51].

In work aimed at incorporating γ-linolenic acid (GLA) into a saturated triglyceride, tristearin was transesterified with GLA using *Rhizopus delemar* lipase as the catalyst [53]. Reactions were carried out at four different compositions, all situated in the hydrocarbon-rich corner of the ternary phase diagram. The surfactant used was an ethoxylated (6 EO) C_{12}-Guerbet alcohol and dodecane was used as the oil component. As seen in Fig. 8, three samples lie in the isotropic L2 region, whereas one sample falls within the two-phase region. The reactions were carried out with stirring. Also shown in Fig. 8, the reaction in the two-phase region (emulsion) is very sluggish compared with the reactions in microemulsions. This set of experiments is a good illustration of the benefits of the much larger oil-water interfacial area of microemulsions compared with emulsions.

III. IMMOBILIZATION OF PROTEINS IN MICROEMULSIONS

A. Attachment to Hydrophilized Surfaces

Immobilization of proteins on solid surfaces can be done in a number of ways. The most commonly used technique utilizes the fact that most proteins often adhere strongly to solid surfaces due to hydrophobic and electrostatic interactions. However, interactions with the underlying surface also cause problems. Strong attraction forces between the molecule and the surface may eventually lead to such a large change in protein conformation that the specific biological activity is lost. Furthermore, long-range attractions by the underlying surface can cause nonspecific adsorption and be the reason for poor specificity in solid-phase immunoassays. In addition,

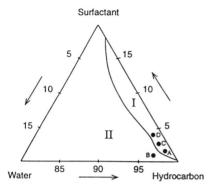

γ- Linoleyl in triglyceride

Composition	Linoleic acid (%)
A	28
B	7
C	29
D	26

FIG. 8 Transesterification of tristearin with γ-linolenic acid using reaction media of slightly different compositions.

although most proteins adhere strongly to solid supports consisting of hydrophobic and charged domains, some molecules eventually desorb.

There has been considerable interest in improving the immobilization technique and different solutions have been proposed. Nonspecific adsorption is often reported to be minimized by the use of "blocking" agents, such as albumin, which compete with other proteins for attachment sites on the plastic surface. Although reduction of nonspecific adsorption can often be attained by this procedure, it is not entirely satisfactory because the problems of poor stability and desorption still remain.

Another approach that is gaining interest is to graft the solid support with chains of a hydrophilic and uncharged material, such as poly(ethylene glycol) (PEG) or a polysaccharide. Such a surface is characterized by low cell and protein adsorption [54,55]. Proteins immobilized to free reactive ends of the hydrophilic polymer interact minimally with the underlying surface. Studies in which proteins have been linked to surfaces via PEG spacers indicate that the biomolecule can be efficiently attached with retained activity and improved stability and without the matrix effects often found with conventional immobilization [56]. Covalent attachment of the protein to the solid support also decreases the risk of desorption. In applications in which proteins are immobilized in devices for extracorporeal treatment, irreversible attachment is necessary, since leakage of foreign proteins into the blood stream may cause anaphylaxis. In addition, covalent coupling offers an advantage, since protein loading can be controlled by varying the amount of activated groups at the polymer surface.

However, covalent attachment of proteins to such hydrophilic noncharged surfaces is not a trivial procedure. The characteristic features of the surface-grafted polymer layer, i.e., hydrophilicity, freedom of electrostatic forces, and rapid motion of the hydrated chains, being valuable in terms of avoiding interaction between the protein and the surface, cause considerable difficulties when it comes to immobilization efficiency. Even with very reactive end groups of the polymer chains, coupling of the biomolecule in solution is rendered difficult by the energy barrier involved in the close approach of a protein to a dense PEG or polysaccharide layer.

If a hydrophilized surface, for instance, a PEG surface, is exposed to a worse than theta solvent, the grafted chains do not reach out into the liquid phase but form a compressed surface layer. (A theta solvent is defined as a solvent in which a polymer solution at low concentration behaves thermodynamically ideally.) In such an environment the PEG chains do not induce repulsion of particles or macromolecules in solution. Also, close approach by a protein to such a surface may be possible because the interaction between the PEG chains and the approaching biomolecule may be attractive. This phenomenon is analogous to steric stabilization of polymer-coated particles, in which it has been shown that when the solvency of the dispersion medium is changed (e.g., by addition of a nonsolvent until the theta point is reached) the particles come in close contact and flocculation occurs [57].

Aliphatic hydrocarbons are examples of worse than theta solvents for PEG. Also microemulsions of W/O type have very low dielectric constants and are nonsolvents for PEG. It has been demonstrated that such microemulsions are excellent media for immobilization of a range of proteins to PEG-grafted surfaces [58]. After coupling, the surface is washed with water, whereby the PEG layer becomes hydrated and protein rejecting. The principle is outlined in Fig. 9.

A variety of proteins have been immobilized to hydrophilized solid surfaces in microemulsions. The technique is of particular interest for solid-phase diagnostics of ELISA type [59]. (ELISA stands for enzyme-linked immunosorbent assay). The ELISA principle is shown in Fig. 10.

Coating of ELISA plates with an immunoreagent (antibody or antigen) is routinely done by simple adsorption from aqueous solution. The reagent usually adheres strongly to the polystyrene plates commonly used, but

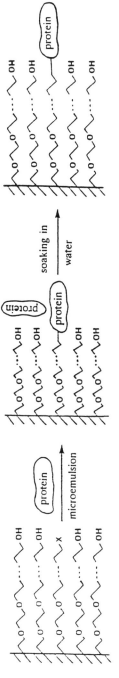

FIG. 9 Immobilization of protein from a microemulsion.

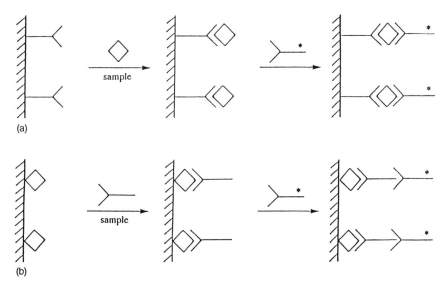

FIG. 10 Two principles of ELISA. In (a) the surface is coated with antibodies against a target antigen. The presence of antigens on the surface is subsequently detected by an enzyme-linked antibody. In (b) the surface is coated with antigens and detection is made by an enzyme-linked antiantibody.

some molecules, particularly those carrying long carbohydrate chains, may eventually desorb. The tendency to desorb is particularly strong when the antibody-antigen complex has been formed on the surface.

The problems associated with the hydrophobic solid phase (i.e., non-specific adsorption, desorption of the antigen-antibody complex, and gradual denaturation of the immunoreagent on the surface) can all be minimized by the technique of covalent binding of the immunoreagent to the PEG surface. This is illustrated schematically in Fig. 11.

Figure 12 shows albumin, immunoglobulin G (IgG), and collagen immobilized on epoxy-functionalized PEG and aldehyde-functionalized polysaccharide surfaces [58]. Immobilization is done from an AOT-based microemulsion and from aqueous buffer. As can be seen, use of a microemulsion as the reaction medium considerably increases the amount of protein immobilized on both PEG and polysaccharide surfaces.

B. Reactivity of Amino Acid Residues

Three factors are believed to govern which amino acid residues of proteins will participate in covalent attachment to a solid support [60]. The relative

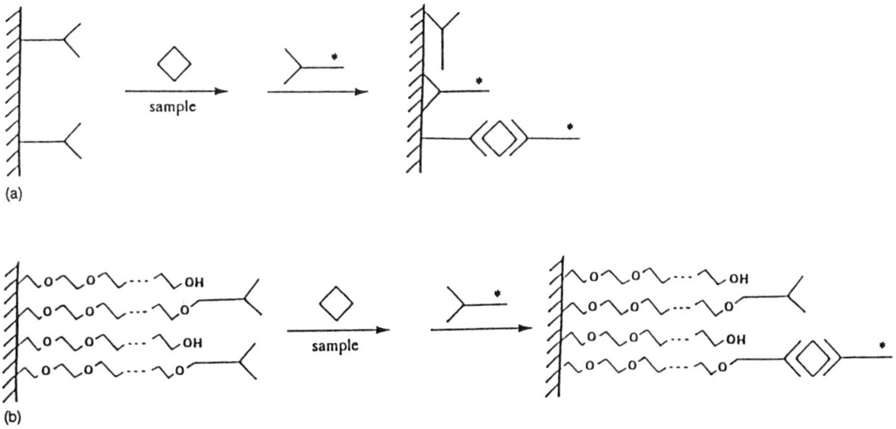

FIG. 11 Path (a) shows a normal ELISA illustrating both nonspecific adsorption of the enzyme-linked antibody and change in conformation of the antibody used for coating. In path (b), where a dense layer of PEG chains is introduced, the tendency for antibodies to interact with the surface is reduced.

concentrations of the amino acids are obviously of importance. Ser, Lys, and Thr, being the most abundant residues, are favored on this account.

The degree of hydrophilicity of the residue is the second factor, the availability of a functional group being dependent on whether it is situated on an amino acid residue that is exposed on the outside of the protein or buried inside the molecule. In aqueous media a protein folds so as to expose its more hydrophilic regions. Hydrophobic residues, such as Tyr, Met, and Thr, are therefore more likely to be internalized than, e.g., Lys and Asp.

The relative reactivity of the amino acid residues is the third factor. It is, of course, not possible to compare accurately residues in terms of reactivity without specifying a particular reaction and reaction conditions. However, for nucleophilic substitution reactions involving amino, thiol, or phenolic hydroxyl groups of the protein, reactions with Lys, Cys, and Tyr are likely to dominate.

If one considers the factors mentioned, relative concentration, hydrophilicity, and reactivity, Lys is predicted to be the most likely coupling residue under aqueous conditions, followed by Cys, Tyr, and His [60]. The relative reactivity of the residues is likely to depend on solvent polarity, since Lys contains an uncharged nucleophile, the ϵ-amino group, whereas Cys and Tyr react in the deprotonated, ionic form. The reaction between

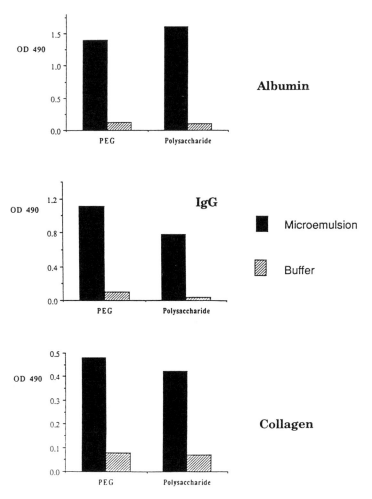

FIG. 12 Albumin, IgG, and collagen immobilized to hydrophilized polystyrene using a microemulsion (shaded) or a buffer (hatched) as the coupling medium.

a reactive electrophile, such as a tresylate (RX), and a nucleophile (Y) proceeds via different types of transition states, depending on whether or not the nucleophile is charged:

$$R-X + Y \longrightarrow Y^{\delta+} \text{----} R \text{----} X^{\delta-} \longrightarrow R-Y + X \quad (1)$$

$$R-X + Y^- \longrightarrow Y^{\delta+} \text{----} R \text{----} X^{\delta-} \longrightarrow R-Y + X^- \quad (2)$$

In reactions of type (1) the reactants are unchanged, but the transition state has built up a charge. Such reactions are aided by polar solvents, which reduce the energy of an ionic transition state. A change from a polar solvent to a nonpolar one thus leads to a reduced state of reaction. In type (2) reactions, on the other hand, the initial charge is dispersed in the transition state, so that the reactions are favored by more nonpolar solvents.

Immobilization via lysine residues is a type (1) reaction, whereas couplings to cystein and tyrosine residues represent reactions of type (2). A change of reaction medium from water to a microemulsion of low polarity is thus likely to reduce attachment via lysine amino groups.

In order to separate the effect on reactivity of amino acid residues from that on availability due to conformational changes when the solvent polarity is decreased, the model studies with free amino acids presented in Fig. 13 were carried out [61]. As can be seen, the effect on relative reactivity is very large, the order of reactivity Lys>Cys>Tyr in aqueous buffer being changed to Cys>Lys>Tyr in the W/O microemulsion. The reactions were carried out by adding tresylated silica particles to solutions of equimolar amounts of the three amino acids lysine, cystein, and tyrosine either in an aqueous buffer of pH 8, 9, or 10 or in a microemulsion based on the same buffers as the polar component. At each reaction condition, e.g., buffer at pH 8, three experiments were performed, all with one radioactive amino acid and two nonlabeled ones. Thus, for each pH and reaction medium "hot" lysine was mixed with "cold" cystein and tyrosine, hot cystein was mixed with cold lysine and tyrosine, and finally hot tyrosine was mixed

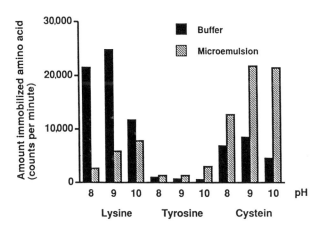

FIG. 13 Immobilization yield of amino acid from mixtures of equimolar amounts of lysine, tyrosine, and cystein. Immobilizations were carried out in microemulsions and in aqueous buffers of pH 8, 9, and 10.

with cold lysine and cystein. This method proved useful for determining the relative reactivities of the three amino acids.

It should be pointed out that the relative reactivities obtained with free amino acids may not be quantitatively transferred into relative reactivities of amino acid residues in proteins. The nucleophilicities of the amino, thiol, and phenolic hydroxyl groups are somewhat different in the amino acid state, i.e., as ^+H_3N—CHR—COO$^-$, and as polypeptides, i.e., as —NH—CHR—CO—. However, the inductive effect on the nucleophilic groups when going from the free acid to the polypeptide should be of the same magnitude for all three residues. It is therefore reasonable to assume that the trends obtained with the free amino acids are also applicable to amino acids residues in proteins.

In the immobilization of proteins a change from water to a W/O microemulsion will probably also suppress coupling via Lys on availability grounds. In the less polar microemulsion milieu the hydrophobic Tyr is likely to be more exposed on the surface, whereas Lys, like other hydrophilic residues, is found more inside the protein.

Thus, immobilization of proteins from W/O microemulsions instead of aqueous buffers is likely to favor coupling via Cys for reactivity reasons and Tyr from protein conformational aspects. The net result will be that attachment via Lys is minimized.

IV. USE OF MICROEMULSIONS IN BIOSEPARATIONS

Use of W/O microemulsions (reversed micelles) for extraction of proteins and other biomolecules is an area that has been explored mainly by Hatton and co-workers. The procedure is based on the observation that different protein molecules exhibit different affinities for properly formulated microemulsions, a property that can be used to achieve a selective separation of a protein of interest from other material formed in an aqueous broth. The topic has been the subject of several reviews [62–65]—also in this book series [66]—and is therefore only conceptually treated here.

Partitioning of a protein between a bulk aqueous phase and a microemulsion depends on parameters related to the aqueous phase, such as pH, ionic strength, and salt type, as well as on the types of solvent and surfactant used in the microemulsion. The majority of work has been done with aliphatic hydrocarbon as the solvent and AOT as the surfactant. pH and ionic strength have been the main parameters used to govern the separation process.

Solution pH affects the protein's net charge and hence its interaction with ionic surfactant head groups. pH can also affect protein conformation, which, in turn, may influence the solubilization properties. If electrostatic interactions are the dominant factor in the solubilization of a protein into a microemulsion, the process should be possible with anionic surfactants only at pH values less than the isoelectric point (pI) of the protein, where electrostatic attractions between the protein and surfactant head groups are favorable. At pH values above the pI, electrostatic repulsions would inhibit protein solubilization. With cationic surfactant the reverse trend would be found. In principle, it should be possible to use the microemulsion extraction technique to separate proteins of varying pI by using aqueous solutions of different pH values, as illustrated in Fig. 14. In most instances this seems to be the case, although the amount of surfactant needed to achieve transition between the solubilization regimes at the pI varies from one protein mixture to another [67,68].

A good example of the use of solution pH to achieve selective solubilization is given in Ref. 67, where a sharp transition in solubilization behavior was found at the pI of three individual proteins, lysozyme, cytochrome c, and ribonuclease A. The pI demarcated the region of no solubilization (pH > pI) from that of almost quantitative transfer to the organic solution (pH < pI). There are also examples in the literature of protein solubilization not being dramatically changed at the protein pI [69]. Probably, electrostatic interactions are normally, but not always, decisive of the solubilization process.

A second parameter of importance to achieving selective extraction of a protein is ionic strength. An increase in ionic strength reduces double-

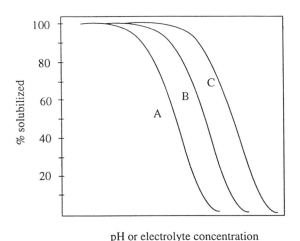

pH or electrolyte concentration

FIG. 14 Schematic illustration of the effect of pH or ionic strength on solubilization of three different proteins, A, B, and C.

layer forces and thus decreases protein-surfactant interactions, inhibiting solubilization of the protein. The electrostatic screening effect also leads to formation of smaller water droplets, since repulsion of surfactant head groups is being reduced. This may lead to size exclusion effects in protein solubilization.

Figure 14 gives a schematic picture of how ionic strength can be used to separate proteins from a mixture by selective solubilization into a microemulsion phase. The procedure has been elegantly demonstrated using the same mixture of lysozyme, cytochrome c, and ribonuclease A as was used in the separation based on differences in solution pH. By intelligent use of the two variables pH and ionic strength, there seems to be a potential for effective separation of many complex protein mixtures.

A protein extracted into a microemulsion can subsequently be recovered by having the microheterogeneous solution come in contact with water, leading to back transfer of the newly isolated product. In studies of the kinetics of the forward and reverse processes, i.e., extraction into and out of the microemulsion, respectively, it has been found that the former transfer is much faster than the latter [70]. In fact, it has been shown that whereas forward transfer of proteins occurs up to three orders of magnitude faster than forward transfer of small solutes, the reverse transfer occurs at rates two to three orders of manitude slower than rates reported for back transfer of small solutes [71]. The facilitated solubilization into microemulsion and the resistance to desolubilization indicate an active role of the biomolecule in the formation of protein-filled reversed micelles. It has been suggested that protein-surfactant interactions constitute a strong driving force for proteins to enter into the microemulsion droplets. For protein positions close to the liquid-liquid interface, charge interactions produce a significant electrostatic force at that interface. The change of interfacial shape resulting from such forces facilitates protein envelopment by the interface as the microemulsion droplet is formed. The magnitude of the electrostatic attraction between a protein particle and the head groups of surfactants aligned at the interface will depend on the pH and ionic strength of the aqueous phase [71].

REFERENCES

1. M. A. Wells, Biochemistry *13*:4937 (1974).
2. P. H. Poon and M. A. Wells, Biochemistry *13*:4928 (1974).
3. R. L. Misiorowsky and M. A. Wells, Biochemistry *13*:4921 (1974).
4. C. Gitler and M. Montal, FEBS Lett. *28*:329 (1972).
5. C. Gitler and M. Montal, Biochem. Biophys. Res. Commun. *6*:1486 (1972).
6. M.-P. Pileni, J. Phys. Chem. *97*:6961 (1993).

7. K. Martinek, A. V. Levashov, N. L. Klyachko, and I. V. Berezin, Dokl. Akad. Nauk SSSR *236*:920 (1977).

8. P. L. Luisi, F. Henninger, and M. Joppich, Biochem. Biophys. Res. Commun. *74*:1384 (1977).

9. K. Martinek, I. V. Berezin, Yu. L. Khmelnitski, N. L. Klyachko, and A. V. Levashov, Biocatalysis *1*:9 (1987).

10. Yu. L. Khmelnitski, A. V. Kabanov, N. L. Klyachko, A. V. Levashov, and K. Martinek, in *Structure and Reactivity in Reverse Micelles* (M. P. Pileni, ed.), Elsevier, Amsterdam, 1989, p. 230.

11. J. W. Shield, H. D. Ferguson, A. S. Bommarius, and T. A. Hatton, Ind. Eng. Chem. Fundam. *25*:603 (1986).

12. R. H. Valivety, P. J. Halling, and A. R. Macrae, in *Biocatalysis in Non-Conventional Media* (J. Tramper, M. H. Vermüe, H. H. Beeftink, and U. von Stockar, eds.), Elsevier, Amsterdam, 1992, p. 549.

13. V. Dorovska-Taran, C. Veeger, and A. J. W. G. Visser, in *Biocatalysis in Non-Conventional Media* (J. Tramper, M. H. Vermüe, H. H. Beeftink, and U. von Stockar, eds.), Elsevier, Amsterdam, 1992, p. 697.

14. R. Bru, A. Sanchez-Ferrer, and F. Garcia-Carmona, Biotechnol. Bioeng. *34*: 304 (1989).

15. V. Papadimitriou, A. Xenakis, and A. E. Evangelopoulos, Colloids Surfaces *B1*:295 (1993).

16. D. Han and J. S. Rhee, Biotechnol. Bioeng. *28*:1250 (1986).

17. K. Martinek, A. V. Levashov, N. Klyachko, Yu. L. Khmelnitski, and I. V. Berezin, Eur. J. Biochem. *155*:453 (1986).

18. M. Gonnelli and G. B. Strambini, J. Phys. Chem. *92*:2854 (1988).

19. P. Walde, D. Han, and P.L. Luisi, Biochemistry *32*:4029 (1993).

20. E. Ruckenstein and P. Karpe, Biotechnol. Lett. *12*:241 (1990).

21. Q. Mao and P. Walde, Biochem. Biophys. Res. Commun. *178*:1105 (1991).

22. C. Laane, S. Borren, R. Hilhorst, and C. Veeger, in *Biocatalysis in Organic Media* (C. Laane, J. Tramper, and M. D. Lilly, eds.), Elsevier, Amsterdam, 1987, p. 65.

23. R. H. Valivety, G. A. Johnston, C. J. Suckling, and P. J. Halling, Biotechnol. Bioeng. *38*:1137 (1991).

24. E. Strika-Alexopoulos, J. Muir, and R. B. Freedman, in *Biocatalysis in Non-Conventional Media* (J. Tramper, M. H. Vermüe, H. H. Beeftink, and U. von Stockar, eds.), Elsevier, Amsterdam, 1992, p. 705.

25. E. Österberg, C. Ristoff, and K. Holmberg, Tenside *25*:293 (1988).

26. M.-B. Stark, P. Skagerlind, K. Holmberg, and J. Carlfors, Colloid Polym. Sci. *268*:384 (1990).

27. P. Skagerlind, M. Jansson, B. Bergenståhl, and K. Hult, J. Chem. Tech. Biotechnol. *54*:277 (1992).

28. P. Skagerlind and K. Holmberg, J. Dispers. Sci. Techn. *15*:317 (1994).

29. G. G. Zampieri, H. Jäckle, and P. L. Luisi, J. Phys. Chem. *90*:1849 (1986).

30. K. J. Levashov, Yu. L. Khmelnitski, N. L. Klyachko, V. Ya Chernyak, and K. Martinek. J. Colloid Interface Sci. *88*:444 (1982).

31. A. M. Brzozowski, U. Derewenda, Z. S. Derewenda, G. G. Dodsson, D. M. Lawson, J. P. Turkenburg, F. Björkling, B. Huge-Jensen, S. A. Patkar, and L. Thim, Nature *351*:491 (1991).
32. M. Bello, D. Thomas, and M.D. Legoy, Biochem. Biophys. Res. Commun. *146*:361 (1987).
33. P. D. I. Fletcher, R. B. Freedman, B. H. Robinson, G. D. Rees, and R. Schomäcker, Biochim. Biophys. Acta *912*:278 (1987).
34. C. P. Singh, D. O. Shah, and K. Holmberg, J. Am. Oil Chem. Soc. *71*:583 (1994).
35. C. P. Singh, P. Skagerlind, K. Holmberg, and D. O. Shah, J. Am. Oil Chem., Soc. *71*:1405 (1994).
36. D. G. Hayes and E. Gulari, Biotechnol. Bioeng. *35*:793 (1990).
37. D. G. Hayes and E. Gulari, Biotechnol. Bioeng. *40*:110 (1992).
38. A. Xenakis, T. P. Valis, and N. Kolisis, Prog. Colloid Polym. Sci. *84*:508 (1991).
39. H. Stamatis, A. Xenakis, U. Menge, and F. N. Kolisis, Biotechnol. Bioeng. *42*:931 (1993).
40. H. Stamatis, A. Xenakis, M. Provelegiou, and F. N. Kolisis, Biotechnol. Bioeng. *42*:103 (1993).
41. S. E. Godtfredsen, O. Kirk, F. Björkling, and L. Bjerre Christensen, Proceedings of the IUPAC-NOST International Symposium on Enzymes in Organic Solvents, New Delhi, 1992.
42. G. D. Rees, M. G. Nascimento, T. R. J. Jenta, and B. H. Robinson, Biochim. Biophys. Acta *1073*:493 (1991).
43. G. D. Rees, T. R. J. Jenta, M. G. Nascimento, M. Catauro, B. H. Robinson, G. R. Stephenson, and R. D. G. Olphert, Indian J. Chem. *32B*:30 (1993).
44. M. G. Nascimento, M. C. Rezende, R. D. Vecchia, P. C. Jesus, and L. M. Z. Aguiar, Tetrahedron Lett. *33*:5891 (1992).
45. K. Holmberg and E. Österberg, J. Am. Oil Chem. Soc. *65*:1544 (1988).
46. D. Han, J. S. Rhee, and S. B. Lee, Biotechnol. Bioeng. *30*:381 (1987).
47. P. D. I. Fletcher, B. H. Robinson, R. B. Freedman, and C. Oldfield, J. Chem. Soc. Faraday Trans. I *81*:2667 (1985).
48. A. Xenakis, T. P. Valis, and F. N. Kolisis, Prog. Colloid Polym. Sci. *79*:88 (1989).
49. Y. Miyake, T. Owari, K. Matsuura, and M. Teramoto, J. Chem. Soc. Faraday Trans. *89*:1993 (1993).
50. K. Holmberg, B. Lassen, and M.-B. Stark, J. Am. Oil Chem. Soc. *66*:1796 (1989).
51. K. Holmberg and E. Österberg, Prog. Colloid Polym. Sci. *74*:98 (1987).
52. E. Österberg, A.-C. Blomström, and K. Holmberg, J. Am. Oil Chem. Soc. *66*: 1330 (1989).
53. A.-C. Blomström, E. Österberg, and K. Holmberg, unpublished work.
54. E. W. Merrill and E. W. Salzman, J. Am. Soc. Artif. Intern. Organs *6*:60 (1983).

55. K. Holmberg, K. Bergström, C. Brink, E. Österberg, F. Tiberg, and J. M. Harris, J. Adhesion Sci. Technol. *7*:503 (1993).

56. J. M. Harris and K. Yoshinaga, J. Bioact. Compatible Polym. *4*:281 (1989).

57. D. H. Napper, J. Colloid Interface Sci. *58*:390 (1977).

58. K. Bergström and K. Holmberg, Colloids Surfaces *63*:273 (1992).

59. V. Thomas, K. Bergström, G. Quash, and K. Holmberg, Colloids Surfaces *A77*:125 (1993).

60. P. A. Srere and K. Uyeda, in *Methods in Enzymology* (K. Mosbach, ed.), Academic Press, New York, 1986, vol. 19, p. 11.

61. K. Holmberg and M.-B. Stark, Colloids Surfaces *47*:211 (1990).

62. K. E. Goklen and T. A. Hatton, Separation Sci. Technol. *22*:831 (1987).

63. K. L. Kadam, Enzyme Microb. Technol. *8*:266 (1986).

64. E. B. Leodidis and T. A. Hatton, in *The Structure, Dynamics and Equilibrium Properties of Colloidal Systems* (D. M. Bloor and E. Wyn-Jones, eds.), Kluwer Academic Publ., Amsterdam, 1990, p. 201.

65. J. M. S. Cabral and M. R. Aires-Barros, in *Recovery Processes of Biological Materials* (J. F. Kennedy and J. M. S. Cabral, eds.), Wiley, Chichester, UK, 1993, p. 247.

66. T. A. Hatton, in *Surfactant-Based Separation Processes* (Surfactant Science Series No. 33), Marcel Dekker, New York, 1989, p. 55.

67. K. E. Goklen, Ph.D. dissertation, Massachusetts Institute of Technology, Cambridge, 1986.

68. J. M. Woll, A. S. Dillon, R. S. Rahaman, and T. A. Hatton, in *Protein Purification: Micro to Macro* (R. Burgess, ed.), Alan R. Liss, New York, 1987.

69. P. L. Luisi, V. E. Mire, H. Jaeckle, and A. Pande, in *Topics in Pharmaceutical Sciences* (D. D. Breimer and P. Speiser, eds.), Elsevier, Amsterdam, 1983, p. 243.

70. M. Dekker, K. Van't Riet, B. H. Bijsterbosch, P. Fijneman, and R. Hilhorst, Chem. Eng. Sci. *45*:2949 (1990).

71. S. R. Dungan, T. Bausch, T. A. Hatton, P. Plucinski, and W. Nitsch, J. Colloid Interface Sci. *145*:33 (1991).

5

Microemulsions in the Pharmaceutical Field: Perspectives and Applications

MARIA ROSA GASCO Dipartimento di Scienza e Tecnologia del Farmaco, Università degli Studi di Torino, Turin, Italy

I. INTRODUCTION

The concept of microemulsions was introduced by Hoar and Schulman in 1943 [1]; a vast literature on these systems has subsequently been developed [2–5].

Microemulsions are transparent, thermodynamically stable dispersions of water and oil, usually stabilized by a surfactant and a cosurfactant. They contain particles smaller than 0.1 μm. Microemulsions are often defined as thermodynamically stable liquid solutions [4,6]; this includes normal micellar solutions, reverse micelles, cores or droplets of water or oil, and, for some systems, even bicontinuous structures, in which neither oil nor water surrounds the other. Analysis of the thermodynamic stability indicates that a microemulsion, being a liquid-liquid dispersed system, consists of two bulk phases separated by an interface region. In this respect microemulsion differ from micellar solutions; there are no direct means, however, of distinguishing between the two states. By contrast, a clear-cut distinction exists between microemulsions and coarse emulsions. The latter are thermodynamically unstable, droplets of their dispersed phase are generally larger than 0.1 μm, and, consequently, their appearance is normally milky rather than transparent.

The stability of microemulsion is a consequence of the ultralow interfacial tension between the oil and water phases. Cosurfactant and surfactant contribute to the low tension; the cosurfactant molecules intercalate between the surfactant molecules at the oil-water interface, thus affecting the curvature of the droplet [2–4]. The molecular structure of the surfactant and cosurfactant, as well as their concentrations, determines the microstructure; the structure of the oil, which may penetrate the interface, also plays a role [7].

Phase diagrams have been used extensively in the study of this phenomenon and have afforded comprehensive knowledge of how to characterize the domains of microemulsions and of other systems (such as liquid crystals and coarse emulsions) [8–10].

As a consequence of the potential advantages of microemulsions over conventional emulsions, interest in their use in the pharmaceutical field as potential drug delivery systems has grown progressively in the past decade.

II. ADVANTAGES OF THE USE OF MICROEMULSIONS AS CARRIERS OF DRUGS

Microemulsions exhibit several properties that are of particular interest in pharmacy:

Their thermodynamic stability allows self-emulsification of the system, whose properties are not dependent on the process followed; the temperature range over which the phases do not separate can be rather wide.

Microemulsions act as supersolvents of drugs (including drugs that are relatively insoluble in both aqueous and hydrophobic solvents), probably as a consequence of the presence of the surfactant and the cosurfactant.

The dispersed phase, lipophilic or hydrophilic [oil-in-water (O/W) or water-in-oil (O/W) microemulsion, respectively], can behave as a potential reservoir of lipophilic or hydrophilic drugs, respectively. The drug will be partitioned between dispersed and continuous phases, and when the system comes into contact with a semipermeable membrane, with skin or mucous membrane, the drug can be transported through the barrier. Drug release with pseudo-zero-order kinetics can be obtained [11], depending on the volume of the dispersed phase, the partition of the drug among interphase and continuous and dispersed phases, and the transport rate of the drug.

The mean diameter of the droplets in microemulsions (considering a droplet microstructure) is below 100 nm. Such a small particle size yields a very large interfacial area, from which the drug can quickly be released into the external phase when in vitro or in vivo absorption takes place, maintaining the concentration in the external phase close to initial levels.

The technology required to prepare microemulsions is simple, because their thermodynamic stability means that no significant energy contribution is required.

Microemulsions can be sterilized by filtration, as the mean diameter of the droplets is below 0.22 μm.

Autoxidation of lipids in O/W microemulsions is lower than in emulsions or micellar solutions as shown in a study of oxidation rates, using linoleic acid as a model molecule [12].

Solid colloidal therapeutic systems can be obtained with both O/W and W/O microemulsions.

Lipophilic and hydrophilic drugs can be carried together in the same microemulsion.

Microemulsions have low viscosity.

The use of microemulsions as delivery systems can improve the efficacy of a drug, allowing the total dose to be reduced and thus minimizing side effects.

Microemulsions may become unstable at high or low temperatures, but their formation is reversible (when the temperature returns to the stability range).

III. FACTORS LIMITING THE USE OF MICROEMULSIONS IN PHARMACY

The limits on the use of microemulsions in the pharmaceutical field arise chiefly from the need for all the components to be acceptable, particularly surfactants and cosurfactants. More studies will have to be done to increase the number of biocompatible components. Other relevant factors are the following:

Thermodynamic stability must be maintained, at least over the range of temperature between 4° and 40°C.

Pressure must be kept reasonably constant during storage.

Salinity may have a great effect on the domains in phase diagrams, as well as on the microemulsion structure itself [13].

The amounts of cosurfactant and surfactant required to form microemulsions are usually higher than those required for emulsions.

IV. MICROEMULSIONS CARRYING DRUGS

A. Formulation

A number of steps are required to formulate a microemulsion for therapeutic use. After selecting the components (usually oil, surfactant, cosurfactant, and hydrophilic medium) and after some preliminary tests, a series of phase diagrams is made to establish the different zones (microemulsions, liquid crystals, coarse emulsion, etc.), usually by physical methods [14] in the absence of the drug. Phase diagrams are typically constructed in three dimensions, either maintaining the fourth component constant or maintaining the ratio between two of the components fixed.

Lecithins are naturally occurring, nontoxic, biocompatible surfactants, and thus the formulation of lecithin-based microemulsions is of considerable pharmaceutical interest. The effects of the nature and concentration of seven short-chain alcohols on the area of existence of microemulsions prepared using water, isopropyl myristate, and lecithin have been investigated [15]; sec-butanol, n-butanol, and isobutanol were found to have the optimum properties, mostly because of their intermediate solubility in water, making possible stable microemulsions at a wide range of lecithin concentrations. Attwood et al. [16] studied the phase properties of an O/W microemulsion, examining the influence of the lecithin/cosurfactant ratio on the area of existence of the O/W microemulsion domain. Figure 1

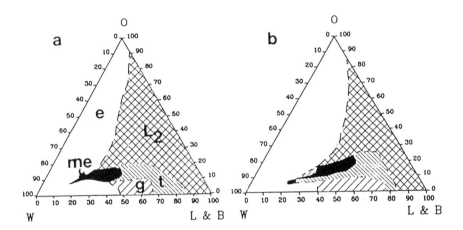

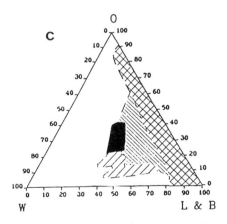

FIG. 1 Partial phase diagrams of the system soya lecithin/butanol/isopropyl myristate/water showing stable oil-in-water microemulsion (me), gel (g), monophasic turbid (t), unstable emulsion (e), and isotropic (L_2) regions, for soya lecithin/butanol weight ratios of (a) 1:0.6, (b) 1:0.45, and (c) 1:0.33. O, isopropyl myristate; W, water; L&B, lecithin + butanol. (From Ref. 16.)

shows partial phase diagrams of the soya lecithin/butanol/isopropyl myristate/water systems, maintaining fixed three different soya lecithin/butanol weight ratios. As the lecithin content increased, with a change in the ratio from 1:0.6 to 1:0.33, the amount of isopropyl myristate in the microemulsion increased from between 8 and 19% by weight (for the 1:0.6 ratio) to between 24 and 42% by weight (for the 1:0.33 ratio). If microemulsions are considered as carriers, the more oil the dispersed phase contains, the greater the opportunity to solubilize a lipophilic drug.

After defining the microemulsion domain, the second step is to define the domain in the presence of the drug; since it is known that addition of a drug to a microemulsion makes the percentages of the components of the mixture vary and may affect the stability of the system, the aim is to establish the maximum amounts that the microemulsions can carry. With a pseudoternary diagram, by keeping fixed the amounts of two components (or more than two in the case of more then five components in the system), the domain in the presence of a drug can be established.

Chemometric procedures provide an effective tool for optimizing pharmaceutical mixtures, as are most dosage forms. These procedures can also be applied to complex systems, such as microemulsions. By statistical methods and with a reduced number of experiments, it is possible to approach suitable drug release as a function of microemulsion formulation or to establish structural parameters that affect the formation of a microemulsion [17,18].

B. Partition of a Drug Among the Phases of a Microemulsion

In predicting the release of a drug from a microemulsion, one of the most important problems is to evaluate its partition among interphase and continuous and dispersed phases; the problem cannot easily be solved, as microemulsions are systems with infinite stability in the range of temperatures studied. In order to know roughly how a drug is partitioned, its partition coefficient (P_{cos}) among oil, water, and cosurfactant, in the ratios present in the microemulsion, can be determined [19]. In order to evaluate the significance of P_{cos}, five drugs with varying lipophilicity (menadione, phenylbutazone, betamethasone, nitrofurazone, and prednisone) were dissolved in an O/W microemulsion (isopropyl myristate, buffer pH 7.0, dioctylsulfosuccinate, butanol) of constant composition. Table 1 shows how the partition coefficients of the drugs differ in the presence and in the absence of the cosurfactant (butanol). Moreover, the logarithms of the coefficients of permeation of the drugs through a hydrophilic membrane were inversely proportional to the logarithms of P_{cos}: the higher the concentration of the

TABLE 1 Partition Coefficients of Drugs

Drug	P_{oct}	P_{IPM}	P_{cos}
Nitrofurazone	1.8 ± 0.1	0.12 ± 0.02	1.5 ± 0.1
Phenylbutazone	5.3 ± 0.2	2.1 ± 0.2	4.5 ± 0.2
Prednisone	29 ± 2	0.30 ± 0.05	6.4 ± 0.2
Betamethasone	95 ± 5	2.3 ± 0.3	30 ± 2
Menadione	160 ± 10	100 ± 8	180 ± 12

P_{oct} for octanol/buffer, P_{IPM} for isopropyl myristate/buffer, P_{cos} for isopropyl myristate/butanol/buffer partition coefficients.
Source: Ref. 19.

drug in the internal phase (reservoir), the lower the amount released over time from the systems (Fig. 2).

Modulation of drug release can be advantageous. This may be achieved in a microemulsion by keeping the amounts of the other components fixed and changing only the amount of the cosurfactant; as a consequence, P_{cos} and permeation coefficients of the drug vary [19].

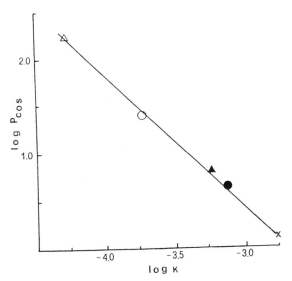

FIG. 2 Logarithm of permeation coefficients (log K_{meas}) of drugs through a hydrophilic membrane versus log P_{cos}. P_{cos} = partition coefficient of drugs among isopropyl myristate, buffer, and butanol. Key: nitrofurazone ($\times$); phenylbutazone ($\bullet$); prednisone ($\blacktriangle$); bethametasone ($\bigcirc$); menadione ($\triangle$). (From Ref. 19.)

C. Increase of the Lipophilicity of Drugs that Are Weak Acids or Bases

Many drugs are weak acids or bases; because of their hydrophilicity and the pH of the external phase, it is often impossible to enrich the dispersed phase of an O/W microemulsion. In many cases the drug is not sufficiently lipophilic to be adequately partitioned in the internal phase. By forming ion pair complexes, it is possible to enhance the lipophilicity of a drug. propranolol, a weak base, has been solubilized in an O/W microemulsion consisting of Tween 60, isopropyl myristate, butanol, and buffer pH 6.5 [11]. The apparent partition coefficient, P_{cos}, of propranolol among isopropyl myristate, butanol, and buffer increased from 1.8 (in the absence of any counterion) to 15.1 in the presence of octanoic acid (1:16.4 drug/octanoic acid molar ratio). Consequently, the permeation coefficient of the drug from the relative microemulsions decreased and the release of propranolol from the microemulsion, in the presence of octanoic acid, followed pseudo-zero-order kinetics.

D. Evaluation of the Influence of Drugs in Microemulsions by Microcalorimetry

Microcalorimetry may be used to evaluate the enthalpy associated with the microemulsification process in the presence and in the absence of drugs [20,21]. The enthalpies obtained by increasing amounts of cosurfactant (butanol) added to a mixture with fixed amounts of the other three components (surfactant, isopropyl myristate, water) can be measured by microcalorimetry.

Two oil-in-water microemulsions with the same oil/water/cosurfactant mixtures but differing in the surfactant and consequently in the percentage amounts of the components were considered. In both cases aliquots of cosurfactant were added until a clear system was formed. The variation in enthalpy, ΔH, and the molar fractions of added butanol below the amount required to form a microemulsion were found to be linearly related. On microemulsion formation, a large variation from linearity (a minimum in endothermicity) was found. It was probably due to the overlap of different phenomena. An exothermic contribution due to the microemulsion formation was superimposed on the endothermic effect of butanol addition to the mixture. The two microemulsions behaved in the same way. The addition of two drugs with different lipophilicities (prednisone and menadione) did not modify the general trend. The only appreciable effect was found with menadione. A lower endothermic value, when lecithin was present in the microemulsion, was found. It is relatable to the weak interactions among components, which would not be easily detected with other techniques.

V.   APPLICATIONS OF MICROEMULSIONS IN THE PHARMACEUTICAL FIELD

Many studies have been performed in vitro and in vivo and various components have been examined. Biocompatibility, which is obviously the first requirement, limits the choice of the components for application in the pharmaceutical field to a limited number of oils, surfactants, and cosurfactants.

A.   Percutaneous Administration

The percutaneous administration route has been extensively studied [22–24]. Drug transport from microemulsions is usually better than that from other ointment, gels, or creams. Systemic medication has also been achieved. An explanation for this facilitated transport is that the drugs are completely dissolved in microemulsions, reaching relatively high concentrations as a consequence of the supersolvent properties of microemulsions, and the dispersed phase can act as a reservoir, making it possible to maintain an almost constant concentration in the continuous phase; thus pseudo-zero-order kinetics can be achieved. Moreover, some components of the microemulsions can operate as enhancers. According to the requirements, the drug properties, and the formulation, the release rate may be controlled.

Friberg [25] states that great attention must be paid to the possibility of water evaporating from a microemulsion after its application onto the skin; usually the vapor pressure of water is higher than those of the other components. A microemulsion is a colloidal dispersion, and when only water is present in the aqueous phase, a preliminary evaluation of the rate of evaporation of the water from the microemulsion after its application to the skin is essential; a consequence of such evaporation could be the formation of a different system consisting of only oil, surfactant, and cosurfactant, which could influence the lipid structure of the stratum corneum of the skin. It is then advisable to use an occlusive patch after applying the microemulsion, even when a humectant, such as propylene glycol, is present in the formulation, to avoid the evaporation of water.

Müller et al. [26] described the percutaneous absorption of radiolabeled arecaidine propylester in rabbits, comparing the drug concentrations in the blood after cutaneous application of a W/O microemulsion, an O/W microemulsion, a microgel, an emulsion, and a lanolin ointment. The highest blood concentrations were obtained for the two microemulsions.

Ziegenmeyer and Führer [27] have compared the release of tetracycline hydrochloride from different formulations. Figure 3 shows the in vitro permeation of the drug through skin membranes versus time for three different formulations. The results show that transport of the drug is significantly better for the microemulsion. Moreover, the lag time is notably lower, probably due to microemulsion components that may have enhancer properties. Franz and Ziegenmeyer [28] obtained a systemic effect with percutaneous administration; tests were conducted on human volunteers using tizanidine and compared with oral administration. Urine concentrations of the unchanged active agent and two of its metabolites were measured; significant percutaneous absorption of tizanidine from the microemulsion was found, which is not possible with other topically administered systems such as emulsions or gels. Labeled alpha-tocopherol, applied to the backs

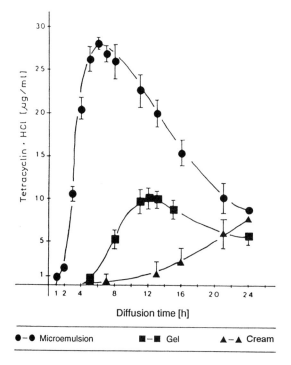

FIG. 3 In vitro permeation of tetracycline hydrochloride through skin membranes from three different formulations: (●) microemulsion; (■) gel; (▲) cream. (From Ref. 27.)

of rats in a microemulsion, W/O emulsion, or Vaseline ointment was transported faster from the microemulsion [29].

Osborne et al. [30] studied the effect of microemulsion composition on the in vitro human skin transport of radiolabeled water from three microemulsions. They characterized a W/O microemulsion region by using a ternary diagram, since their system had only three components (water/ octanol/dioctylsulfosuccinate). The system with a 58:42 dioctylsulfosuccinate/octanol weight ratio can incorporate over 70% water. Therefore a 58:42 fixed weight ratio of surfactant to octanol was maintained and water concentrations of 15, 35, and 68% by weight were used, resulting in clear systems. Most of the water in the 15% water microemulsions is bound to the surfactant head groups and is not available for transport across the skin; thus the transport of water from that microemulsion was lower than the transport from water alone. For the microemulsions with a higher water content, transport of water was enhanced. Table 2 gives the in vitro transdermal flux values of water and the lag times for microemulsions containing different amounts of water: the flux of water from the 68% water microemulsion is more than 10 times that from the 15% water microemulsion. These results were confirmed in a study of the transport of glucose, as a model hydrophilic molecule, from the same three microemulsions across human cadaver skin [31]; it was shown that both 35 and 68% water microemulsions caused enhanced transport but no transport was discernible for the 15% water microemulsion.

Müller and Kleinebudde [32,33] have studied the influence of in situ formation of microemulsions on the transdermal delivery of drugs. The phase diagrams (or pseudoternary phase diagrams) of the four-component

TABLE 2 In Vitro Transdermal Flux Values and Lag Times for the Microemulsion Formed by Addition of Water to a 58:42 Weight Ratio of Dioctylsulfosuccinate/octanol

	Flux (mg cm^{-2} h^{-1})	Lag time (h)	Corr.[a]
Pure water	0.8	2.1	0.9997
15% water microemulsion	0.3	−1.8	0.9968
35% water microemulsion	1.6	4.2	0.9996
68% water microemulsion	4.0	5.1	0.9968

[a]Linear regression correlation coefficient based on data collected from 11–24 h for the 15 and 35% microemulsions, 13–24 h for the 68% water microemulsion, and 8–19 h for water.
Source: Ref. 31.

system (Tween 85/Pluronic L31/isopropyl palmitate/water) were investigated, and the microemulsion region was characterized; water-free systems were then selected for preparation in the microemulsion domain; indomethacin and β-blockers of different polarities and solubilities in oil were used as model drugs. In the basic water-free system, the saturation solubility of the drugs increased with increased drug lipophilicity and as the melting point decreased. After application of the water-free system onto the skin, the water uptake from the stratum corneum forms the microemulsion and the water content decreases the solubility of the drug, leading to supersaturated systems and thereby promoting drug absorption. Tests on the relation between the saturation solubility of drugs and water content, however, showed that only drugs which the authors indicated being nonpolar relative to the microemulsion became supersaturated. In contrast, semipolar substances, such as the β-blocker atenolol, exhibited an increase in saturation solubility with increasing water content. The pharmacodynamic effect of skin application of water-free microemulsion bases saturated with a series of β-blockers, covered by an occlusive patch, was evaluated using rabbits as suitable in vivo models [34]. Occlusion leads to water uptake from the skin due to hydratation, changing the water-free system to a microemulsion. Carazolol applied dermally is therapeutically equivalent to about 1% of the dose given intravenously. Figure 4 shows the influence of bupranolol concentration on transdermal absorption and consequently on the pharmacodynamic effect. The most marked increase in effect with time was obtained with a near-saturated preparation (7.65%).

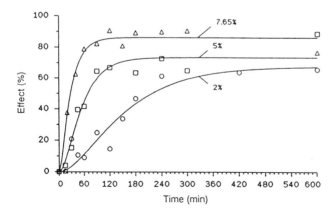

FIG. 4 Effect vs. time curves after dermal application of microemulsions containing 7.65, 5, and 2% bupranolol, respectively (dose: 2 mg kg^{-1}; fitted curves, mean of three rabbits). (From Ref. 34.)

The pharmacodynamic effect of the model drug bupranolol, applied as a water-free microemulsion base in vivo to rabbits, was examined as a function of the solubility in vitro of bupranolol with respect to the water content [35]. The in situ formation of supersaturated microemulsions from applied water-free bases enhances the bupranolol flux through the skin. The pharmacodynamic effects obtained with the water-free bases applied in vivo can be correlated with the decline of the solubility of bupranolol vs. water curves in vitro; the higher the solubility, the lower the pharmacodynamic effect.

Azelaic acid applied percutaneously has been shown to have some therapeutic effect on lentigo maligna; high amounts are required to be absorbed. Administration of azelaic acid is not feasible by the oral or the parenteral route, as it is quickly biotransformed. An O/W microemulsion carrying azelaic acid was selected on the basis of the apparent partition coefficient P_{cos}, the in vitro permeation coefficient, and the amount of drug solubilized; the system was then thickened [36]. Figure 5 shows the fluxes of azelaic acid through hairless mouse skin from the microemulsion (6.4% of azelaic acid) and from a gel (15% suspension of acelaic acid). The gel was selected because it contained several components of the microemulsion

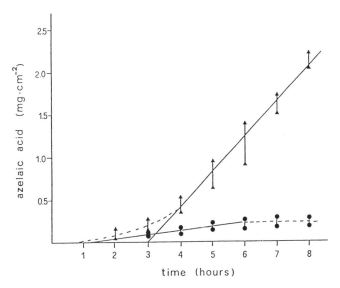

FIG. 5 Permeation profiles of azelaic acid through hairless mouse skin from viscosized microemulsion (▲) and from gel (●). Vertical bars indicate SD. (From Ref. 36.)

[37]. The flux of azelaic acid from the microemulsion, at steady state, is about 10 times higher than from the gel. The thickened microemulsion gave positive results in vivo [38]. Complete remission of the lesions was achieved in all cases of lentigo maligna treated in the last 8 years [39].

Luisi et al. [40] studied W/O microemulsion gels; they were obtained by starting from a reverse micellar solution of lecithin in an organic solvent and adding a small amount of water; the gels are isotropic, thermoreversible, and optically transparent. They can be used as a matrix for the transdermal delivery of drugs [41]; scopolamine release from a microemulsion gel was seven times higher after 75 h than from a scopolamine solution at the same concentration. Transdermal delivery of various drugs has been studied in vitro, using microemulsion gels as the matrix. Isopropyl palmitate was studied as a constituent of microemulsion gels [42]. Isopropyl palmitate incorporation in isolated stratum corneum of human skin was measured by Fourier transform–infrared (FT–IR) spectroscopy; it is considered that modifications of the barrier function of the stratum corneum, due to the interactions of the components of the matrix gel with the skin, could induce more efficient drug transport.

B. Ocular Administration

Drugs intended to treat eye diseases are essentially delivered topically. O/W microemulsions have been investigated for ocular administration, to dissolve poorly soluble drugs, to increase absorption, and to prolong release time.

Siebenbrodt and Keipert [43] developed and characterized lecithin–Tween 80 based microemulsions, which dissolve some poorly soluble drugs such as atropine, chloramphenicol, and indomethacin, often used in local ocular therapy. Three drugs were solubilized in therapeutically relevant concentrations (0.5%) in microemulsions, and low physiological irritation and elevated and prolonged in vitro release were observed [44].

With the aim of enhancing the amounts of drug transported through the cornea and limiting transport through the conjunctiva, timolol, a β-blocker used in glaucoma therapy, has been dissolved in a lecithin-based O/W microemulsion [45]. Timolol was rendered lipophilic by ion pair formation with octanoate. Aliquots of timolol in solution, as an ion pair in solution, and as an ion pair in microemulsion were instilled in rabbits. Figure 6 shows the amounts of timolol in aqueous humor versus time; the areas under the curve for timolol in aqueous humour, after administration of the microemulsion and the ion pair solution, were 3.5 and 4.2 times higher, respectively, than that observed for timolol alone; even if the best result was obtained for the solution containing timolol as an ion-pair, absorption

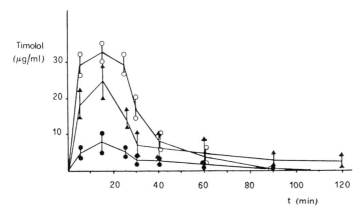

FIG. 6 Aqueous humour concentration-time profiles following multiple instillations in rabbits' eyes: (●) timolol alone; (○) timolol as an ion pair in solution; (▲) timolol as an ion pair in microemulsion. (From Ref. 45.)

times lengthened for timolol in microemulsion. It is probably that the tiny nanodroplets remained on the cornea for some time and acted as a microreservoir of timolol, appreciably prolonging the absorption time.

C. Peroral Administration

The recent advances in biotechnology permit rather high amounts of peptides and proteins to be produced. However physicochemical and biological properties of polypeptides, such as short half-life, biodegradability and conformational stability, cause considerable design difficulties in their formulation, particularly for peroral administration. Studies concerning the protection of biodegradable drugs from the biological environment when they are administered by the peroral route in the form of microemulsions consider primarily peptides and small proteins. Absorption in the small intestine has been extensively studied; Ritschel et al. [46] studied the absorption of cyclosporine, which is a potent, neutral, cyclic endekapeptide, an immunosuppressant drug widely used in transplants, whose bioavailability after peroral administration is very poor. Two W/O microemulsions were administered to rats perorally. It was found that for one of them, the absolute and relative bioavailability is better than that of commercially available solutions.

More recently, Drewe et al. [47] enhanced the oral absorption of cyclosporine by administering hard gelatin capsules containing two O/W microemulsions (slow and fast release) and a solid micellar solution of cyclosporine to healthy male volunteers. Absorption increased on average by 45% for the solid micellar solution and 49% for the faster releasing microemulsion, compared to the reference soft gelatin capsule. No differences were noted between the reference and the slow-releasing microemulsion (Fig. 7). The improved conditions for drug absorption may be related to the greater solubility of the lipophilic cyclosporine in the fast-releasing microemulsion and the micellar gelified solution, giving instantaneous absorption. By incorporating the drug in a microemulsion concentrate, a reduction of inter- and intraindividual variability in cyclosporine pharmacokinetics has been obtained [48]. Following oral administration, the concentrate forms a microemulsion in aqueous fluids, and cyclosporine is rapidly available for absorption, providing a more predictable concentration-time profile than the currently marketed formulations. Moreover, the influence of a fat-rich meal on the pharmacokinetics of cyclosporine

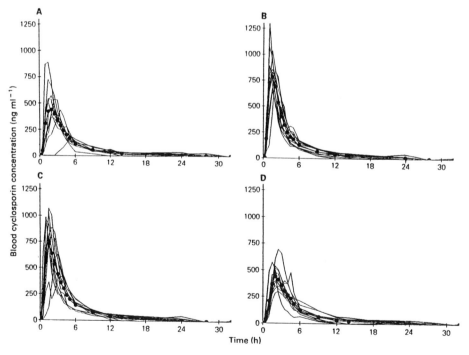

FIG. 7 Individual (——) and mean (-●-●-●-) blood cyclosporine concentrations after a single oral administration of 3 × 50 mg of (A) cyclosporine standard formulation; (B) fast-release microemulsion; (C) solid micellar solution; (D) slow-release microemulsion. (From Ref. 47.)

absorption has been shown to be less pronounced for the microemulsion concentrate than for other commercial mixtures [49].

Significant intraduodenal enhancement of a model molecule, calcein, and of a peptide has been found in rats after administration of a W/O microemulsion [50]. A mixture of two nonionic surfactants at low and high hydrophile-lipophile balance (HLB), with no cosurfactant, was used for the microemulsion. Although the small intestine has the greatest absorbing capacity, its high peptidase content has a negative effect on delivery of peptide drugs. The release of peptides from a carrier system to the colon, where enzyme activity is lower, can improve their absorption. Gelled microemulsions have been considered as carriers for the colonic release of insulin [51]; they were filled into gelatin capsules hardened with formaldehyde and coated with polymers, in order to obtain pH-dependent and time-controlled release of insulin. The capsule was programmed to dissolve in the colon, as tested with a model molecule. The experiment was carried out in beagle dogs, and the reduction of the blood glucose concentration was evaluated. The colonic release of insulin from the capsule dosage form gave a reduction of 18% of the glucose level, which reached 21% in the presence of an enzyme inhibitor. The reference intravenous insulin gave a glucose reduction of 65% but a short duration of the effect (about 2 h); in the other cases the duration was 6–8 h.

D. Parenteral Administration

1. O/W Microemulsions

O/W microemulsions are used mainly as carriers of lipophilic drugs in order to attain prolonged release and to administer parenterally lipophilic substances that are not soluble in water. They can be administered intravenously, intramuscularly, or subcutaneously. Various patents concerning the vectorization of fluorocarbons, calcium antagonists, steroids, and other lipophilic drugs have already been reported [52]. O/W microemulsions have been injected into rats with the aim of targeting very lipophilic drugs [53] into reticuloendothelial system (RES) tissues (liver and spleen). The results indicate that the higher the partition coefficient of the drug, the better the targeting; an octanol-water partition coefficient above 10^8 is required to deliver the drug effectively to RES tissues.

2. W/O Microemulsions

W/O microemulsions can be used for subcutaneous and intramuscular administration; they may be indicated for parenteral administration of hydrophilic drugs with the aim of obtaining prolonged release.

D-Trp-6-LH-RH is an analogue of the peptidic hormone luteinizing hormone–releasing hormone (LH–RH); its half-life is 2 h 15 min. The effect of administration of the hormone is to decrease testosterone. After a single intramuscular injection of D-Trp-6-LH-RH contained in a phospholipid-based microemulsion in rats (3 mg/kg), the testosterone level decreased (Fig. 8). The level of testosterone in plasma between 10 and 20 days after the injection was lower than that observed after daily injection of 100 μg/kg of the drug in solution [54].

A microemulsion (surfactant egg lecithin) containing 0.5 mg/mL insulin, was administered subcutaneously in rabbits [55]. The results demonstrated that some pharmacokinetic parameters were modified compared to those found with a solution: the half-life of insulin was 1.6 h for the solution and reached 4.3 h for the microemulsion; t_{max} values were 0.7 and 1.8 h, respectively.

The behavior of a microemulsion at the injection site was investigated by observing the release of a radiodiagnostic pertechnetate from a W/O microemulsion and an aqueous solution after subcutaneous administration in rabbits, imaging the administration sites with a gamma camera [56]. Disappearance of pertechnetate in aqueous solution from the injection site

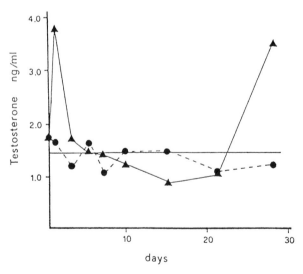

FIG. 8 Plot of mean plasma concentration-time curves of testosterone in rats, following a single intramuscular (i.m.) injection (3 mg/kg) of D-Trp-6-LH-RH in microemulsion (▲) and daily i.m. injection (100 μg/kg) of buffered solution (●). (From Ref. 54.)

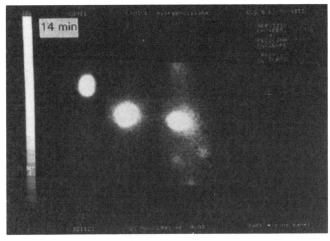

(a)

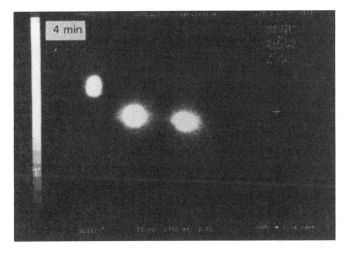

(b)

FIG. 9 Temporal sequences of images at 4, 14, 28, 50, and 66 min after injection. Left to right: reference source of radioactivity, rabbit injected with microemulsion, rabbit injected with solution. (From Ref. 56.)

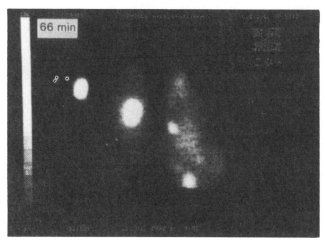

(c)

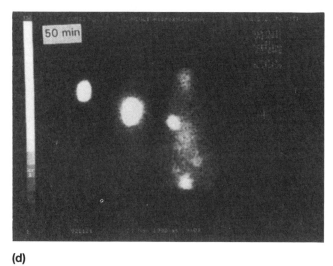

(d)

FIG. 9 Continued

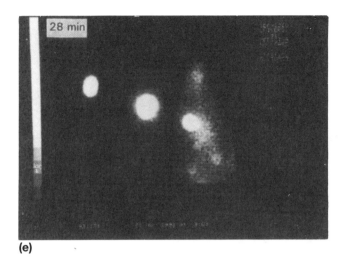

(e)

was about 10 times faster than that of pertechnetate in the microemulsion (Fig. 9). The kinetics followed was of the first order, as usual in subcutaneous injection.

W/O microemulsions may be indicated for the parenteral administration of short-half-life hydrophilic drugs. Specifically, peptides and small proteins could be administered parenterally in W/O microemulsions, with advantages such as protection of the molecule from the biological environment, prolonged release, and reduction of the drug's side effects.

VI. SOLID COLLOIDAL THERAPEUTIC SYSTEMS OBTAINED FROM MICROEMULSIONS

When a droplet microstructure is present, microemulsions offer a system of preformed nanodroplets that can be utilized to prepare different solid therapeutic colloidal systems.

A. From Water-in-Oil Microemulsions

Colloidal drug carriers are potential tools for achieving site-specific drug delivery. Polyalkylcyanoacrylate nanoparticles have been investigated as potential lisosomotropic carriers of drugs; they are prepared by emulsion polymerization, in which droplets of the insoluble monomers are emulsified

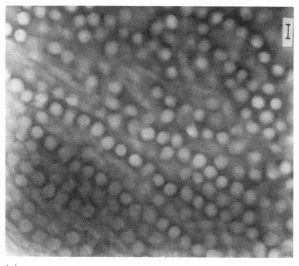

(a)

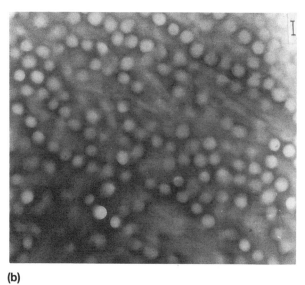

(b)

FIG. 10 Photomicrographs of lipospheres. Bar = 100 nm.

in the aqueous phase [57]. Samples of the nanoparticles showed that the inner structure was a highly porous matrix. The acute toxicity of poly-alkylcyanoacrylate nanoparticles was shown to be rather low and the toxicity of doxorubicin adsorbed onto such particles was markedly reduced [58]. Nanocapsules are prepared by dissolving the monomer alkylcyanoacrylate and a lipophilic drug in a lipidic phase that is slowly injected into an aqueous solution of a nonionic surfactant [59]. The most significant advantage of the use of polyalkylcyanoacrylates is their biodegradability.

From W/O microemulsions, nanocapsules of polyalkylcyanoacrylate can be prepared [60]. On slow addition of the lipophilic monomer to the microemulsion, an in situ nucleophilic polymerization takes place, yielding colloidal nanocapsules. The reaction also occurs in waterless microemulsions, with a nucleophilic component used for the dispersed phase. When a hydrophilic drug, such as doxorubicin, is dissolved in the aqueous dispersed phase, a large amount of the drug (about 10%) can be encapsulated [61], as shown in vitro and in vivo in rabbits [62]. Nanocapsules can function as microreservoirs of hydrophilic drugs.

B.   From Oil-in-Water Microemulsions

Many studies have been carried out on lipophilic carriers for the delivery of drugs, particularly with the aim of decreasing systemic side effects and of targeting them to the lymphatic system. Liposomes are suitable drug carriers [63]; lipid microspheres are liquid carriers of lipophilic drugs [64] obtained by dissolving the drugs in lecithin-based emulsions. Neither of these carriers is solid. Lipid nanopellets, consisting of lipids and surfactants, were described by Speiser [65], but particles over 10 μm were present. More recently, solid lipid nanoparticles were produced by high-pressure homogenization of melted lipids in water with or without surfactant [66,67].

Solid colloidal lipophilic systems can be obtained from warm O/W microemulsions. Components with relatively low melting points (60–70°F) (such as triglycerides and alcanoic acids) were chosen as the oil phase. They were melted and the mixture of the other components, in predetermined amounts, previously warmed to the melting temperature of the oil, was then added. Warm clear systems were obtained by mild mixing. The microemulsions were then dispersed in cold water (2–3°F) by mechanical mixing, yielding liposBorne [65]. They were washed by diafiltration and then freeze-dried. A colloidal dispersion was obtained by redispersion in water. The liposBorne have a diameter between 50 and 200 nm depending on the components and conditions.

Drugs can be dissolved in the oil phase of the warm microemulsions and on cooling are randomly distributed in the lipospheres. Figure 10 shows photomicrographs of lipospheres. Their regularity shows that the lipospheres may be considered solidified droplets of the hot microemulsion.

Drugs with high lipophilicity and other drugs that are weak bases or acids may be incorporated into lipospheres [69,70]; the latter drugs can be incorporated as ion pairs, reaching concentrations of 7–8%. Depending on the lipophilicity of the drug, release from lipospheres may be very slow. Hydrophilic peptides can also be incorporated into lipospheres [71] and have very slow and prolonged release in vitro.

VII. CONCLUSIONS

Microemulsions appear to be a promising delivery system for various routes of administration, such as percutaneous, peroral, ocular, and parenteral. They can offer several advantages, such as enhanced absorption of drugs, modulation of the kinetics of drug release, and decreased toxicity. Solid colloidal therapeutic systems may also be obtained from microemulsions.

REFERENCES

1. T. P. Hoar and J. H. Schulman, Nature *152*:102 (1943).
2. L. M. Prince, ed., *Microemulsions: Theory and Practice,* Academic Press, London, 1977.
3. K. Shinoda and S. E. Friberg, *Emulsions and Solubilization,* Wiley, New York, 1974.
4. H. L. Rosano and M. Clausse, eds., *Microemulsion Systems*, Surfactant Series, vol. 24, Marcel Dekker, New York, 1987.
5. S. E. Friberg and P. Bothorel, eds., *Microemulsions: Structure and Dynamics*, CRC Press, Boca Raton, FL, 1987.
6. I. Danielsson and B. Lindman, Colloid Surface *3*:391 (1981).
7. K. Shinoda, M. Araki, A. Sadaghiani, A. Khan, and B. Lindman, J. Phys. Chem. *95*:989 (1991).
8. H. L. Rosano, J. Soc. Cosmet. Chem. *25*:609 (1974).
9. M. Kahlweit, E. Lessner, and R. Strey, J. Phys. Chem. *87*:5032 (1983).
10. H. Kunieda and K. Shinoda, J. Colloid Interface Sci. *107*:107 (1985).
11. M. R. Gasco, M. E. Carlotti, and M. Trotta, Int. J. Cosmet. Sci. *10*:263 (1988).
12. M. E. Carlotti, M. R. Gasco, M. Trotta, and S. Morel, J. Soc. Cosmet. Chem. *42*:285 (1991).
13. P. Guering and B. Lindman, Langmuir *1*:464 (1985).
14. H. N. Bhargava, A. Narurkar, and L. M. Lieb, Pharm. Technol. *11*:46 (1987).
15. R. Aboofazeli and M. J. Lawrence, Int. J. Pharm. *93*:161 (1993).

16. D. Attwood, C. Mallon, and C. J. Taylor, Int. J. Pharm. *84*:R5 (1992).
17. F. Pattarino, E. Marengo, M. R. Gasco, and R. Carpignano, Int. J. Pharm. *91*: 157 (1993).
18. M. Gallarate, F. Pattarino, E. Marengo, M. R. Gasco, STP Pharma Sci. *3*:413 (1993).
19. M. Trotta, M. R. Gasco, and S. Morcl, J. Controlled Release *10*:237 (1989).
20. B. Fubini, M. R. Gasco, and M. Gallarate, Int. J. Pharm. *42*:19 (1988).
21. B. Fubini, M. R. Gasco, and M. Gallarate, Int. J. Pharm. *50*:213 (1989).
22. F. Fevrier, M. F. Bobin, C. Lafforgue, M. C. Martini, STP Pharma Sci. *1*:60 (1991).
23. A. Jayakrishnan, K. Kalaiarasi, and D. O. Shah, J. Soc. Cosmet. Chem. *34*: 335 (1983).
24. G. Vuleta and M. Stupar, Arh. Farm. *35*:79 (1985).
25. E. Friberg, J. Soc. Cosmet. Chem. *41*:155 (1990).
26. B. W. Müller, K. J. Franzky, and C. J. Kolln, Eur. Pat. Appl. 0-152-945 (1985).
27. J. Ziegenmeyer and C. Führer, Acta Pharm. Technol. *26*:273 (1980).
28. J. Franz and J. Ziegenmeyer, DE 3,212,053 (1982).
29. M. C. Martini, M. F. Bobin, H. Flandin, F. Caillaud, and J. Cotte, J. Pharm. Belg *39*:348 (1984).
30. D. W. Osborne, A. J. I. Ward, and K. J. O'Neill, Drug Dev. Ind. Pharmacy *14*:1203 (1988).
31. D. W. Osborne, A. J. I. Ward, and K. J. O'Neill, J. Pharm. Pharmacol. *43*: 451 (1991).
32. B. W. Müller and P. Kleinbudde, Pharm. Ind. *50*:370 (1988).
33. B. W. Müller and P. Kleinbudde, Pharm. Ind. *50*:1301 (1988).
34. J. Kemken, A. Ziegler, and B. W. Müller, J. Pharm. Pharmacol. *43*:679 (1991).
35. J. Kemken, A. Ziegler, and B. W. Müller, Pharm. Res. *9*:554 (1992).
36. M. R. Gasco, M. Gallarate, and F. Pattarino, Int. J. Pharm. *69*:193 (1991).
37. U. Maru, P. Michaud, J. Garrigue, J. Oustrin, and R. Ruffiac, J. Pharm. Belg. *37*:207 (1982).
38. M. R. Gasco, M. Gallarate, F. Pattarino, and G. Zina, Proc. Int. Symp. Control. Rel. *17*:419 (1990).
39. S. Colona, M. G. Bernengo, F. Pattarino, and M. R. Gasco, Eur. J. Pharm. Brupharm. 1996, 11S, 42nd Congress of the International Association for Pharmaceutical Technology. Mainz, March 7–9, 1996.
40. P. L. Luisi, R. Scartazzini, G. Haering, and P. Schurtenberger, Colloid Polym. Sci. *268*:356 (1990).
41. P. Walde, H. Willimann, C. Nastruzzi, R. Scartazzini, P. Schurtenberger, and P. L. Luisi, Proc. Int. Symp. Control. Rel. Bioact. Mater. *17*:421 (1990).
42. P. Walde, F. Dreher, and P. L. Luisi, Proc. Int. Symp. Control. Rel. Bioact. Mater. *20*:406 (1993).
43. I. Siebenbrodt and S. Keipert, Pharmazie *46*:435 (1991).
44. I. Siebenbrodt and S. Keipert, Eur. J. Pharm. Biopharm. *39*:25 (1993).
45. M. R. Gasco, M. Gallarate, M. Trotta, L. Bauchiero, E. Gremmo, and O. Chiappero, J. Pharm. Biomed. Anal. *7*:433 (1989).

46. W. A. Ritschel, S. Adolph, G. B. Ritschel, and T. Schroeder, Methods Find. Exp. Clin. Pharmacol. *12*:127 (1990).
47. J. Drewe, R. Meier, J. Vonderscher, D. Kiss, U. Posanski, T. Kissel, and K. Gyr, Br. J. Clin. Pharmacol. *34*:60 (1992).
48. J. M. Kovarik, E. A. Mueller, J. B. van Bree, W. Tetzloff, and K. Kutz, J. Pharm. Sci. *83*:444 (1994).
49. E. A. Mueller, J. M. Kovarik, J. B. van Bree, J. Grevel, P. W. Luecker, and K. Kutz, Pharm. Res. *11*:151 (1994).
50. P. P. Constantidines, J. P. Scalart, C. Lancaster, J. Marcello, G. Marks, H. Ellens, and P. Smith, Proc. Int. Symp. Control. Rel. Bioact. Mater. *20*:184 (1993).
51. M. E. K. Kraeling and W. A. Ritschel, Methods Find. Exp. Pharmacol. *14*:199 (1992).
52. S. Keipert, I. Siebenbrodt, F. Luders, and M. Bornschein, Pharmazie *44*:433 (1989).
53. T. Kakutani, Y. Nishihara, K. Takahashi, and K. Kirano, Proc. Int. Symp. Control. Rel. Bioact. Mater. *18*:359 (1991).
54. M. R. Gasco, F. Pattarino, and F. Lattanzi, Int. J. Pharm. *62*:119 (1990).
55. M. R. Gasco, S. Morel, E. Tonso, and I. Viano, Proc. Int. Symp. Control. Rel. Bioact. Mater. *19*:502 (1992).
56. M. Bellò, D. Colangelo, M. R. Gasco, F. Maranetto, S. Morel, V. Podio, G. L. Turco, and I. Viano, J. Pharm. Pharmacol. *46*:508 (1994).
57. P. Couvreur, M. Roland, and P. Speiser, U.S. Patent 4,329,332 (1982).
58. P. Couvreau, B. Kante, L. Grislain, M. Roland, and P. Speiser, J. Pharm. Sci. *72*:790 (1982).
59. N. Al-Khouri-Fallouh, L. Roblot-Treupel, H. Fessi, J. P. Devissaguet, and F. Puisieux, Int. J. Pharm. *28*:125 (1986).
60. M. R. Gasco and M. Trotta, Int. J. Pharm. *29*:267 (1986).
61. R. Carpignano, M. R. Gasco, and S. Morel, Pharm. Acta Helv. *66*:28 (1991).
62. M. R. Gasco, S. Morel, M. Trotta, and I. Viano, Pharm. Acta Helv. *66*:47 (1991).
63. G. Gregoriadis, ed., *Liposome Technology,* vol. 3, CRC Press, Boca Raton, FL, 1984.
64. Y. Mizushima, Drugs Exp. Clin. Res. *11*:595 (1985).
65. P. Speiser, Eur. Patent O 167 825 (1990).
66. K. Westesen, B. Siekmann, and M. H. J. Koch, Int. J. Pharm. *93*:189 (1993).
67. R. H. Müller, C. Schwarz, W. Mehnert, and J. C. Lucks, Proc. Int. Symp. Control. Rel. Bioact. Mater. *20*:480 (1993).
68. M. R. Gasco and S. Morel, Il Farmaco *45*:1127 (1990).
69. R. Cavalli, O. Caputo, and M. R. Gasco, Int. J. Pharm. *89*:R9 (1993).
70. R. Cavalli, M. R. Gasco, and S. Morel, STP Pharma Sci. *2*:514 (1992).
71. S. Morel, R. Cavalli, and M. R. Gasco, Int. J. Pharm. *105*:R1 (1994).

6

Solubilization of Drugs in Microemulsions

MARIA JOSÉ GARCÍA-CELMA   Departamento de Farmacia, Facultad de Farmacia, Universidad de Barcelona, Barcelona, Spain

I. INTRODUCTION

The design and development of new drug delivery systems to enhance the effectiveness of existing drugs have an important role in pharmaceutical research. It is well known that drug efficacy can be limited by poor aqueous solubility. It is also known that the side effects of some drugs are the result of their poor solubility. The ability to increase aqueous solubility can help to improve the therapeutic efficacy of a drug and allow a reduction in the total dose needed, minimizing toxic side effects [1].

Solubilization by surfactants is one of the most studied techniques for dissolving drugs. Among the various surfactant systems of interest in pharmacy, microemulsions have attracted considerable attention in recent years. Microemulsions are clear dispersions usually obtained by mixing oil (generally a nonpolar solvent), water, and a surfactant(s). They possess interesting physicochemical properties, namely transparency, low viscosity, thermodynamic stability, high solubilization power, and low interfacial tensions [2–4]. Because of these specific properties microemulsions can be useful as new pharmaceutical dosage forms for overcoming solubility problems [5,6]. The thermodynamic stability of microemulsions, as opposed to the thermodynamic instability of other disperse systems such as emulsions or suspensions, is an important consideration in providing a guarantee of accurate dosage, particularly when small amounts of drugs are to be delivered.

Some authors have developed microemulsions as possible therapeutic systems to allow sustained or controlled drug release for topical, transdermal, oral, rectal, and parenteral administration [6]. Microemulsions can be used to deliver a combination of drugs of varying lipophilicity as well as to prepare oral dosage forms of drugs whose bioavailability is hindered by chemical instability or their hydrophobic nature and low aqueous solubility [7]. However, the use of microemulsions as drug delivery systems involves limiting factors, such as the high surfactant content in most of the formulations reported in the literature, the potential toxicity of several components of model microemulsions (surfactants, cosurfactants) and possible adverse effects of these materials on the body, and the concomitant solubilization of other ingredients in the formulation with consequent alterations in stability and effectiveness. To make these surfactant systems pharmaceutically and biologically more useful, it is necessary to explore the feasibility of formulating microemulsions using commercially available nontoxic and safe ingredients.

Several physical techniques can be employed in the characterization of microemulsions, including photon correlation spectroscopy (light scattering), small-angle X-ray scattering, small-angle neutron scattering, electron microscopy, phase diagrams, and conductivity and viscosity determinations [8–12]. Each technique contributes to the understanding of the microstructure of the microemulsion system. The characterization of microemulsions requires the use of several of these techniques rather than only one method. Problems may arise in the interpretation of scattering data because of strong interparticle interactions, particularly when systems have a high particle volume fraction. The complexity of the characterization as well as the high cost of these techniques may also cause difficulties in industrial application of microemulsions.

The object of this chapter is to review drug solubilization in microemulsions, mainly with nontoxic surfactants and biocompatible solvents suitable for pharmaceutical formulations, with emphasis on the effect of these systems on the chemical stability and bioavailability of drugs. This chapter provides several examples that can be useful for application to new and unstudied drugs. It is difficult to understand the complex relationships between the components and their measured properties and to assess how various levels of the components (surfactant, cosurfactant, water, and oil) affect particle size, solubilizing capability, and other properties. Although our current state of knowledge does not allow exact prediction of physicochemical characteristics and pharmacokinetics from the chemical structure of the drug and microemulsion composition, the correlation developed for several molecules can be a valuable contribution.

II. MICROEMULSIONS AS DRUG DELIVERY SYSTEMS

A. Antineoplasics

One of the major goals of anticancer therapy is to obtain constant and prolonged concentrations of the active compounds in the body, particularly at the tumor lesion level [13]. Efficient delivery of antitumor agents is a key step in optimizing the in vivo activity of compounds displaying in vitro growth inhibition of tumor cell lines.

Inclusion of lipid-soluble antineoplasic agents in phospholipid and cholesteryl ester microemulsions and evaluation of the system against in vitro murine leukemias under a variety of conditions were investigated by Halbert et al. [14]. Low-density lipoproteins (LDLs) have targeting potential because the dividing cells require large quantities of cholesterol for cell membrane synthesis, so a system with LDLs could deliver cytotoxin to cancer cells, providing a specific drug delivery system. Lipid-soluble cytotoxic agents were mixed with LDLs and incorporated in microemulsions. Drug incorporation was related to compounds' physicochemical properties and, in serum-free media, the microemulsions exhibited growth inhibitory effects.

Doxorubicin is a hydrophilic drug widely used for its antitumor activity. Gasco et al. [15] studied the release behavior of doxorubicin from oil-in-water (O/W) and water-in-oil (W/O) microemulsions containing the surfactant Aerosol OT, polysorbate 80, or lecithin and the influence of the surfactant on the partition of the drug between the continuous and dispersed phases. Most of the drug should have been dissolved in the aqueous phase of the three microemulsions assayed, but no release could be noted from the O/W microemulsions and very slow release occurred from the W/O microemulsion. The influence of the surfactant on the lack of diffusion of doxorubicin was attributed to the formation of complexes between the drug and the surfactant, increasing the lipophilicity of the drug. Furthermore, the capability of the dispersed phase of W/O microemulsions to function as a reservoir for doxorubicin and idarubicin was investigated [16]. The diffusion of the drugs from aqueous solutions to W/O microemulsions containing lecithin as the surfactant and their accumulation in the dispersed phase indicated a higher diffusion rate for idarubicin than for doxorubicin. The higher lipophilicity of idarubicin allowed the formation of complexes with lecithin that were more stable than the ones formed by doxorubicin.

The acute toxicity of doxorubicin can be significantly reduced by binding to nanoparticles [17]. Polymethylcyanoacrylate nanoparticles were obtained by polymerization at the interphase of a nonaqueous water-in-oil microemulsion and this method allowed the encapsulation of doxorubicin dissolved in the dispersed phase, with incorporation of about 10% of the doxorubicin and a yield of about 95% [18].

Topical and transdermal delivery of antitumor drugs could be a powerful strategy for reducing drug toxicity and, at the same time, restricting the therapeutic effects to specific areas of targeted tissues. Therefore, experimental systems leading to high concentrations of the active agent at the cutaneous and subcutaneous level could be of great interest for novel approaches and specially for the chemotherapeutic treatment and prevention of neoplasic skin diseases. This approach could be particularly indicated in the treatment of melanoma tumor lesions and micrometastases, which, because of their spreading in the body, represent a difficult target for surgical practice. Transdermal lecithin microemulsion gels containing a tetrabenzamidine derivative (0.2 mg/mL) were prepared and studied for antitumor activity in tumor-bearing mice [19]. Application of these microemulsion gels at the tumor lesion level reduced the tumor mass in mice as shown in Fig. 1.

B. Peptide Drugs

The labile nature and rapid systemic elimination of peptides and proteins require their formulation in carrier systems that protect them from metabolic degradation and prolong their plasma half-lives. Microemulsions have been investigated as vehicles for peroral and rectal administration of peptides. In general, peptides are poorly absorbed from the gastrointestinal

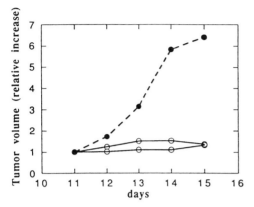

FIG. 1 In vitro antitumor activity of the bromo derivative of tetra-*p*-amidinophenoxy neopentane (TAPP-Br) administered by topical application of TAPP-Br–containing lecithin gels. Kinetics of the increase of tumor mass developed in one control mouse (●) and in two nude mice treated with TAPP-Br–containing lecithin gel (○), 10–16 days after the injection of tumorigenic (FH06T1-1) cells. (From Ref. 19.)

tract and usually need extensive formulation research to improve bioavailability. Cyclosporine is an immunosuppressant agent that has a markedly variable absolute bioavailability, ranging from 1 to 95%, and pharmacokinetics following oral administration [20]. A major source of variability appears to be related to poor absorption, due mainly to its relatively high molecular weight and low solubility in aqueous fluids. Oil-in-water microemulsions were proposed by Ritschel et al. [21] as carriers for the oral administration of cyclosporine. In order to optimize cyclosporine oral absorption, it has been incorporated in a preconcentrate microemulsion, and after oral administration it immediately forms a microemulsion in aqueous fluids [22,23]. When this cyclosporine microemulsion formulation was administered by the oral route, the absorption rate and systemic availability of cyclosporine were greater than for a standard cyclosporine formulation at all dose levels investigated and the variability in the pharmacokinetic parameters may be reduced [24–26]. Microemulsion formulations can be administered orally in soft gelatin capsules, in solution with water, or in solution with fruit juice. Cyclosporine administered in microemulsions was immediately available for absorption and absorbed more rapidly and more efficiently than the same amount of drug administered in other dosage form. The new microemulsion formulations may have a considerable impact on the management of liver transplant recipients in reducing the requirement for intravenous cyclosporine with its attendant side effects and in accelerating discharge from hospital [27].

A water-in-oil microemulsion in which the aqueous phase contained insulin and the oil phase contained lecithin, nonesterified fatty acids, and cholesterol in critical proportions was administered to three diabetic patients [28]. All three patients showed a substantial reduction in blood glucose. However, the hypoglycemic effect of an intraperitoneal injection of insulin in a microemulsion formulation that had been stored for 1–3 months was studied in mice [29], and after administration of the microemulsion plasma glucose levels increased significantly instead of decreasing. It was concluded that the microemulsion may have a significant glucogenic effect of its own and the potency of insulin may decrease on prolonged storage in the microemulsion; insulin does not retain its potency longer than 14–28 days. The results obtained cannot be extrapolated to other animals, formulations, or routes of administration. In a later study, Ritschel [30] investigated the gastrointestinal absorption of three peptides (insulin, vassopressin, and cyclosporine) from oil-in-water microemulsion formulations (Fig. 2). Improvements in the bioavailability of these peptides were found to be dependent on droplet size, digestibility of the lipid used, and type of lipid and surfactant compounds.

Luteinizing hormone–releasing hormone (LH-RH) activates the release of the pituitary hormones that control reproductive development. A study was conducted to prolong the action of an LH-RH analogue by using a water-in-oil microemulsion containing natural and biocompatible compounds for parenteral administration [31]. Although the analogue has a half-life longer than that of LH-RH (2 h compared with 15 min), prolonged delivery systems could offer the advantage of maintaining controlled levels of peptide over an extended period, thus reducing the frequency of administration. The behavior of the LH-RH analogue dissolved in the internal phase of the microemulsion was evaluated both in vitro and in vivo in rats. The results suggested that prolonged action and protection of biodegradable molecules such as this LH-RH analogue could be achieved by using water-in-oil microemulsions administered parenterally.

C. Sympatholytics

Transdermal administration of drugs has advantages in therapy, but it is limited by poor penetration of drug through the stratum corneum [32]. The absorption of drugs can be enhanced by increasing the thermodynamic activity of the drug in the vehicle, which can be expressed as relative solubility (current concentration in the vehicle/concentration of saturated

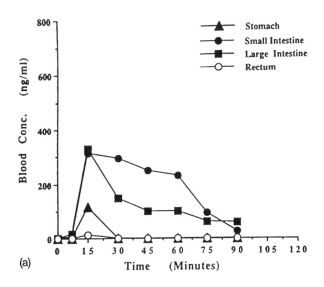

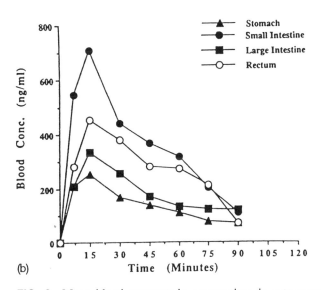

FIG. 2 Mean blood vasopressin versus time in rats upon administration of (a) vasopressin aqueous solution and (b) vasopressin-containing microemulsion into various ligated, gastrointestinal segments. (From Ref. 30.)

vehicle) [33,34]. Absorption rates should be very high from supersaturated vehicles (RS > 1), but supersaturated systems cannot be stored because of their physical instability and must be formed in situ by application of systems that become supersaturated during the application period, such as microemulsions. Water uptake from the occluded skin could change a water-free microemulsion base into a microemulsion, and the increasing content of water would decrease the solubility of apolar drugs. This leads to in situ formation of a supersaturated microemulsion that has a particularly high absorption rate because of the enhanced diffusion pressure of the drug. A saturated solution of the model drugs bupranolol and timolol in a water-free microemulsion base was applied to a clipped area of the dorsal skin of rabbits with an occlusive patch [35]. Evaluations were made in comparison to matrix patches containing bupranolol or timolol. In matrix systems the drug was suspended and penetration requires its dissolution, whereas microemulsions were saturated or supersaturated solutions and no dissolution process is required for diffusion. Faster increasing effects and higher maxima for both drugs were found after application in microemulsions compared with matrix patches as shown in Fig. 3. The pharmacodynamic effects of bupranolol microemulsion systems in rabbits following topical application were correlated with solubility in vitro to determine the effects of supersaturation [36]. The in situ formation of supersaturated microemulsions from microemulsion bases enhanced drug flux through the skin.

The use of microemulsions as ocular drug delivery systems has been proposed to achieve sustained release of a drug applied to the cornea and higher penetration into the deeper layers of the cornea and the aqueous humor. Microemulsions appear to be potentially interesting ophthalmic vehicles for β-blockers, such as timolol or levobunolol hydrochloride, used for topical treatment of increased intraocular pressure in patients with chronic open glaucoma or ocular hypertension. The bioavailability and absorption of timolol alone, as an ion pair solution and as a microemulsion, following topical administration to rabbits' eyes as determined by thin-layer chromatography (TLC) were investigated [37]. Both the availability and absorption were higher for the ion pair solution and microemulsion than for the drug alone. In vitro and in vivo studies of the release of timolol from different ophtalmic formulations were reported [38]. Prolonged persistence in aqueous humor and better transcorneal penetration were achieved with microemulsions and no side effects were described. In another study lipospheres containing timolol maleate, as ion pairs to increase its lipophilicity, were prepared from microemulsions and characterized in vitro [39].

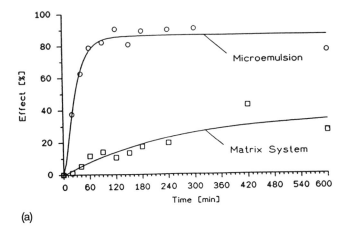

(a)

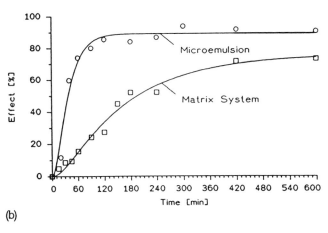

(b)

FIG. 3 Pharmacodynamic effect of transdermal (a) bupranolol and (b) timolol in rabbits. (From Ref. 35.)

An oil-in-water microemulsion containing levobunolol hydrochloride coupled to octanoic acid to increase its lipophilicity and improve its ocular penetration was studied [40]. Studies of permeation through an artificial membrane suggested a possible reservoir effect of the dispersed phase of the microemulsion.

It is well known that when propranolol, a β-blocker, is administered orally, the bioavailability is low and varies with individuals because of the first-pass effect caused by the liver [41]. Transdermal drug delivery systems have been applied to some drugs in an effort to decrease the frequency of drug administration and to increase patients' compliance. The availability of a long-acting transdermal form of propranolol could facilitate compliance and would avoid the first-pass mechanism after oral administration. Oil-in-water microemulsions containing propranolol were studied [42]. The lipophilicity of propranolol was enhanced by formation of lipophilic ion pairs to obtain a dispersed phase that could act as a reservoir; octanoic acid was used as a counterion. The diffusion rates of propranolol from microemulsions through a hydrophilic membrane decreased as the concentration of octanoic acid increased. These microemulsions should yield complete dissolution of the drug between the dispersed and continuous phase, sustained release of propranolol due to the reservoir effect of the dispersed phase, and the possibility of modulating the release of the drug by varying the concentration of the counterion.

D. Local Anesthetics

Incorporating local anesthetic free bases in microemulsions could improve their solubility, penetration across the skin, and pharmacological effect. Lidocaine was solubilized in a microemulsion composed of water, and isopropyl myristate and stabilized by low-toxicity ethoxylated sorbitan esters [43]. The amount of lidocaine dissolved in the microemulsion was proportional to the isopropyl myristate content and was limited to its solubility in this solvent. The addition of lidocaine lowered the phase inversion temperature of the system and increased the temperature range for microemulsion stability.

Microemulsions and miniemulsions obtained in the system water/caster oil ethoxylated derivatives/isopropyl myristate allowed solubilization of therapeutic concentrations of benzocaine, lidocaine, and tetracaine free bases [44]. The solubilities of benzocaine, lidocaine, and tetracaine in microemulsions with a water content of 95% were 10, 5, and 72 times higher, respectively, than their solubilities in water. Because of the solubility properties of these free bases, it was assumed that they are solubilized in an oil-surfactant microenvironment. Light scattering studies of microemulsions with solubilized drugs showed that the droplets were larger than those corresponding to microemulsions without the drugs. This confirmed the hypothesis that the drugs were solubilized inside the microdroplets.

Although the use of solutions as drug delivery systems has significant advantages, the use of aqueous solutions can increase the chemical reactivity and possibly drug unstability. Solutions of tetracaine in water showed degradation products after 24 h, whereas tetracaine solubilized in microemulsions with a high water content was not degradated after storage for 1 month [44].

The local anesthetic effect of 0.1% heptacaine formulated in water-in-oil microemulsion bases applied to the skin of rabbits and dogs was studied [45,46]. An aqueous solution of 0.1% heptacaine used as the standard had no effect, whereas all the evaluated microemulsion vehicles exerted effects.

E. Steroids

Steroids are poorly water-soluble drugs that have wide pharmacological applications, and there is a need for solutions of these compounds for topical and parenteral administration.

The solubilization of hydrocortisone in microemulsions was found to be six times higher than that of isopropanol [5]. The level of incorporation of several steroids (testosterone, testosterone enanthate, testosterone propionate, progesterone, and medroxyprogesterone acetate) in a nonionic surfactant micellar system and in oil-in-water microemulsions was studied (Table 1) [47]. In all cases, drug incorporation increased with increasing surfactant content. The increase in solubilization in the microemulsion systems compared with the micelles was greatest at low surfactant concentrations. It was concluded that the drug-carrying improvement of an O/W microemulsion over a micellar system depends on the solubility of the drug in the dispersed oil phase and is significant only for very lipophilic drugs.

The reservoir effect of oil-in-water microemulsions for prednisone, a lipophilic drug, was examined by permeability studies [48]. Three microemulsions containing a fixed amount of lecithin as the surfactant and different percentages of butanol as the cosurfactant were used. The permeability constants obtained with microemulsions were four to five times lower than those obtained with solutions. The amount of alcohol influenced the diffusion rate of the drug. Similar results were obtained with solutions and oil-in-water microemulsions containing prednisone, betamethasone, and the other three drugs with different lipophilicities [49].

F. Anxiolytics

Benzodiazepines are lipophilic drugs that have several side effects when administered by the parenteral route, probably associated with precipitation of the drug in aqueous solutions. In order to overcome this problem, different formulations for the transdermal administration of benzodiazepines have been investigated. Trotta et al. [50] studied in vitro skin permeation of diazepam from fluid and viscosized O/W microemulsions. An amount

TABLE 1 Solubilization of Several Steroids in a Nonionic Surfactant Micellar System and in O/W Microemulsions

Drug	Brij 96 (% w/w)	Drug incorporation (% w/v ± SD)	
		Micellar system	Microemulsion
Testosterone	10	0.172 ± 0.015	0.223 ± 0.025
	15	0.218 ± 0.021	0.230 ± 0.049
	20	0.306 ± 0.024	0.297 ± 0.015
Testosterone propionate	10	0.290 ± 0.041	0.402 ± 0.035[a]
	15	0.351 ± 0.072	0.531 ± 0.056[a]
	20	0.447 ± 0.098	0.656 ± 0.080[a]
Testosterone enanthate	10	3.50 ± 0.36	5.72 ± 0.34[a]
	15	4.56 ± 0.43	5.89 ± 0.43[a]
	20	5.24 ± 0.59	6.19 ± 0.47[a]
Progesterone	15	0.629 ± 0.067	0.789 ± 0.036[a]
	20	0.906 ± 0.19	0.911 ± 0.038
Medroxyprogesterone acetate	15	0.511 ± 0.075	0.544 ± 0.053
	20	0.689 ± 0.032	0.748 ± 0.057

[a]Indicates a significant difference ($P < .05$) between uptake into the microemulsion and micellar systems at the same concentration of surfactant.
Source: Ref. 47.

of 30 mg/g of drug could be solubilized. The composition of the external phase affected the release of the drug. Pseudo-zero-order kinetics was achieved over the permeation experiments.

G. Anti-infective Drugs

The treatment of cellular infections with chemotherapeutic agents is often impeded because of the inability of these agents to penetrate infected cells. Drug carriers may be useful anti-infective delivery systems leading a favorable lipid-water distribution coefficient that allows the accumulation of sufficient drug in the interior of the infected cells [51]. A number of anti-infective drugs have been solubilized in microemulsions. For most of these drugs only solubilization studies and in vitro drug release have been reported.

Ziegenmeyer and Führer [52] described an increased rate of diffusion of tetracycline hydrochloride, a hydrosoluble compound, into the skin when formulated in a water-in-oil microemulsion. Diffusion from the microemulsion was found to be in the range of 5 to 6 h, compared with an approximate peak of 12 h from a gel and a peak of more than 24 h from a cream.

Solubilization of antifungal drugs (clotrimazole, ciclopirox olamine, and econazole nitrate) in ternary water/nonionic surfactant/oil systems was studied [53]. The solubility of these antifungal drugs ranged from practically insoluble to slightly soluble in water as well as in most of the oils used in pharmaceutical formulations. The maximum drug solubilization values were obtained with water/polysorbate 80/oil systems. The results showed that it is possible solubilize 1% w/w of antifungal agent in suitable topical microemulsions with a water content higher than 50% w/w. Solubilization of higher concentrations of econazole nitrate was achieved in a system composed of water/N^α-lauroyl-L-arginine-methyl-ester hydrochloride (LAM)/2-phenoxyethanol, and 1% of the drug was solubilized in compositions with a water content higher than 90% w/w as shown in Table 2 [54,55]. In vitro studies showed the efficacy of some of these formulations against selected microorganisms.

H. Vitamins

Fat-soluble vitamins have been solubilized in aqueous media using microemulsions. These vitamins are often more stable when solubilized in a microemulsion than when moderately dissolved in vegetable oil; in a microemulsion they remain solubilized throughout the shelf life of the product.

TABLE 2 Solubilization of Econazole Nitrate in the Water/
N^α-Lauroyl-L-arginine-methyl-ester hydrochloride (LAM)/2-
Phenoxyethanol (2-PhE) System

2-PhE/LAM ratio (w/w)	Water (%)	Maximum econazole nitrate solubilization (%)	
		25°C	40°C
85/15	10	22.8	28.8
85/15	90	0.1	0.4
30/70	96	1.3	1.5
50/50	65	9.1	11.7
50/50	96	1.1	1.1
70/30	25	30.5	32.0
70/30	90	0.9	0.9
25/75	25	17.6	22.6

Menadione was solubilized in the internal phase of two oil-in-water microemulsions having the same oil/water/cosurfactant system (isopropyl myristate, water, 1-butanol) but differing in the surfactant (Aerosol OT or egg lecithin) [56]. Nonaqueous microemulsions consisting of four biocompatible substances (lecithin as surfactant, taurodeoxycholic acid as cosurfactant, ethyl oleate as disperse phase, and 1,2-propylene glycol as continuous phase) were studied to assess their performance as carriers for the oral administration of retinol and to determine the role of the components in the release through a hydrophilic membrane of this model lipophilic drug [57]. The results emphasized the effect of the cosurfactant and oil phase of the system on drug permeation behavior from the waterless microemulsion and confirmed the reservoir effect of the internal phase of the microemulsion.

Ascorbic acid and the naturally occurring α-tocopherol have been used as antioxidants in food systems and pharmaceutical formulations. It has been shown that the oxidation of fats at the air-water interface is related to the molecular conformation, and antioxidant effects in emulsions can be improved by localization of the antioxidant in the interfacial film [58]. The effect of ascorbic acid, a water-soluble antioxidant, and α-tocopherol, an oil-soluble antioxidant, on fat oxidation was studied in a microemulsion consisting of soybean oil, sunflower oil monoglycerides, and water. It was observed that the presence of ascorbic acid in the water aggregates gave a pronounced reduction rate and, even after storage at 40°C for 100 days, only minor oxidation of the oil was observed. A control emulsion without ascorbic acid was highly oxidized at 100 days. Both antioxidants were solubilized in the same microemulsion; however, the protection against oxidation by ascorbic acid was insufficient when α-tocopherol was added to the oil for storage times above approximately 50 days [59]. The oxidation of fish oil was decreased if the oil was formulated as a microemulsion with ascorbate and α-tocopherol used as antioxidants [60].

Percutaneous absorption of α-tocopherol through Sprague-Dawley rat skin was greatly enhanced by use of either an O/W or a W/O microemulsion preparation. The vitamin was delivered predominantly to the epidermis, and undue accumulation of this vitamin in the organs other than the skin was avoided. A cream and a lotion product that contained the same amount of vitamin had resulted in excessive accumulation of the vitamin in organs such as the liver, body fats, and muscle [61].

I. Anti-inflammatory Drugs

Specific targeting of anti-inflammatory drugs may reduce the side effects generally associated with systemic administration of these drugs and could open up new possibilities for the therapy of chronic rheumatic diseases. The physical and chemical stability of an O/W microemulsion of butibufen, a lipophilic nonsteroideal anti-inflammatory agent, was studied under various conditions. Butibufen degradation was so small that the variation in concentration could not be detected before 1 year, despite the high sensitivity of the analytical method used [62].

Indomethacin has been incorporated in self-microemulsifying drug delivery systems (SMEDDS). These formulations are mixtures of oil, nonionic surfactant, and cosurfactant that are solid at room temperature but self-emulsify into water at 37°C with moderate stirring. One feature of these mixtures is their ability to form a microemulsion when exposed to gastrointestinal fluids. This behavior makes them good candidates for oral or rectal delivery of lipophilic or slightly soluble drugs. Farah et al. [63] studied the in vitro dissolution kinetics of indomethacin from SMEDDS in a gastric medium as a function of composition (Fig. 4). The dissolution kinetics of indomethacin were substantially improved compared with the kinetics when the drug was in a powder or marketed form, and they concluded that the higher the HLB value of the surfactant and cosurfactant mixture, the higher the dissolution rate. These authors investigated the release of indomethacin from premicroemulsified suppositories [64]. Indomethacin was released in vitro from these formulations twice as fast as from a classical suppository formula. The results obtained in vivo with rabbits showed an improvement of absorption parameters.

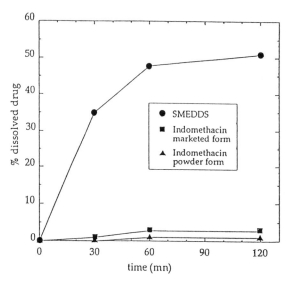

FIG. 4 Kinetics of indomethacin dissolution in vitro at pH 1.2. (From Ref. 63.)

J. Dermatological Products

Several studies have been carried out to compare the bioavailability of drugs in microemulsions and other pharmaceutical dosage forms. The percutaneous absorption of tyrosine, a stimulator in the biosynthesis of melanine, incorporated in liquid crystal, emulsion, and microemulsion formulations was studied in vitro [65]. The microemulsion and liquid crystals enhanced tyrosine penetration through epidermis compared with the emulsion. It was concluded that, compared with an emulsion and liquid crystals, a microemulsion of tyrosine has the best transepidermal penetration with the least irritation.

Many dermatological products may require a good skin reservoir of the therapeutic agents to attain their desirable clinical effects. Azelaic acid (nonadienoic acid), topically applied, has beneficial effects in some hyperpigmentary disorders. This drug was solubilized in a water-in-oil microemulsion [66]. The advantages of the use of microemulsions as carriers for azelaic acid are that the acid can be dissolved and not suspended in the topical system and there is a reservoir of the acid in the disperse phase. The necessity of achieving and maintaining intralesional concentrations of the diacid for a long period is a limiting factor. High rates of transport through a lipophilic membrane were obtained, but the microemulsions could not be applied on the skin because of their fluidity. Gelification of microemulsions has been carried out in order to improve the possibilities of application. Gelification is usually achieved by gelifiant incorporation with different methods: (1) to the aqueous components, (2) to the oily components, (3) to the prepared microemulsion, or (4) incorporation of one gelifiant to the aqueous components and another one to the oily components. Gelifiant dispersion requires additional mixing energy. In this case, a viscosized microemulsion was obtained by adding Carbopol 934 and the viscosization of a microemulsion and permeation of azelaic acid through hairless mouse skin were studied. The percentage of azelaic acid transported from the microemulsion was several times higher than from a gel.

Furthermore, the transport of azelaic acid in a topical microemulsion through hairless mouse skin was studied [67]. The microemulsion containing the highest concentration of drug, partly dissolved and partly suspended, showed better transport through the skin than the microemulsion with a lower concentration of drug but completely dissolved. The reservoir of drug in the microemulsion was then increased by adding suspended drug.

A topical formulation capable of producing a good skin reservoir for a sun protection agent is highly desirable because it can maintain an untraviolet absorbing agent in the epidermis for a long time. A sunscreen agent, octyl dimethyl PABA, and a skin moisturizing agent, cetyl alcohol, were used as model molecules to study the permeation characteristics of a microemulsion delivery system [68]. A water-in-oil microemulsion, a cream, and a lotion formulation were employed to deliver the alcohol and the sunscreen in vitro using human and hairless mouse skin. The delivery efficiencies were compared in terms of rate and depth of product penetration into human or hairless mouse skin. The advantage of using microemulsions to achieve deeper and faster penetration of the permeating compounds was clearly demonstrated. The microemulsion had the ability to deliver into the skin two to six times faster and at least twice as much as two macroemulsions, a cream and a lotion.

An oily sunscreen, 2-ethylhexyl p-methoxycinnamate, was solubilized in several microemulsions [69]. The maximum solubilization values were achieved in the pseudoternary water/nonionic surfactant mixture/isopropyl myristate and propyleneglycol/cationic surfactant mixture/benzyl alcohol systems as shown in Fig. 5, in microemulsions near the liquid crystalline phase area.

Cosmetic formulations usually contain several active principles incorporated in a suitable vehicle. Solubilization of lipophilic and hydrophilic products (tocopherol acetate and mandelic acid) in water-in-oil microemulsions was investigated [70]. The degree of solubilization was found to

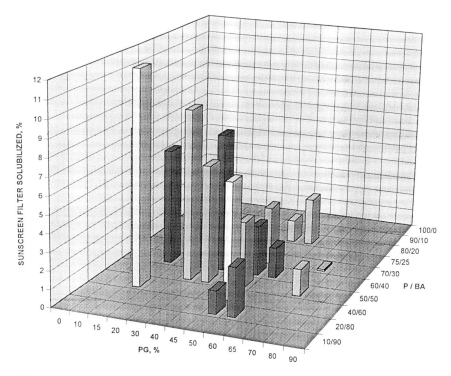

FIG. 5 Solubilization of 2-ethylhexyl *p*-methoxycinnamate in the propylene glycol (PG)/phospholipid EFA (P)/benzyl alcohol (BA) system.

be dependent on composition variables. Whereas tocopherol acetate could be solubilized in all the microemulsions tested, mandelic acid could not be solubilized in formulations with a water concentration higher than 45% w/w. Solubilization of a mixture of both antiaging compounds could be achieved in microemulsions with a low surfactant content.

K. Oxygen

Certain fluorocarbon oils are able to store oxygen and release it in the presence of carbon dioxide; however, for physiological applications, the oil must be made miscible with blood. A novel microemulsion system containing a fluorinated oil and polyethoxylated (20) sorbitan monooleate (Montanox 80) was characterized for its usefulness as a blood substitute (Table 3) [71,72]. Oxygen absorption by the system was comparable to that of blood. The system was composed of small aggregates that pass

TABLE 3 Characteristics of Several Microemulsions of the System Water/Montanox 80/C_8F_{17}—CH_2—CH=CH—C_4H_9

No.	Composition of microemulsions (wt %)			Viscosity (cp) ± 0.1 cp	Micelle radius (Å) ± 1 Å	Oxygen absorption (mL/100 mL)	
	Montanox 80	C_8F_{17}—CH_2—CH= CH—C_4H_9	H_2O			Measured (±2 mL/100 mL)	Calculated
B	2.6	48.7	48.7	70.0	—	23	17.0
C	5.5	16.6	77.8	1.1	64	32	5.4
D	7.15	7.15	85.7	1.1	36	9	2.3
E	1.3	9.1	89.6	0.9	43	21	2.8

Source: Ref. 72.

through the capillaries and can be compatible with blood. No toxic effects of the fluorinated oil were observed following intraperitoneal injection in rats.

III. CONCLUSIONS

Microemulsions represent an alternative to classical formulations as drug delivery systems and may present new possibilities for therapy that have not yet been investigated. They are potential colloidal carriers for the targeting of drugs to specific areas, and their solubilizing capability combined with the possibility of controlling drug release would not only improve drug efficacy but also reduce unwanted toxic side effects. Optimization of microemulsion formulations by choosing surfactants and cosurfactants suitable for specific routes of administration will be a challenge in future years.

REFERENCES

1. S. H. Yalkowsky, ed., *Techniques of Solubilization of Drugs*, Marcel Dekker, New York, 1981.
2. L. M. Prince, *Microemulsions. Theory and Practice*, Academic Press, New York, 1977.
3. S. E. Friberg and P. B. Bothorel, *Microemulsions: Structure and Dynamics*, CRC Press, Boca Raton, FL, 1987.
4. I. Danielsson and B. Lindman, Colloids Surf. *3*:391 (1981).
5. A. Jayakrishnan, K. Kalaiarasi, and D. O. Shah, J. Soc. Cosmet. Chem. *34*: 355–350 (1983).
6. H. N. Bhargava, A. Narurkar, and L. M. Lieb, Pharmaceut. Technol. *11*:46–52 (1987).
7. D. Attwood, in *Colloidal Drug Delivery Systems* (J. Kreuter, ed.), Marcel Dekker, New York, 1994, pp. 31–71.
8. D. J. Cebula, D. Y. Myers, and R. H. Ottewill, J. Chem Soc. Faraday Trans. 1. *77*:2585 (1981).
9. R. C. Baker, A. T. Florence, Th. F. Tadros, and R. M. Wood, J. Colloid Interface Sci. *100*:311 (1984).
10. R. C. Baker, A. T. Florence, R. H. Ottewill, and Th. F. Tadros, J. Colloid Interface Sci. *100*:332 (1984).
11. I. S. Barnes, S. T. Hyde, B. W. Ninham, P. J. Derian, M. Drifford, and T. N. Zemb, J. Phys. Chem. *92*:2286 (1988).
12. P. K. Vinson, J. G. Sheeban, W. G. Miller, L. E. Scriven, and H. T. Davis, J. Phys. Chem. *95*:2546 (1991).
13. R. Langer, Science *249*:1527–1533 (1990).
14. G. W. Halbert, J. F. B. Stuart, and A. T. Florence, Int. J. Pharm. *21*(Sep): 219–232 (1984).

15. M. R. Gasco, F. Pattarino, and I. Voltani, Farmaco. Ed. Prat. *43*(1):3–12 (1988).
16. F. Pattarino, M. R. Gasco, and M. Trotta, Farmaco *44*:339–344 (1989).
17. P. Couvreur, L. Grislain, V. Lenaerts, F. Brasseur, P. Guiot, and A. Biernacki, in *Polymeric, Nanoparticles and Microspheres* (P. Guiot and P. Couvreur, eds.), CRC Press, Boca Raton, FL, 1986, p. 27.
18. R. Carpignano, M. R. Gasco, and S. Morel, Pharm. Acta Helv. *66*(1):28–32 (1991).
19. C. Nastruzzi and R. Gambari, J. Controlled Release *29*(Feb):53–62 (1994).
20. B. D. Kahan, M. Ried, and J. Newburger, Transplant Proc. *15*:446 (1983).
21. W. A. Ritschel, S. Adolph, G. B. Ritschel, and T. Schroeder, Meth. Find. Exp. Clin. Pharmacol. *12*(2):127–134 (1990).
22. J. Drewe, R. Meier, J. Vonderscher, D. Kiss, K. Gyr, et al., Br. J. Clin. Pharmacol. *34*(Jul):60–64 (1992).
23. J. M. Kovarik, E. A. Mueller, A. Johnston, G. Hitzenberger, and K. Kutz, Pharmacotherapy *13*(6):613–617 (1993).
24. A. K. Trull, K. K. Tan, J. Uttridge, T. Bauer, N. V. Jamieson, et al., Lancet *341*(Feb 13):433 (1993).
25. J. M. Kovarik, E. A. Mueller, J. B. Van Bree, W. Tetzloff, and K. Kutz, J. Pharm. Sci. *83*(Mar):444–446 (1994).
26. E. A. Mueller, J. M. Kovarik, J. B. Van Bree, W. Tetzloff, and K. Kutz, Pharm. Res. *11*(Feb):301–304 (1994).
27. I. S. Sketris, V. McAlister, and M. R. Wright, Ann. Pharmacother. *28*(Jul–Aug):962–963 (1994).
28. Y. W. Cho and M. Flynn, Lancet *2*:1518–1519 (1989).
29. D. G. Patel, W. A. Ritschel, P. Chalasani, and S. Rao, J. Pharm. Sci. *80*(Jun): 613–614 (1991).
30. W. A. Ritschel, Meth. Find. Exp. Clin. Pharmacol. *13*:205 (1991).
31. M. R. Gasco, F. Pattarino, and F. Lattanzi, Int. J. Pharm. *62*(Jul 31):119–123 (1990).
32. T. B. Fitzpatrick, *Dermatology in General Medicine*, McGraw-Hill, New York, 1971.
33. R. Brandau and B. H. Lippold, eds., *Dermal and transdermal absorption*, Wissenschaftl. Verlagsges, Stuttgart, 1982.
34. M. F. Coldman et al., J. Pharm. Sci. *58*:1098–1102 (1969).
35. J. Kemken, A. Ziegler, and B. W. Müller, Meth. Find. Exp. Clin. Pharmacol. *13*(5):361–365 (1991).
36. J. Kemken, A. Ziegler, and B. W. Muller, Pharm. Res. *9*(Apr):554–558 (1992).
37. M. R. Gasco, M. Gallarate, M. Trotta, L. Bauchiero, O. Chiappero, et al., J. Pharm. Biomed. Anal. *7*(4):433–439 (1989).
38. B. Boles Carenini, E. Gremmo, B. Brogliatti, and M. R. Gasco, New Trends Ophthalmol. 4(3):177–179 (1989).
39. M. R. Gasco, R. Cavalli, and M. E. Carlotti, Pharmazie *47*(Feb):119–121 (1992).

40. M. Gallarate, M. R. Gasco, M. Trotta, P. Chetoni, and M. F. Saettone, Int. J. Pharm. *100*(Nov 8):219–225 (1993).
41. P. A. Routledge and D. G. Shand, Clin. Pharmacokinet. *4*:73 (1979).
42. M. R. Gasco, M. E. Carlotti, and M. Trotta, Int. J. Cosmet. Sci. *10*:263–269 (1988).
43. J. Carlfors, I. Blute, and V. Schmidt, J. Disper. Sci. Tech. *12*(5–6):467–482 (1991).
44. M. I. Delgado, C. Solans, and M. J. García-Celma, Pharmaceutical Technology Conference, Barcelona, 1995.
45. M. Zabka and M. Benkova, Farm-Obz. *61*(Aug):413–426 (1992).
46. M. Zabka and M. Benkova, Cesk.-Farm. *42*(Apr):170–172 (1993).
47. C. Malcolmson and M. J. Lawrence, J. Pharm. Pharmacol. *45*(Feb):141–143 (1993).
48. M. R. Gasco, M. Gallarate, and F. Pattarino, Il Farmaco. Ed. Pr. *43*(10): 325–330 (1988).
49. M. Trotta, M. R. Gasco, and S. Morel, J. Controlled Release *10*:237–243 (1989).
50. M. Trotta, M. R. Gasco, and F. Pattarino, Proceedings of the 10th Pharmaceutical Technology Conference, Bologna, Italy, 1991, pp. 388–404.
51. A. Trouet and P. Tulkens, in *The Future of Antibiotherapy and Antibiotic Research* (L. Ninet, P. E. Bost, D. H. Bouanchand, and J. Florent, eds.), Academic Press, London, 1981, pp. 337–349.
52. J. Ziegenmeyer and C. Fuehrer. Acta Pharm. Technol. *26*(4):273 (1980).
53. M. J. García-Celma, N. Azemar, M. A. Pes, and C. Solans, Solubilization of antifungal drugs in water/POE (20) sorbitan monooleate/oil systems. Int. J. Pharm. *105*:77–81 (1994).
54. M. A. Pes, M. R. Infante, M. J. García-Celma, B. Bricio, and C. Solans, VIIth International Conference on Surface and Colloid Science, Compiegne, 1991.
55. M. J. García-Celma, B. Bricio, M. A. Pes, M. R. Infante, and C. Solans, Symposium on Topical Administration of Drugs, Stockholm, 1992.
56. B. Fubini, M. R. Gasco, and M. Gallarate, Int. J. Pharm. *50*(Mar 15):213–217 (1989).
57. F. Pattarino, E. Marengo, M. R. Gasco, and R. Carpignano, Int. J. Pharm. *91*(Apr 26):157–165 (1993).
58. C. Ruben and K. Larsson, J. Dispers. Sci. Techn. *6*:213 (1985).
59. L. Moberger, K. Larsson, W. Buchheim, and H. Timmen, J. Dispers. Sci. Tech. *8*(Jun):207–215 (1987).
60. M. Jakobsson and B. Sivik, J. Disper. Sci. Tech. *15*(5):611–619 (1994).
61. M. C. Martini, M. F. Bobin, H. Flandin, F. Caillaud, and J. Cotte, J. Pharm. Belg. *39*(6):348–354 (1984).
62. L. Gonzalez-Tavares, P. Sanz-Saiz, M. J. Perez de la Cruz, M. A. Camacho, and J. L. Martin, STP Pharma. Sci. *1*(3):195–199 (1991).
63. N. Farah, M. de Taddeo, J. P. Laforêt, and J. Denis, AAPS Annual Meeting, Orlando, FL, 1993.

64. N. Farah, J. Denis, J. P. Laforêt, and J. Joachim, Pharmaceutical Technology Conference, Strasbourg, 1994.
65. F. Fevrier, M. F. Bobin, C. Lafforgue, and M. C. Martini, STP Pharma Sci. *1*(1):60–63 (1991).
66. M. R. Gasco, M. Gallarate, and F. Pattarino, Int. J. Pharm. *69*(Mar 20): 193–196 (1991).
67. F. Pattarino, M. E. Carlotti, and M. R. Gasco, Pharmazie *49*(Jan):72–73 (1994).
68. E. E. Linn, R. C. Pohland, and T. K. Byrd, Drug Dev. Ind. Pharm. *16*(6): 899–920 (1990).
69. M. J. García-Celma, A. Porta, N. Azemar, and C. Solans, VIIIth European Colloid and Interface Society Conference, Montpellier, 1994, p. 285.
70. M. J. García-Celma, N. Azemar, I. Carrera, and C. Solans, Preprints of the IFSCC International Conference, vol. II, Platja d'Aro, Girona, 1993, pp. 235–243.
71. C. Cecutti, I. Rico, A. Lattes, A. Novelli, A. Rico, G. Marion, A. Graciaa, and J. Lachaise, Eur. J. Med. Chem. *24*:485–492 (1989).
72. C. Cecutti, A. Novelli, I. Rico, and A. Lattes, J. Disper. Sci. Tech. *11*(2): 115–123 (1990).

7
Microemulsions in Foods: Properties and Applications

STEPHANIE R. DUNGAN   Departments of Food Science and Technology, and Chemical Engineering and Materials Science, University of California, Davis, Davis, California

I. INTRODUCTION

The coexistence of oil and water within many foods has made the study of emulsions an important one for food scientists and food processors for several decades. In an emulsion, small (0.1 to 10 μm) droplets of oil or water are stabilized within the second fluid by the presence of surfactants at the surface of the droplet. In foods, these surfactants may be naturally occurring compounds such as monoglycerides, phospholipids, or proteins or may be other substances approved as food additives. Because emulsions are thermodynamically unstable systems, their properties depend on the methods used in their formation, and any emulsion formed will eventually separate into bulk water and oil phases over time. Hence much of the study of food emulsions has been directed toward understanding how to form emulsions with desirable and reproducible properties and how to stabilize such emulsions for a time comparable to the shelf life of the food.

Although microemulsions have received relatively little attention in the food science literature, they may have distinct advantages over emulsions as vehicles for contacting oil and water phases within foods. Like an emulsion, a microemulsion is generally considered to be droplets of one type of fluid encased in a surfactant shell and dispersed throughout a second, immiscible fluid (Fig. 1). However, microemulsions are thermodynamically stable phases, and as such they can be formed with defined properties that depend only on thermodynamic conditions rather than on the kinetics of

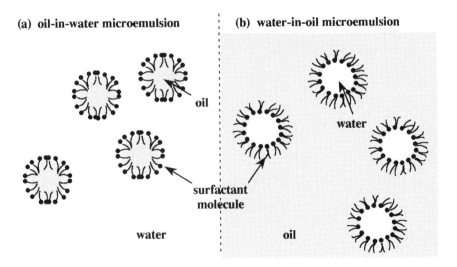

FIG. 1 Schematic of microemulsion systems.

formation. Further, the microemulsion formed is generally a homogeneous and well-characterized system. For example, it is possible to create microemulsions consisting of a fairly monodisperse system of spherical droplets, in contrast to the generally broad distribution of droplet sizes found within an emulsion. As their name implies, microemulsion droplets are also considerably smaller than emulsion drops, ranging from one to tens of nanometers in diameter. The equilibrium nature of microemulsions enables them to remain stable as long as conditions around them are unchanged, and thus they are unaffected by the destabilization mechanisms (creaming, flocculation, and coalescence) that so strongly affect emulsion stability. On the other hand, by changing temperature or composition to values at which the microemulsion is no longer energetically favorable, one may destabilize the system in a controlled fashion. Hence microemulsions could potentially bring unique advantages to foods, providing systems that are reproducible and well characterized, are stable indefinitely, but also can be destabilized at will.

However, these advantages of microemulsion systems do not come without a price. The thermodynamic stability of these phases means that they can be formed only in specific ranges of temperature, pressure, and composition. Thus, the food processor does not have complete freedom in designing his or her system. Knowing what conditions will allow a microemulsion to be established requires information about the phase diagram for the mixture of interest. In general, the ingredients of the mixture must be specifically chosen in order to achieve a microemulsion with desired characteristics.

Hence a major thrust of research in the area of food microemulsions is the investigation of aggregates formed by edible surfactants within oil/water systems. Section II reviews current information on water-in-oil and oil-in-water microemulsions formed using food-grade materials. The section also includes some studies of more complex food mixtures such as liquid crystals and gels. The list of all of these types of systems is still quite limited, and many of the researchers discussed in this section have examined mixtures in which one component is nonedible—e.g., a food-grade surfactant mixed with water and a nonedible oil, or food oils combined with nonedible surfactants. Clearly more research is needed to discover other mixtures that will produce microemulsions or liquid crystals using food-grade components, with work on mixtures of surfactants being one promising area for future study.

In this discussion it is important to keep in mind a semantic issue associated with the definition of a microemulsion. Micelles, the aggregates formed by surfactant molecules in water, are able to incorporate solute molecules within their interior in a process known as solubilization. For example, it is possible to solubilize hydrophobic solutes such as oil molecules within micelles formed within an aqueous phase. Clearly if the quantity of oil incorporated within the micelle is considerable, one can refer to this aggregate as a microemulsion droplet rather than as a micelle within which oil is solubilized. It is not clear at which point solubilization ends and microemulsion formation begins. Hence in this chapter we use a rather broad definition of a microemulsion that includes surfactant aggregates containing only small quantities of oil or water.

The potential for using microemulsions in foods to provide a stable mixture of oil and water phases is mentioned above. However, microemulsions have also attracted attention for their use in other novel applications such as component solubilization, enhanced reaction efficiency, and extraction techniques. These applications have considerable potential in the area of food technology. Section III examines possibilities for incorporating food ingredients such as flavors, preservatives, and vitamins within micelles and microemulsions. Such an approach can lead to greatly enhanced quantities of these ingredients within a food and can also have important effects on the release rate of those components and on their stability to degradation. Section IV provides a discussion of reactions carried out within a microemulsion medium, for which the reaction rate and efficacy can be greatly enhanced by the presence of the surfactant aggregates. Finally, Sec. V explores the potential of microemulsions for extracting food components from a complex mixture, in a separation method that is potentially both selective and cost-effective.

As should be clear from the following sections, microemulsions have a host of promising applications in the food industry, and their potential in this area has only begun to be explored. In addition to the applications discussed in this chapter, there will doubtless be other improvements to foods made possible through the use of these types of mixtures. Because of the complexity of food systems and the importance of sensory perception in food technology, it is often difficult to predict how changes in a food composition will alter factors such as flavor, texture, and processing effectiveness. Only through further exploration of microemulsions made using food-grade components will we be able to realize the possibilities of this technology.

II. FORMATION OF MICROEMULSIONS BY FOOD-GRADE COMPONENTS

A key determinant of the type of assembly amphiphilic molecules will form at equilibrium is the "geometry" of the surfactant [1]. Within a micellar aggregate, for example, one observes that the hydrophobic tail groups of the surfactant molecules must fit within a relatively confined core of the aggregate, whereas the distance between head groups is significantly less constrained. Thus we expect that surfactant molecules with relatively bulky head groups and nonbulky tails will have the easiest time forming a micelle. Considering next an oil-in-water microemulsion droplet, the presence of a second solvent in the droplet core means that the tails of the surfactant are less constrained than they are in the micellar aggregate. Thus surfactants with a more sterically hindered tail group will be likely to form such microemulsions. In a water-in-oil droplet the situation is reversed—here the tails need to be widely spaced and the heads are constrained. Hence surfactants with small head groups and large tail groups favor this form of microemulsion.

In practice there are two ways in which the surfactant geometry required for microemulsion formation can be achieved. First, a surfactant with an intrinsically bulky tail group can be selected for the system, often by using surfactants with two tail groups. Alternatively the added bulk needed in the tail region can be achieved by adding to the system a second long-chain molecule, known as a cosurfactant. These cosurfactants are typically alcohols, which reside between the tails of the surfactant within the microemulsion droplet. These cosurfactants are polar enough to be surface active, but their hydrocarbon structure allows them to increase the bulkiness of the tail group region.

A. Water-in-Oil Microemulsions

Phospholipids comprise a collection of ubiquitous food-grade surfactants whose double-tailed structure makes them promising for use in forming microemulsions (see Fig. 2). Thus it is no surprise that these molecules

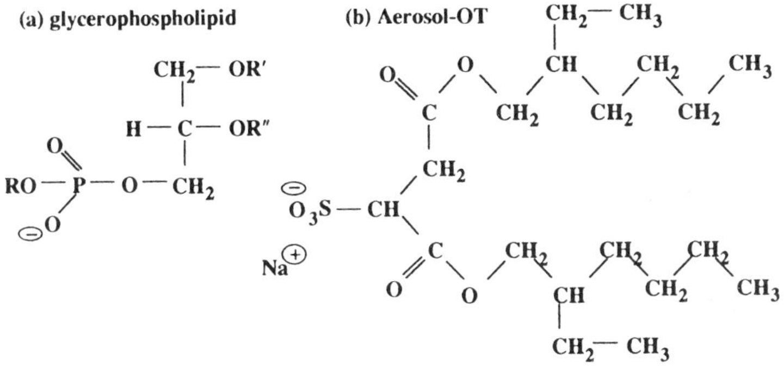

FIG. 2 Double-tailed surfactant molecules.

have been extensively studied with regard to the formation of water-in-oil microemulsions. Phospholipids can have a variety of different structures, resulting from variations in the R, R', and R" groups indicated in Fig. 2. Of these, phosphatidylcholines, also known as lecithin, have been the most thoroughly investigated. Dilinoleyl phosphatidylcholine has been shown using phosphorus-31 nuclear magnetic resonance (^{31}P-NMR) to form water-in-oil microemulsions with chlorobenzene as a solvent. ^{31}P-NMR and proton NMR (^{1}H-NMR) have been used to demonstrate that dipalmitoyl phosphatidylcholine forms water-in-oil microemulsions at 52°C in benzene [2,3], chlorobenzene [2], or o-dichlorobenzene [4]. Both these surfactants appear to form droplets containing up to approximately 20 molecules of water for every phosphatidylcholine molecule. This molar ratio of water to surfactant is conventionally referred to as the w_o value.

Phospholipids extracted from foods have been investigated as well—these extracts actually contain a mixture of different phospholipid types. Phosphatidylcholine from egg forms water-in-oil microemulsions in benzene [5–10], o-dichlorobenzene [2], diethyl ether [11], carbon tetrachloride [8,9], and cyclohexane [8,9], as shown by light scattering, electron microscopy, ^{31}P-NMR and ^{1}H-NMR, infrared spectrometry, and viscosity and diffusion measurements. Shervani and co-workers [10] also demonstrated egg phosphatidylcholine microemulsion formation within octane and dodecane. The water content in these microemulsions ranges from w_o values of 0 to 46. Finally, soybean phosphatidylcholine forms a microemulsion with a w_o value of 16 in a hexane solvent at 4°C, as determined by small-angle neutron scattering [12].

Studies indicate that the water-in-oil microemulsions formed from lecithin contain monodisperse spherical droplets [8] on the order of 20 nm in radius. It is anticipated that the size of these droplets will increase with increasing w_o values, but no clear correlation between the two has been established. At high water contents the droplets begin to change shape, and a shift to a clear viscous gel phase is observed in many systems. This gel contains arrays of water tubules within the organic external phase [13,14]. Such gel systems are discussed in more detail below. It has also been determined that characteristics such as the mobility and dielectric properties of the water within the microemulsion droplets change as w_o varies, with the incorporated water becoming more mobile and more similar to bulk water as w_o increases.

The amount of water that can be incorporated per mole of surfactant within these phosphatidylcholine microemulsions depends on the type of phospholipid, the type of organic solvent, and temperature. The presence of solutes such as salts within the aqueous phase may also affect the value

of w_o [15]. Shervani et al. [10] have shown that for egg lecithin microemulsions, the amount of water per mole of surfactant increases with increasing temperature above about 10°C—below that temperature the microemulsion does not form (see Fig. 3). This work also indicates that water
solubility increases as the polarity of the organic phase increases. The
amount of water that can be incorporated at room temperature for a
benzene-continuous system is 19 moles of water per mole of surfactant.
This value translates to 0.44 g of water per gram of lecithin. Finally, significantly higher temperatures are needed with a pure dipalmitoyl
phosphatidylcholine/benzene/water system to achieve similar values of w_o.

In addition to these studies of phospholipid-based microemulsions, some
work has been done in which a cosurfactant is used in combination with
the phospholipid. It is expected that the cosurfactant will effectively add
to the bulkiness of the two tails in the phospholipid, thus enhancing the
likelihood of water-in-oil microemulsion formation. Shinoda et al. [16]
explored the phase behavior of soybean phosphatidylcholine/water/
hexadecane systems when either ethanol or propanol is used as a cosurfactant. They found that water-in-oil microemulsions were formed at low
weight percentages of the alcohol relative to the surfactant. At 3 wt%
lecithin the addition of propanol increased the water solubilized within the
microemulsion, whereas at 1 wt% lecithin the cosurfactant had little effect
on w_o. Shervani and co-workers [10] added cholesterol to an egg lecithin/
oil/water system and found that it likewise acts as a cosurfactant. As shown
in Fig. 3, the amount of water solubilized within the microemulsion was
increased in the presence of cholesterol. Further, the addition of cholesterol
enabled the system to incorporate more water without forming gels—in
other words, it expanded the microemulsion region in the phase diagram
and decreased the size of the gel region.

Figure 2 shows a second double-tailed surfactant, known as sodium
bis(2-ethylhexyl)sulfosuccinate or Aerosol OT (AOT). The two bulky tails
coupled with the small head group of this surfactant causes it readily to
form water-in-oil microemulsions. As a consequence, a number of applications for such microemulsions have been carried out using AOT, resulting
in significant information about its microemulsion-forming properties. AOT
is also approved as a food additive in the United States, and consequently
the extensive information we have on this surfactant could be useful in the
area of food science.

AOT-based microemulsions contain approximately monodisperse and
spherical droplets of water, whose size is linearly related to the amount of
water per mole of surfactant incorporated into the aqueous phase [17,18].
These microemulsions can solubilize large quantities of water, yielding w_o

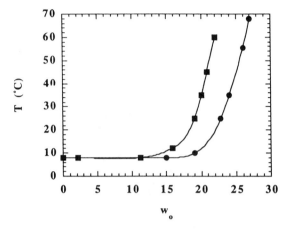

FIG. 3 Water solubilization as a function of temperature for an egg lecithin/benzene/water microemulsion (■) without cholesterol; (●) with cholesterol.

values up to 60 (corresponding to 2.5 g of water per gram of surfactant) and higher. The amount of water that can be incorporated is largely dependent on the salt concentration of the aqueous phase, with lower salt concentrations resulting in greater water solubility. As is the case for lecithin-based systems, higher temperatures result in more extensive water incorporation. Figure 4 shows a schematic of the phase diagram for AOT in isooctane [19].

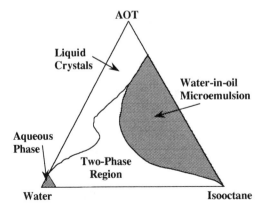

FIG. 4 Phase diagram of AOT/isooctane/water system. (Adapted from Ref. 19.)

Because AOT is not very palatable, there is a desire to find other microemulsions with high water solubility that are not based solely on this surfactant. Studies have shown [20,21] that AOT can be combined with the food-grade surfactant sorbitan monostearate or with the similar surfactant sorbitan monolaurate to yield a microemulsion with considerable capability for water solubilization. The results of Osborne and co-workers [21] demonstrate that increasing the amount of sorbitan monolaurate increases the solubility of water in a hexadecane microemulsion system up to certain level (40 wt% sorbitan monolaurate). At higher fractions of sorbitan monolaurate the water solubility decreases but is still significant. Similar qualitative results would be anticipated for sorbitan monostearate.

The water-in-oil microemulsion studies discussed above focus only on non-food-grade oils, and translating these results to more polar triglyceride solvents may prove difficult. There have been only three studies focusing on completely food-grade water-in-oil microemulsions. Larsson [22] found that a thermodynamically stable oil continuous phase could be formed with water, sunflower oil monoglycerides, and soybean oil. X-ray diffraction results suggest that some type of lamellar structure was formed in these phases, rather than the more typical spherical microemulsion droplets. These microemulsions could incorporate 15% water by weight.

El-Nokaly and co-workers [23] surveyed several industrially available surfactants to determine their effectiveness in creating a water-in-triglyceride microemulsion. Using soybean oil as the continuous phase, they tested several glycerides and polyglycerol esters of fatty acids. Four of these surfactants appeared to form an oil-continuous microemulsion: polyglycerol linoleate, polyglycerol oleate, the monoglyceride mixture AM#505, and polyoxyethylene sorbitol oleate. Polyglycerol linoleate was the most efficient at solubilizing water, yielding a w_o value of 4 moles of water per mole of surfactant. The authors believe the relative effectiveness of this surfactant is due to the comparable size and extensibility of the head and tail groups in the molecule, together with a "kink" due to unsaturation in the tail group that should increase its bulkiness.

Still more recently, polyoxyethylene (40) sorbitol hexaoleate was used with an ethanol cosurfactant to incorporate water in soybean oil [24]. The amount of water solubilized increased as the ratio of cosurfactant to surfactant increased, and increasing temperature likewise increased water solubilization. A constant ratio of 1.5 g of surfactant per gram of oil was used; this high concentration of surfactant was able to solubilize 40% water by weight at room temperature.

In addition to these studies of completely edible microemulsion phases, other research has utilized non-food-grade surfactants and cosurfactants to incorporate water within edible oils. Paul and coworkers [25] used Triton

X-100 plus butanol to solubilize water in saffola oil. Lastly, monoglycerides and acetic acids of monoglycerides were used with butanol, isopropanol, and hexanol to incorporate water within Canola oil [26].

B. Oil-in-Water Microemulsions

As little as has been done researching the formation of water-in-oil microemulsions from food-grade materials, even less has been accomplished in the area of oil-in-water systems. Early work by Ekman and Lundberg [27] found that aqueous solutions of egg lecithin could solubilize up to 15% by weight triolein, with similar results found in dipalmitoyl phosphatidylcholine solutions. Burns and Roberts [28] and Lin et al. [29] examined the solubilization of the triglyceride tributyrin within diheptanoyl phosphatidylcholine micelles. It was found that at very low amounts of triglyceride, the oil forms a rodlike mixed micelle with the lecithin. Above a w_o value of 0.18, the surfactant aggregate becomes spherical and the oil forms a continuous core at the center of the aggregate. Small-angle neutron scattering results indicate that the size of these spherical aggregates increases linearly with the fraction of oil incorporated. At room temperature a maximum of 0.36 mole of tributyrin may be solubilized per mole of lecithin.

More extensive oil solubilization within lecithin-based microemulsions was obtained by Shinoda et al. [16]. This study considered the formation of oil-in-water microemulsions as well as water-in-oil phases. The relative flexibility of phospholipids to form both of these types of phases is due to the comparable size of their head and tail groups, a property that also makes them ideally suited for forming the bilayer structure found in cell membranes. In contrast to the lecithin-based water-in-oil microemulsions mentioned above, however, a cosurfactant is needed to form oil-continuous phases from these molecules. Shinoda et al. [16] found that at least 14 wt% propanol in water was needed to form a hexadecane-in-water microemulsion with soybean lecithin as a surfactant. As was the case for water-in-oil microemulsions, at 1 wt% lecithin the amount of oil solubilized is fairly insensitive to propanol concentration, whereas with 3 wt% lecithin the oil solubilized decreases as the propanol concentration increases. Quite extensive quantities of hexadecane—at least 38% by volume—could be solubilized in these systems. Similar results were obtained using ethanol as the cosurfactant.

In addition to these studies of lecithins, some success has been achieved with solubilizing edible oils in other food-grade surfactants. In an unpublished report, Treptow [30] obtained a oil-in-water microemulsion using peppermint oil and Tween 20 (see Fig. 5). Tween 20 together with the surfactant G1045 (ICI, Inc.) could also be used to solubilize up to 10%

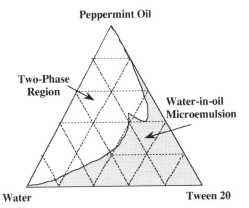

FIG. 5 Phase diagram of Tween 20/peppermint oil/water system. (Adapted from Ref. 23.)

soybean oil in a water-continuous microemulsion [30]. Monocaprylin and sodium xylene sulfonate have been shown to create tricaprylin microemulsions in which up to 15 wt% of the oil was incorporated [31]. Finally, Wolf and Havekotte [32] discuss a wide range of oil-in-water microemulsions formed from monoglycerides, sodium palmitate, AOT, or Tween 60. These microemulsions, which were formed using only edible oils, contain large quantities of an alcohol, with weight ratios of alcohol to water ranging from 2 to 17. Propylene glycol is the primary cosurfactant used.

There has been some recent work utilizing edible compounds as the oil constituents in oil-in-water microemulsions, in which the surfactant is not food grade. Paul and co-workers [25] report the phase behavior of a mixture containing saffola oil together with Triton X100, butanol, and water; they observed a two-phase water continuous region under certain conditions. Sodium dodecyl sulfate and pentanol were also shown solubilize significant amounts of vegetable oils in water by Romero et al. [33], with the resulting microemulsion being used as a medium for aniline detection in vegetable oils.

C. Other Surfactant Structures

In addition to forming discrete droplets of one fluid within another, which is the structure conventionally associated with microemulsions, mixtures of oil, water, and surfactant can form thermodynamically stable phases containing other, more complex structures. These phases can often be useful in the same applications as are proposed for microemulsions. In the food area, lecithin-based phase structures are the best studied. Luisi and co-workers [13,34] and Shervani et al. [10,35] found that by increasing the water content of lecithin water-in-oil microemulsions a gel structure was formed. It has been speculated [10] that addition of considerable amounts of water to the microemulsion system starts to distort the spherical water droplets into asymmetric shapes, resulting in an observable increase in the fluid viscosity [14]. These shapes eventually evolve into a tubular network structure within a gel, a structure that has been seen using small-angle neutron scattering and quasielastic light scattering [13,14,36].

Another structure often seen in surfactant-containing systems is that of a bicontinuous solution. As its name suggests, in a bicontinuous phase regions of oil and water intertwine to create two separate but continuous domains of fluid separated by surfactant interface. Such a phase is generally formed when the concentrations of the two solvent fluids are comparable. In a study of the effect of cosurfactants on lecithin-based mixtures, Shinoda et al. [16] found that bicontinuous microemulsions were formed at cosurfactant (propanol) concentrations in the range of 11 to 14 wt% in water. At lower propanol concentrations an oil-continuous microemulsion was formed; at higher concentrations a water-continuous microemulsion was formed.

The gel and bicontinuous surfactant structures discussed here have many of the same advantages as microemulsions containing spherical droplets: they effectively make contact between oil and water solutions in a thermodynamically stable form. Thus they lend themselves to several applications in the food area, some of which are discussed in the following sections. However, these complex structures, often generally referred to as liquid crystals, impart another advantage to the food processor—they can greatly enhance the stability of more traditional food emulsions. Consider the phase diagram shown in Fig. 6a. The region marked W + D + O represents a bicontinuous phase coexisting with an aqueous and an oil phase—this is known as a three-phase or Winsor III system. Because the bicontinuous phase has a density intermediate between those of the oil and water phases, it forms at the interface between those phases, as shown in Fig. 6b. If we now consider the same system in emulsified form, we recognize that in such an emulsion, droplets of oil contained within an aqueous phase are coated by an interfacial layer consisting of the bicontinuous phase. This region, which includes a mixture of both solvents plus surfactant, can provide a thick coating that stabilizes the emulsion droplets to coalescence.

As discussed by Friberg and Solans [37], the reasons for this added stability in the presence of a liquid crystal interfacial layer are threefold.

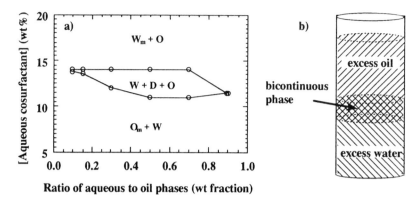

FIG. 6 Formation of a bicontinuous phase. (a) Phase diagram of soy bean lecithin/hexadecane/water system. (Adapted from Ref. 16.) D represents the bicontinuous middle phase, W_m and O_m represent water-continuous and oil-continuous microemulsion phases, and W and O represent the excess water and oil phases. (b) Three-phase system.

First, this layer has a much higher surface viscosity than a single surfactant monolayer and thus increases the hydrodynamic resistance to droplet coalescence. Second, the layer also resists deformation and hence should prevent the droplets from fusing. Finally, the complex composition of this liquid crystal layer alters the van der Waals interaction between the droplets in such a way as to decrease the likelihood of coalescence. The added stability these phases can impart will apply to foams as well as to emulsions.

Although the possibility of forming these liquid crystal coatings about emulsion particles has long been recognized [31,38], Friberg and Solans [37] have noted that relatively little has been done to explore the connection between surfactant aggregation structures and emulsion stability. The effect of phospholipid phase behavior on emulsion stability was studied by Rydhag and Wilton [39], who found that surfactant concentrations above a critical value yielded high emulsion stability, corresponding to the presence of liquid crystal multilayers approximately 8 to 10 nm in thickness. Only 3 to 6% surfactant based on the oil phase is needed to achieve such an emulsion. Krog et al. [40] use polarized microscopy to demonstrate the presence of lecithin liquid crystals in a salad dressing emulsion. With respect to other surfactants, Tween 40 with sorbitan monopalmitate was used to increase the stability of a sunflower oil emulsion, in which the presence of liquid crystals was verified by electron microscopy [41,42].

III. SOLUBILIZATION WITHIN MICROEMULSIONS

One of the most obvious applications for microemulsions in foods is to facilitate the addition of food ingredients. This can be as simple as replacing an oil-in-water emulsion with an oil-in-water microemulsion to achieve greater food stability; alternatively, components other than water or oil can be solubilized within the microemulsion. Flavors, preservatives, and nutrients that are poorly soluble in water can be incorporated in a water-based food by solubilizing the component within surfactant aggregates. Similarly, water-in-oil microemulsions can be used to combine water-soluble substances with edible oils. In both cases microemulsions can provide a well-controlled, highly stable medium for the combination of these ingredients.

Microemulsions can provide many advantages for the incorporation of food ingredients. Flavors, colors, and aromas that are fat soluble may be lacking when oils are removed in low-fat foods—microemulsions could provide a lower calorie alternative for reintroducing those missing sensory elements. The transparent quality of microemulsions can provide a host of new options for altering the appearance of food, as well as being a useful quality when employing these phases in food analysis. The small droplet sizes in microemulsions provide excellent contact between the lipid and aqueous phases—particularly advantageous when solubilizing perservatives within the droplets. There is also some indication that microemulsions or micelles may protect solubilized components from unwanted degradative reactions [43,44]. Finally, the stability characteristics of microemulsions could be a real benefit during food processing and storage.

A. Flavors, Aromas, Dyes

The hydrophobic nature of many flavors and aromas makes them excellent candidates for incorporation within microemulsions. However, there have been just a few studies considering flavor solubilization within these surfactant systems. Slocum et al. [45] found that greatly enhanced solubility of a range of flavor elements, including alcohols, 2-ketones, 1-alcohols, and ethyl esters, could be achieved through solubilization in a variety of surfactants. Of these, Tween 20 was the only food-grade surfactant considered, but it was able to solubilize larger quatities of flavors than the three other surfactants studied. Up to 3 moles of flavor could be solubilized per mole of surfactant—a quite attractive result considering the very small quantities of flavor components that are generally needed in foods.

Considerable amounts of peppermint oil can be solubilized within an aqueous phase through incorporation in a Tween 20–based microemulsion, as described in an unpublished report by Treptow [30] (see Fig. 5). Also, several of the microemulsions discussed by Wolf and Havekotte [32] contained essential oils as their organic phase (see discussion in Sec. II). More extensive have been the studies of dyes (see, for example, Refs. 46–48), although food dyes or colorings have not been considered explicitly. Note that in some of these studies the flavor is incorporated directly in the micelle, in a process that can be considered micellar solubilization or microemulsion formation [30], depending on the relative amount of the solute. The cited dye studies, on the other hand, consider a microemulsion containing an oil as the discontinuous phase and then solubilize the dye within that oil-based system. Note also that flavor incorporation is not limited to oil-soluble flavors: El-Nokaly et al. [23] report the successful incorporation of water-soluble flavors within food-grade water-in-oil microemulsions.

One aspect of flavor incorporation that makes this area particularly fascinating for study is the question of how microemulsion solubilization will affect flavor release. It is believed that the size distribution of emulsions droplets will influence the perception of flavors incorporated within the droplet, and thus it is not unreasonable to expect an effect due to microemulsion solubilization on sensory outcomes. Labows [49] has used headspace gas chromatography to measure the effect of micelle solubilization of flavors on the release of those substances into the air. Micelles and microemulsions may also play a role in the release of flavors from emulsified systems. Studies in our laboratory [50–52] have shown that micelles within the aqueous continuum can enhance the transport of oils from emulsion droplets through the water phase. If transport through the aqueous phase is a necessary component of flavor release, the presence of micelles or microemulsion droplets coexisting with emulsified systems could have a significant effect on flavor perception.

B. Vitamins and Preservatives

Microemulsions also have potential for enhancing the solubility of hydrophobic vitamins or other nutrients within water-based foods (or water-soluble nutrients within oil-based foods). The pharmaceutical and medical literature is replete with studies of enhanced micellar delivery of vitamins, in particular vitamin E, vitamin K_1, and β-carotene [53–58]. Chiu and Yang [54] further claim that solubilization within nonionic surfactants protects vitamin E from oxidative degradation. Water-soluble vitamins could similarly be incorporated within a water-in-oil microemulsion [59]. This approach could be taken with a variety of food preservatives, thereby using the microemulsion system to expose more effectively the preservative to the component upon which it acts.

Many of the vitamins mentioned above have important antioxidant effects, and microemulsions may provide a way to contact such agents with the substrates they are designed to protect. The in vivo effectiveness of the antioxidants vitamins E and C has been shown to be enhanced in micellar solutions, with important implications for cancer prevention [55,56,60,61]. Osborne and co-workers studied the effect of vitamin E (α-tocopherol) on lipid peroxidation within a pentaethylene glycol n-dodecyl ether liquid crystalline phase. Finally, Han et al. [59] used water-in-oil microemulsions to enhance the effectiveness of ascorbic acid in preventing the lipid oxidation of a sardine oil or soybean oil continuous phase. Lecithin was used as the surfactant. Figure 7 shows the dramatic effect of the ascorbic acid solubilized within the microemulsion droplets on reducing oil oxidation, relative to that occurring in the absence of any antioxidant or in the presence of oil-soluble antioxidants. Interestingly, this study also found that adding both ascorbic acid within the water droplets and δ-tocopherol to the oil phase reduced the oxidation even further—the two antioxidants appeared to have a synergistic effect. This finding illustrates the important benefits a microemulsion system can impart to foods by providing good contact between water-soluble and oil-soluble substances.

IV. MICROEMULSIONS AS REACTION MEDIA

The ability to incorporate solutes within the droplets of a microemulsion raises the possibility of using microemulsions as reaction media, particu-

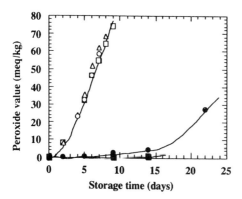

FIG. 7   Effect of solubilized antioxidants on the oxidation of fish oil. Experiments measure the oil oxidation within a water-in-oil microemulsion containing (○) no antioxidant, (□) δ-tocopherol solubilized in the oil phase, (△) rosemary extract solubilized in the oil phase, (●) ascorbic acid solubilized within the water droplets, and (■) ascorbic acid solubilized within the water droplets and δ-tocopherol solubilized in the oil phase. (Adapted from Ref. 59.)

larly in situations where the reaction involves interactions between water-soluble and oil-soluble substances. A particularly fruitful application of this idea has been to carry out enzymatic reactions using water-in-oil microemulsions (see Sec. II in Chapter 4). Here the enzyme is solubilized within the water droplets contained within the microemulsion, where it can access an organic substrate more effectively than if the reaction is carried out in an aqueous phase. Proteins solubilized in this manner are often found to retain their conformation and activity.

Much of the literature (see Refs. 59, 62, and 63 for early studies) on enzymatic reactions within these microemulsions has focused on the surfactant AOT (see Fig. 2b), because the microemulsions formed using this surfactant are the best characterized. However, a number of researchers have also used phospholipids to form their microemulsion phases (see Walde et al. [64] for a review). Since both of these surfactants are approved for food use, they should provide appropriate media for reactions carried out in the food industry. Although in a microemulsion-based process the surfactant can be recycled and therefore should not remain with the product, it is of course preferable to use food-grade surfactants in case any trace amounts are found with the synthesized material.

A number of enzymes have been encapsulated in microemulsions made from both surfactants, including cytochrome c, lysozyme, ribonuclease A, trypsin, α-chymotrypsin, phosphatases, and various lipases and phospholipases. The enzyme activity in these systems can be far greater than that in aqueous solution, a phenomenon sometimes called "superactivity." Although this enhanced activity has sometimes been ascribed to a more favorable conformation of the protein within the microemulsion droplet, these effects may be due largely to the complex partitioning of substrates and products between the water pool in the droplet, the oil phase, and the droplet interface. In any case, this enhanced activity makes enzymatic reactions within water-in-oil emulsions a promising technology. Research has indicated that it is also possible to carry out enzymatic reactions using more complex phases, including cubic phases [65] and bicontinuous phases [66,67].

The study of lipase solubilization within water-in-oil microemulsions may be a particularly fruitful one for food scientists. Lipase hydrolysis rates in microemulsions containing a variety of food oils have been determined for systems containing AOT [68–72] and phosphatidylcholine [11,73–75]. In microemulsions made from either of these surfactants, enzyme activity is seen to be highest at intermediate ratios of water to amount of surfactant. As shown in Fig. 8, very low w_o values reduce the enzyme activity, whereas as the water content increases to high values the activity approaches that in aqueous solution. The optimal water content value will

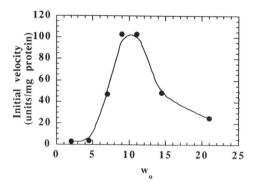

FIG. 8 Effect of water content on lipase activity in a water-in-oil microemulsion. (Adapted from Ref. 73.)

depend on the type of surfactant, as will many other properties of the system. For example, Stark et al. [72] compared lipase activity in AOT-based microemulsions with that in microemulsions based on the nonionic surfactant pentaethylene glycol monododecyl ether and found that the AOT-based system performed considerably better. This result is attributed to the much bulkier head group of the nonionic surfactant, which may prevent the enzyme from easily accessing the interfacial region. Differences were also observed by Peng and Luisi [76] when using various phospholipids. They found that increasing the chain length of the alkyl group in the molecule increases its ability to solubilize water, with a concomitant effect on the protein activity.

The study by Chen and Pai [73] is a direct application of microemulsion hydrolysis to a food process—that of hydrolyzing milk fat by lipase to produce butterlike or cheeselike flavors. In this work the authors solubilize *Candida cylindracea* lipase in a soybean lecithin–based water-in-butteroil microemulsion. They found that the enzyme displayed less sensitivity to pH and to temperature than an enzyme used in an emulsified system (the conventional approach to milkfat hydrolysis). The enzyme activity also increased with increasing surfactant concentration, a result attributed to the accompanying increase in interfacial area. The much larger interfacial areas present in the microemulsion system than in a conventional emulsified system had the result that much higher enzyme concentrations could be used before the system reached saturation.

Enzymatic reactions are not the only bioreactions that may be carried out in water-in-oil microemulsions. Research by Walde [77] demonstrates the potential of microemulsions in food analysis involving enzymatic reactions. In this work phenol red, a fatty acid indicator, is solubilized within water microemulsion droplets incorporated in food oils. This microemulsion system allows for the simple spectrophotometric determination of fatty acid content of vegetable oils. There are also reports in the literature that entire cells can be solubilized within these water-in-oil microemulsions and remain viable [78–82]. Microemulsions based on Tween, a mixture of phospholipids known as asolectin, or AOT have proved to be successful media for solubilization of the cells studied, which include *Escherichia coli*, yeast, and plant cells. Incorporation of these large bodies within a microemulsion raises the possibility of microbiological reactions carried out within organic media.

Oil-in-water microemulsions may also have a beneficial effect on reaction kinetics. As discussed in the preceding section, poorly water-soluble reactants can have their aqueous solubility dramatically increased through incorporation within oil microemulsion droplets, and this can obviously have a beneficial effect on reactions carried out within an aqueous phase. Microemulsions have advantages over oil-in-water emulsions as reaction media because of their stability and because their much smaller droplet size results in far better contact between reactants or catalysts interacting across the oil-water interface. These microemulsion systems can have other influences as well: Das et al. [83] found that the rate of inversion of cane sugar was higher in an oil-in-water microemulsion than in water alone. Here the effect is apparently due to the decreased polarity of the microemulsion medium, which appears to enhance the reaction kinetics of this food reaction.

V. USE OF MICROEMULSIONS IN EXTRACTION TECHNIQUES

As discussed in the preceding sections, the ability of microemulsions to incorporate solutes within their discontinuous phase enables these mixtures to be used to enhance the solubility of food ingredients or to carry out reactions in novel media. In some cases, however, the solute not only is soluble within the microemulsion droplet, it actually prefers that phase to that of a bulk solvent. In other words, a polar solute may actually preferentially solubilize within a water-in-oil microcmulsion droplet over an aqueous solution, or a hydrophobic solute may prefer the oil microemulsion droplets to a bulk oil phase. In such situations it is possible to use the microemulsion phase to extract solutes from a bulk solution. Such an approach has been studied for separating proteins and other biomolecules from an aqueous solution, using a water-in-oil microemulsion. For hydrophobic solutes the focus has been primarily on using micelles to extract solutes from oil, although again the difference between a micelle and a microemulsion droplet is blurred once significant amounts of solute become incorporated within the micelle.

A. Protein Extraction

The possibility of using water-in-oil microemulsions to extract protein molecules from an aqueous phase was demonstrated by Göklen and Hatton [84,85] and by others [86,87]. In this approach, an aqueous protein-containing phase is contacted with a water-in-oil microemulsion. Given the right conditions, the protein then preferentially solubilizes within the microemulsion phase, at the which time the two phases can be separated. Recovery of the protein may be accomplished by contacting the protein-containing microemulsion with a fresh aqueous solution under conditions favoring the transfer of protein out of the aqueous phase. This process has some nice features in that the microemulsion phase can be a gentle solvent for extracting the protein without altering its enzymatic or functional properties, and yet the process can be readily scaled up using conventional liquid-liquid extraction technology. The dependence of solubility on protein properties also makes selective separations a possibility.

Clearly, to utilize this approach effectively one needs an understof the dependence of protein solubility in the microemulsion on the system properties (see Cabral and Aires-Barros [88] for a review). With the negatively charged AOT surfactant most commonly studied thus far, electrostatics are found to play a major role in the extent of transfer [85,89,90]. At pH values below the isoelectric point, the protein has a net positive charge that facilitates its transfer into the microemulsion droplet, where it can interact attractively with the surfactant head groups. Hence we see in Fig. 9 the extent of protein transfer increasing as pH decreases. By screening the electrostatic interactions between the charged protein and surfactant, the addition of salt tends to mitigate this effect, and Fig. 9 likewise shows a decrease in protein transfer as salt concentration increases. pH and ionic strength also influence the rate of transfer of the protein into or out of the microemulsion phase [91,92], with rates of forward transfer observed to be much higher than that of the back extraction step.

Surfactant concentration and the amount of water within the microemulsion phase also affect the amount of protein transferred [89,90]. Figure 10 shows that the amount of protein solubilized increases significantly as the water content rises above a certain amount. Surfactant concentration also has an effect on protein partitioning for a constant ratio of water to surfactant, particularly at low surfactant concentrations and low pH values.

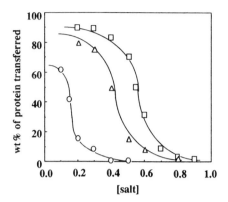

FIG. 9 Weight percent of α-chymotrypsin transferred into an AOT-based water-in-isooctane microemulsion as a function of the aqueous pH and salt concentration. (○) pH 9.2; (△) pH 8.2; (□) pH 6.4.

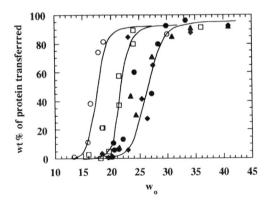

FIG. 10 Weight percent of α-chymotrypsin transferred into an AOT-based water-in-isooctane microemulsion as a function of the water content and surfactant concentration in the organic phase. Aqueous pH is 6.4. (○) [AOT] = 0.025 M; (□) [AOT] = 0.05 M; (♦) [AOT] = 0.1 M; (▲) [AOT] = 0.2 M; (●) [AOT] = 0.4 M.

Protein extraction using water-in-oil microemulsions has been studied in several applications promising to those in the food industry. Extracellular enzymes were effectively recovered from a fermentation broth, leaving other components of the broth behind [93]. Microemulsion extraction is also a useful approach for the recovery of intracellular enzymes, because the contact of surfactant solutions with cells can disrupt cell membranes, thereby liberating the protein. Giovenco et al. [94] utilized such a technique for simultaneously removing proteins from cells and incorporating them within the microemulsion phase. In these studies, a protein of interest can not only be concentrated but also selectively purified from other proteins in the solution. Göklen and Hatton [85] demonstrated this fact by separating cytochrome c from lysozyme using an AOT-based microemulsion. Use of biospecific ligands can enhance the degree of separation [95–97]. This ability to separate and concentrate proteins from a protein mixture could have promise for the separation of whey protein mixtures to obtain individual proteins with much higher value. Work is under way in our laboratory to apply such a technology to whey protein purification. Finally, it is worth noting that biomolecules other than proteins, such as amino acids, peptides, and nucleic acids, have been separated using water-in-oil microemulsions [98,99], expanding the possibilities for food applications.

B. Hydrophobic Solute Extraction

Outside food technology, there has been substantial interest in taking advantage of the solubilizing properties of micelles to accomplish novel separations of water-insoluble compounds. Using an aqueous micellar solution to extract hydrophobic substances has considerable environmental and economic advantages over organic solvent extraction. This approach could likewise be used to effect food separations—to remove undesirable flavors or nutrients from foods or to extract and purify valuable food components such as flavors and aromas. Such a method has been successfully applied to extract cholesterol from butteroil [100,101]. The cholesterol preferentially partitions from the butteroil bulk phase into micelles formed from a food-grade material known as Quillija saponins. Our laboratory is now working to connect the partitioning behavior of the cholesterol in this system to the micellar properties of the saponins, information that is currently unavailable in the literature.

It may be possible in many cases to use oil-in-water microemulsions to carry out extractions now performed using micellar phases. Hence, instead of incorporating the solute within a surfactant micellar aggregate, the solute would partition into the oil-filled droplets present within a microemulsion. Advantages of using microemulsions directly could include a larger capacity for the material within the relatively larger droplets of the microemulsion and an additional degree of control over the extent of partitioning through one's choice of the interior oil phase. Such an approach was taken by Chhatre et al. [102] to extract nicotine from tobacco waste. These researchers contacted a nicotine-containing kerosene phase with a kerosene-in-sulfuric acid solution microemulsion. Cetyl trimethyl ammonium bromide, sodium dodecyl sulfate, and Triton X-100 were used with pentanol to create the microemulsion phase. The microemulsion was used to extract the nicotine from the bulk kerosene, at which point the nicotine could react with the sulfuric acid solution to create nicotine sulfate, the desired product. Hence extraction and reaction can be accomplished in the same step, and the excellent contact between the microemulsion droplets and the aqueous phase allows high reaction rates.

C. Process Development

The use of microemulsion phases for extraction purposes allows the separation to be carried out using conventional liquid-liquid extraction equipment. Such equipment includes spray columns, centrifugal extractors, and mixer-settler units [86,88,100]. In a more novel configuration, the microemulsion phase is contacted with the feed liquid across a hollow-fiber membrane. Such an approach has been used by Dahuron and Cussler [103] to extract proteins into a water-in-oil microemulsion. Hurter and Hatton [104] also used a hollow-fiber membrane device to extract hydrophobic solutes from an aqueous phase into a block copolymer micellar phase; in this case the membrane serves to retain the micelles within the extracting phase.

Calvert et al. [105] have proposed incorporating micelles within an aqueous hydrogel and using such a material to extract hydrophobic solutes. The gel is contacted with the feed solution, thereby allowing the solute to diffuse through the gel interior and into the micelles, where they are retained by the gel fibers. A similar approach could be taken with an oil-in-water microemulsion, in which case the microemulsion droplet would be trapped by the gel matrix. This gel configuration presents interesting possibilities for combining extraction with chemical reactions, by coimmobilizing enzymes or other catalysts within the gel.

A gel-based material has also been developed for a water-in-oil microemulsion phase. Here the gel is created by solubilizing gelatin within the aqueous pools of the microemulsion droplets. The gelatin then links the droplets together across an organic-continuous phase, creating an organogel [106,107]. Such a gel could be utilized for extraction purposes by contacting it with an aqueous feed containing hydrophilic solutes.

VI. CONCLUSIONS

Research on the science and application of microemulsion systems is considerably more extensive in fields outside food science and technology. Yet many of the applications discussed in this chapter—enhancing the solubility of ingredients using microemulsions, utilizing microemulsions as reaction media, and effecting extractions with microemulsion phases—could yield promising new technologies within the food industry. One of the primary barriers to the application of microemulsions to food problems has been the need for more information on microemulsions created from food-grade components. Forming microemulsions within polar oils such as triglycerides may pose a particular challenge. Further progress in this area may be possible by examining mixtures of surfactants, in order to discover surfactant geometries that work for a particular system [16,21].

Existing research on microemulsions applied to food problems has pointed the way to novel and promising technologies. Microemulsions are potentially safe, highly effective, scalable, and low-cost systems in which to accomplish such goals as enhancing the antioxidant effects of ascorbic acid [59], effecting hydrolysis of milkfat [73], or incorporating larger aqueous quantities of vitamin E [54]. Such qualities should be very appealing in the food industry. With a great deal of promise and such intriguing intellectual challenges to be met, investigation into microemulsion applications presents a rich area for food science research.

REFERENCES

1. J. N. Israelachvili, D. J. Mitchell, and B. W. Ninham, J. Chem. Soc. Faraday Trans. II 72:1525 (1976).
2. L. R. C. Barclay, J. M. MacNeil, J. A. van Kessel, B. J. Forrest, N. A. Porter, L. S. Lehman, K. J. Smith, and J. C. Ellington, J. Am Chem. Soc. 106:6740 (1984).
3. S.-T. Chen and C. S. Springer, Chem. Phys. Lipids 23:23 (1979).
4. P. H. Elworthy and D. S. McIntosh, Kolloid Z. Zeitschr. Polym. 195:27 (1964).
5. C. A. Boicelli, F. Conti, M. Giomini, and A. M. Giuliani, Chem. Phys. Lett. 89:490 (1982).
6. J. B. Davenport and L. R. Fisher, Chem. Phys. Lipids 14:275 (1975).
7. P. H. Elworthy and D. S. McIntosh, J. Phys. Chem. 68:3448 (1964).
8. V. V. Kumar, C. Kumar, and P. Raghunathan, J. Colloid Interface Sci. 99: 315 (1984).
9. V. V. Kumar, P. T. Manoharan, and P. Raghunathan, J. Biosci. 4:440 (1982).
10. Z. Shervani, A. Maitra, T. K. Jain, and Dinesh, Colloids Surf. 60:161 (1991).

11. P. H. Poon and M. A. Wells, Biochemistry *13*:4928 (1974).
12. V. R. Ramakrishnan, A. Darszon, and M. Montal, J. Biol. Chem. *258*:4857 (1983).
13. P. L. Luisi, R. Scartazzini, G. Haering, and P. Schurtenberger, Colloid Polym. Sci. *268*:356 (1990).
14. P. Schurtenberger, R. Scartazzini, L. J. Magid, M. E. Leser, and P. L. Luisi, J. Phys. Chem. *94*:3695 (1990).
15. C. A. Boicelli, F. Conti, M. Giomini, and A. M. Giuliani, Gazz. Chim. Ital. *113*:573 (1983).
16. K. Shinoda, M. Araki, A. Sadaghiani, A. Khan, and B. Lindman, J. Phys. Chem. *95*:989 (1991).
17. H. F. Eicke and J. Rehak, Helv. Chim. Acta *59*:(1976).
18. M. Zulauf and H. F. Eicke, J. Phys. Chem. *83*:480 (1979).
19. B. Tamamushi and N. Watanabe, Colloid Polym. Sci. *258*:174 (1980).
20. K. A. Johnson and D. O. Shah, J. Colloid Interface Sci. *107*:269 (1985).
21. D. W. Osborne, C. V. Pesheck, and R. J. Chipman, in *Microemulsions and Emulsions in Foods* (M. El-Nokaly and D. Cornell, eds.), American Chemical Society, Washington, DC, 1991, pp. 62–79.
22. K. Larsson, in *Microemulsions and Emulsions in Foods* (M. El-Nokaly and D. Cornell, eds.), American Chemical Society, Washington, DC, 1991, pp. 44–50.
23. M. El-Nokaly, G. Hiler, and J. McGrady, in *Microemulsions and Emulsions in Foods* (M. El-Nokaly and D. Cornell, eds.), American Chemical Society, Washington, DC, 1991, pp. 26–43.
24. R. F. Joubran, D. G. Cornell, and N. Parris, Colloids Surfaces A: Physicochem. Eng. Aspects *80*:153 (1993).
25. B. K. Paul, M. L. Das, D. C. Mukherjee, and S. P. Moulik, Ind. J. Chem. *30A*:328 (1991).
26. A. M. Vesala, J. B. Rosenholm, and S. Laiho, J. Am. Oil Chem. Soc. *62*:1379 (1985).
27. S. Ekman and B. Lundberg, Acta Chem. Scand. B *32*:197 (1978).
28. R. A. Burns and M. F. Roberts, J. Biol. Chem. *256*:2716 (1981).
29. T.-L. Lin, S.-H. Chen, N. E. Gabriel, and M. F. Roberts, J. Phys. Chem. *94*:855 (1990).
30. R. S. Treptow, Research and Development Report, The Procter & Gamble Company, June 1, 1971.
31. S. E. Friberg and L. Rydhag, Kolloid-Z. Z. Polym. *244*:233 (1971).
32. P. A. Wolf and M. J. Havekotte, U.S. Patent 4,835,002 (May 30, 1989).
33. J. S. E. Romero, E. F. S. Alfonso, M. C. G. Alvarez-Coque, and G. R. Ramos, Anal. Chim. Acta *235*:317 (1990).
34. R. Scartazzini and P. L. Luisi, J. Phys. Chem. *92*:829 (1988).
35. Z. Shervani, T. K. Jain, and A. Maitra, Colloid Polym. Sci. *269*:720 (1991).
36. P. Schurtenberger, L. J. Magid, P. Lindner, and P. L. Luisi, Prog. Colloid Polym. Sci. *94*:3695 (1993).
37. S. Friberg and C. Solans, Langmuir *2*:121 (1986).

38. S. Friberg, L. Mandell, and M. Larsson, J. Colloid Interface Sci. *29*:155 (1969).
39. L. Rydhag and I. Wilton, J. Am. Oil Chem. Soc. *58*:830 (1981).
40. N. Krog, N. M. Barfod, and R. M. Sanchez, J. Dispers. Sci. Tech. *10*:483 (1989).
41. N. Pilpel and M. E. Rabbani, J. Colloid Interface Sci. *119*:550 (1987).
42. N. Pilpel and M. E. Rabbani, J. Colloid Interface Sci. *122*:266 (1988).
43. H. Krasowska, Int. J. Pharm. *4*:89 (1979).
44. M. E. Moro, J. Novillo-Fertrell, M. M. Velazquez, and L. J. Rodriguez, J. Pharm. Sci. *80*:459 (1991).
45. S. A. Slocum, A. Kilara, and R. Nagarajan, in *Flavors and Off-Flavors* (G. Charalambous, ed.), Elsevier Science Publishers, Amsterdam, 1989, pp. 233–247.
46. N. S. Dixit and R. A. Mackay, J. Am. Chem. Soc. *105*:2928 (1983).
47. S. E. Friberg, T. Young, E. Barni, and M. Croucher, J. Dispers. Sci. Tech. *13*:611 (1992).
48. K. R. Wormuth, L. A. Cadwell, and E. W. Kaler, Langmuir *6*:1035 (1990).
49. J. N. Labows, J. Am. Oil Chem. Soc. *69*:34 (1992).
50. D. J. McClements and S. R. Dungan, J. Phys. Chem. *97*:7304 (1993).
51. D. J. McClements and S. R. Dungan, Colloids Surfaces A: Physicochem. Eng. Aspects, *104*:127 (1995).
52. D. J. McClements, S. R. Dungan, J. B. German, and J. E. Kinsella, Food Hydrocolloids *6*:415 (1992).
53. L. M. Canfield, T. A. Fritz, and T. E. Tarara, Methods Enzymol. *189*:418 (1990).
54. Y. C. Chiu and W. L. Yang, Colloids Surfaces *63*:311 (1992).
55. Z. L. Liu, Z. X. Han, P. Chen, and Y. C. Liu, Chem. Phys. Lipids *56*:73 (1990).
56. Z. L. Liu, L. J. Wang, and Y. C. Liu, Sci. China, B *34*:787 (1991).
57. G. Schubiger, O. Tonz, J. Gruter, and M. J. Shearer, J. Pediatr Gastroenterol. Nutr. *16*:435 (1993).
58. M. J. Winn, P. M. White, A. K. Scott, S. K. Pratt, and B. K. Park, J. Pharm. Pharmacol. *41*:257 (1989).
59. D. Han, O. S. Yi, and H. K. Shin, J. Food Sci. *55*:247 (1990).
60. Z. L. Liu, Z. X. Han, L. M. Wu, P. Chen, and Y. C. Liu, Sci. China B *32*:937 (1989).
61. K. Mukai, M. Nishimura, and S. Kikuchi, J. Biol. Chem. *266*:274 (1991).
62. S. Barbaric and P. L. Luisi, J. Am. Chem. Soc. *103*:4239 (1981).
63. P. D. I. Fletcher, B. H. Robinson, R. B. Freedman, and C. Oldfield, J. Chem. Soc. Faraday Trans. I *81*:2667 (1985).
64. P. Walde, A. M. Giuliani, C. A. Biocelli, and P. L. Luisi, Chem. Phys. Lipids *53*:265 (1990).
65. B. Ericsson, K. Larsson, and K. Fontell, Biochim. Biophys. Acta *729*:23 (1983).

66. K. M. Larsson, P. Adlercreutz, and B. Mattiasson, Biotechnol. Bioeng. *36*: 135 (1990).
67. C. Sonesson and K. Holmberg, J. Colloid Interface Sci. *141*:239 (1991).
68. D. Han and J. S. Rhee, Biotechnol. Bioeng. *28*:1250 (1986).
69. K. Holmberg and E. Osterberg, J. Am. Oil Chem. Soc. *65*:1544 (1988).
70. T. Kim and K. Chung, Enzyme Microb. Tech. *11*:528 (1989).
71. E. A. Malakhova, B. I. Kurganov, A. V. Levashov, I. V. Berzin, and K. Martinek, Dokl. Akad. Nauk SSSR *270*:474 (1983).
72. M.-B. Stark, P. Skagerlind, K. Holmberg, and J. Carlfors, Colloid Polym. Sci. *268*:384 (1990).
73. J.-P. Chen and H. Pai, J. Food Sci. *56*:234 (1991).
74. R. L. Misiorowski and M. A. Wells, Biochemistry *13*:4921 (1974).
75. S. Morita, H. Narita, T. Matoba, and M. Kito, J. Am. Oil Chem. Soc. *61*: 1571 (1984).
76. Q. Peng and P. L. Luisi, Eur. J. Biochem. *188*:471 (1990).
77. P. Walde, J. Am. Oil Chem. Soc. *67*:110 (1990).
78. N. W. Fadnavis, N. P. Reddy, and U. T. Bhalerao, J. Org. Chem. *54*:3218 (1989).
79. T. Haag, T. Arslan, and D. Seebach, Chimia *43*:351 (1989).
80. G. Haering, F. Meussdoerffer, and P. L. Luisi, Biophys. Res. Commun. *127*: 911 (1985).
81. A. Hochkoeppler and P. L. Luisi, Biotechnol. Bioeng. *37*:918 (1991).
82. N. Pfammatter, A. A. Guadalupe, and P. L. Luisi, Biochem. Biophys. Res. Commun. *161*:1244 (1989).
83. M. L. Das, P. K. Bhattacharya, and S. P. Moulik, Langmuir *6*:1591 (1990).
84. K. Göklen and T. A. Hatton, Biotech. Prog. *1*:69 (1985).
85. K. Göklen and T. A. Hatton, Sepn. Sci. Tech. *22*:831 (1987).
86. M. Dekker, K. V. Riet, S. R. Weijers, J. W. A. Baltussen, C. Laane, and B. H. Bijsterbosch, Chem. Eng. J. *33*:B27 (1986).
87. E. M. Leser, G. Wei, P. L. Luisi, and M. Maestro, Biochem. Biophys. Res. Commun. *135*:629 (1986).
88. J. M. S. Cabral and M. R. Aires-Barros, in *Recovery Processes for Biological Materials* (J. F. Kennedy and J. M. S. Cabral, eds.), Wiley, New York, 1993, pp. 247–271.
89. P. D. I. Fletcher and D. Parrott, J. Chem. Soc. Faraday Trans. I *84*:1131 (1988).
90. B. Kelley, R. S. Rahaman, and T. A. Hatton, in *Analytical Chemistry in Organized Media: Reversed Micelles* (W. L. Hinze, ed.), JAI Press, Greenwich, 1991, pp. 123–142.
91. S. R. Dungan, T. Bausch, P. Plucinski, W. Nitsch, and T. A. Hatton, J. Colloid Interface Sci. *145*:33 (1991).
92. S. R. Dungan and T. A. Hatton, J. Colloid Interface Sci. *164*:200 (1993).
93. R. S. Rahaman, J. Y. Chee, J. M. A. Cabral, and T. A. Hatton, Biotech. Prog. *4*:218 (1988).

94. S. Giovenco, F. Verheggen, and C. Laane, Enz. Microb. Technol. *9*:470 (1987).
95. R. W. Coughlin and J. B. Baclaski, Biotech. Prog. *6*:307 (1990).
96. B. D. Kelley, D. I. C. Wang, and T. A. Hatton, Biotechnol. Bioeng. *42*:1199 (1993).
97. J. M. Woll, T. A. Hatton, and M. L. Yarmush, Biotech. Prog. *5*:57 (1989).
98. T. A. Hatton, in *Ordered Media in Chemical Separations* (W. L. Hinze and D. W. Armstrong, eds.), American Chemical Society, Washington, DC, 1987, pp. 170–183.
99. P. L. Luisi, V. E. Imre, H. Kaeckle, and H. Pande, in *Topics in Pharmaceutical Sciences* (D. D. Breimer and P. Speiser, eds.), Elsevier, Amsterdam, 1983, pp. 243–254.
100. E. Sundfeld, J. M. Krochta, and T. Richardson, J. Food Process Eng. *16*:207 (1993).
101. E. Sundfeld, S. Yun, J. M. Krochta, and T. Richardson, J. Food Process Eng. *16*:191 (1993).
102. A. S. Chhatre, N. K. Yadav, and B. D. Kulkarni, Sep. Sci. Tech. *28*:1465 (1993).
103. L. Dahuron and E. L. Cussler, AIChE J. *34*:130 (1988).
104. P. N. Hurter and T. A. Hatton, Langmuir *8*:1291 (1992).
105. T. L. Calvert, R. J. Phillips, and S. R. Dungan, AIChE J. *40*:1449 (1994).
106. C. Quellet and H. F. Eicke, Chimia *40*:233 (1986).
107. G. Haering and P. L. Luisi, J. Phys. Chem. *90*:5892 (1986).

8
Microemulsions in Cosmetics

HIDEO NAKAJIMA Basic Research Laboratories, Shiseido Research
Center, Yokohama, Japan

I. INTRODUCTION

Schulman found that a transparent or translucent dispersion, which gives
no phase separation, is spontaneously generated by adding a polar ingre-
dient such as an alcohol (cosurfactant) to an ordinary coarse emulsion with
an ionic surfactant. He called it a "microemulsion" [1] and considered that
such a system is a colloidal dispersion including spherical droplets of extremely small size [1]. However, it was not clarified whether Schulman's
microemulsion is thermodynamically stable or unstable, i.e., a solubilization (single-phase) system. Thereafter, it was confirmed from the studies
of the phase diagrams that the microemulsions are thermodynamically stable and single-phase solubilized solutions [2–4]. Furthermore, it was clar-

ified that similar dispersions are also present in oil-water-nonionic surfactant systems [4,5]. A microemulsion in a strict sense has been defined [6–8] as a single-phase system of water, oil, and amphiphile(s), which is transparent or translucent and a thermodynamically stable isotropic solution. Regarding single-phase microemulsions, many basic studies have been reported from the standpoint of the analysis of the phase diagrams and thermodynamic consideration.

On the other hand, the term microemulsion in a broad sense includes fairly stable but thermodynamically unstable dispersions that are transparent or translucent [7,8]. However, until recently, only a few studies of thermodynamically unstable microemulsions were reported [9,10]. It is not clear whether microemulsions described in practical reports such as patents are thermodynamically stable or unstable systems. Some of them can be presumed to be thermodynamically unstable systems. If there is a true emulsion system having two phases in which the droplet size is extremely small and the system looks transparent or translucent, it can be regarded as a microemulsion in a broad sense. An ultrafine emulsion has been studied [11–13] that has an appearance similar to that of a single-phase microemulsion and has a droplet size of 10 to 100 nm. The ultrafine emulsion is a true emulsion including two phases and is thermodynamically unstable, but a markedly stable ultrafine emulsion with a long shelf life can be obtained. Accordingly, they can be considered equivalent to microemulsions in a broad sense.

Single-phase microemulsions will be summarized below from the standpoint of applications in cosmetics and medical products. Next, ultrafine emulsions (thermodynamically unstable microemulsions) will be discussed in detail and applications in cosmetics and medical products will be introduced.

II. SINGLE-PHASE MICROEMULSIONS

As described above, microemulsions in a strict sense are transparent or translucent single-phase systems of oil, water, and amphiphile in which large swollen micelles are dispersed. It is well known that a larger amount of surfactant is usually required to form single-phase microemulsions compared with macroemulsions. In cosmetic applications, it is often very important to minimize the amount of surfactant for cost and safety reasons. Many surfactants are irritating to the skin when used in high concentrations. Therefore, there is a problem of safety in the application of microemulsions to the skin. It is also caused by the utilization of medium-chain alcohols as cosurfactants and of hydrocarbons having a relatively small carbon number as oils. Taking these aspects into consideration, the preparation methods for microemulsions and the possibility of applications to cosmetics will be stated in the following.

A. Conditions of Formation of Microemulsions

An important subject in the preparation of microemulsions is to attain large solubilization or to mix oil and water completely with less surfactant (including cosurfactants). The ultimate condition for the formation of microemulsions can be readily understood from the phase diagrams of water/nonionic surfactant/oil systems as described by Kunieda and Shinoda [14] and Shinoda [15]. It is well known that the hydrophile-lipophile balance (HLB) of nonionic surfactants depends on temperature. Figure 1 [15] shows the phase diagram of the water/$C_{12}H_{25}O(C_2H_4O)_5H$/$C_{14}H_{30}$ system. The nonionic surfactant shows hydrophilicity at a low temperature and forms oil-swollen micelles in the water phase (W_m). The HLB of the nonionic surfactant shifts toward lipophilic with increasing temperature and the amount of the solubilized oil increases and then reaches its maximum because the surfactant is apt to have balanced hydrophile-lipophile properties. On the other hand, the nonionic surfactant is lipophilic at a high

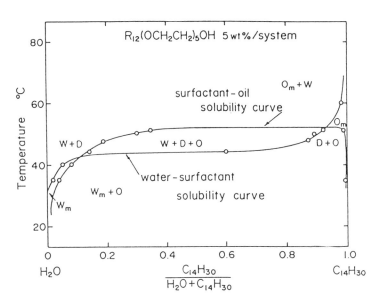

FIG. 1 Phase diagram of H_2O-$C_{14}H_{30}$ containing 5 wt%/system of $R_{12}(OCH_2CH_2)_5OH$ as a function of temperature. Abscissa is the weight fraction of $C_{14}H_{30}$ in solvents. D = surfactant. (From Ref. 14.)

temperature and forms water-swollen reverse micelles in the oil phase (O_m). The HLB of the surfactant shifts toward hydrophilic with decreasing temperature. The amount of solubilized water increases because the HLB of the surfactant is apt to balance, and then the amount of solubilized water reaches its maximum. Thus, the amount of solubilized substances changes remarkably according to the change in the HLB of the surfactant. The above two phases (Wm, Om) can be regarded as oil-in-water (O/W) and water-in-oil (W/O) microemulsions, respectively, in which the swollen micelles with oil or water are dispersed in the continuous phase. When the HLB of the surfactant in the system is just balanced, the surfactant phase (D) appears, in which oil and water are dissolved. The surfactant phase is also referred to as a bicontinuous microemulsion [8,14]. As mentioned above, it is clear that the HLB of surfactant (or mixture of surfactants) in the system is an important factor in the solubilization of oil and water with surfactant.

The amount of solubilized oil or water is also dependent on the structure of the surfactant itself. Figure 2 [11] shows the effect of the surfactant structure on the solubilization of hexadecane in aqueous solutions of pure nonionic surfactants, $C_{12}H_{24}O(C_2H_4O)_6H$, $CH_{14}H_{29}O(C_2H_4O)_7H$, and $C_{16}H_{33}O(C_2H_4O)_8H$. Although all of them have similar hydrophile-lipo-

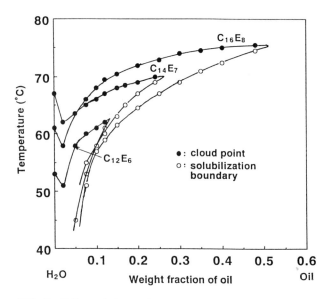

FIG. 2 Effect of the surfactant structure on the solubilization of oil in nonionic surfactant/hexadecane/H_2O system.

phile properties, the amount of solubilized oil increases with increasing molecular size of the surfactant. A surfactant with a longer alkyl chain can solubilize a larger amount of oil at the same HLB of surfactant [11,16,17]. Furthermore, the amount of solubilized oil depends on the structure of the oil. Figure 3 shows the effect of the structure of oil on the solubilization of alkane in aqueous solutions of pure nonionic surfactants. The amount of solubilized oil differs depending on the structure of oil and the amount of hexadecane is larger than that of 2,6,10,15,19,23-hexamethyltetracosane (squalane, C_{30}), although the surfactants have the same alkyl chain and similar HLBs in the system. The amount of solubilized alkane increases with decreasing molecular volume of alkane.

The consideration of nonionic surfactant microemulsions described above is also applicable to microemulsions formed by ionic surfactants. A typical microemulsion of an ionic surfactant is prepared by incorporating a medium-chain alcohol as the cosurfactant. The HLBs of ionic surfactants are usually too hydrophilic. Because the addition of alcohols increases the solubility of ionic surfactants in oil and shifts the HLB of surfactant in the system toward lipophilic, the hydophile-lipophile property of ionic surfactants could be balanced to form microemulsions. It may be also possible to use a relatively lipophilic surfactant as the cosurfactant in place of medium-chain alcohols. If an ionic surfactant having the balanced hydophile-

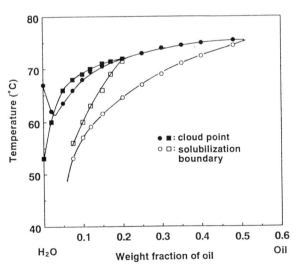

FIG. 3 Effect of temperature on the solubilization of oil in $C_{16}E_7$/squalane/H_2O system (□) and $C_{16}E_8$/hexadecane/H_2O system (○).

lipophile property can be obtained, ionic surfactant microemulsions may be formed without a cosurfactant. These results are shown later in detail.

Mathis et al. [18] reported a fluorocarbon oil microemulsion using a nonionic surfactant with a fluorocarbon chain as a lipophilic group. The microemulsions are formed by changing the temperature and balancing the HLB of the surfactants.

From the foregoing results and other studies in this book, it can be understood that balancing the HLB of surfactants in systems is the ultimately important factor for the formation of single-phase microemulsions.

B. Preparation of Microemulsions

Appropriate selection of surfactants and cosurfactants is a key factor in the successful formulation of microemulsions. There are literally thousands of surfactants available on the market, but not all of them are suitable for use in cosmetics. The selection of surfactants and cosurfactants is also strongly dependent on their safety for the skin.

Utilization of nonionic surfactants is usually desirable from the viewpoint of safety to the human body. However, as shown in Figs. 1, 2, and 3, the temperature range in which the microemulsions form is undesirably narrow because the HLB of nonionic surfactants of polyoxyethylene derivatives is strongly dependent on temperature. This is a serious problem in formulations for cosmetic products. Jayakrishinan et al. [19] reported pharmaceutically acceptable formulations of W/O microemulsions in which Brij 35 and Aracel 186 are employed as nonionic surfactants and isopropanol is incorporated as a cosurfactant. The microemulsions were stable up to 70°C without causing phase separation.

On the other hand, since the HLB of ionic surfactants is scarcely dependent on temperature, ionic surfactant microemulsions can be formed over a wide temperature range. But they have a safety problem because of the utilization of medium-chain alcohols as cosurfactants.

Sagitani and Frieberg [20] reported that the W/O ionic surfactant microemulsions with hexadecane are prepared by using a lipophilic nonionic surfactant as a cosurfactant in place of a medium-chain alcohol. Sodium alkyl sulfate is used as a primary surfactant and tetraethyleneglycol monododecyl ether is used as a secondary surfactant. Rosano and Carallo [21] reported the preparation of microemulsions utilizing alkyl dimethyl amine oxide as the cosurfactant. These microemulsions consist of a sodium alkyl sulfate (C_{12}, C_{14}, or C_{16}), an alkane (C_8, C_{10}, C_{12}, C_{14}, or C_{16}), an alkyl dimethyl amine oxide (C_{12}, C_{14}, or C_{16}), and water.

Shinoda et al. [17,22] studied the formation of microemulsions using ionic surfactants. Figure 4 [17] shows the phase diagram of the water/

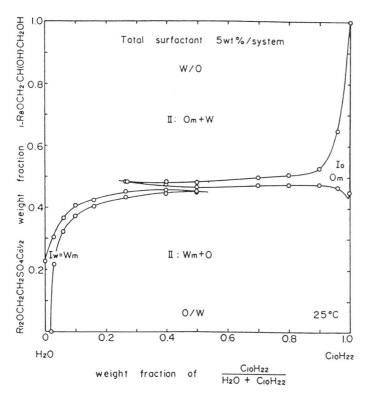

FIG. 4 Phase diagram of $H_2O/C_{12}H_{25}OCH_2CH_2SO_4Ca_{1/2}$/glycerol mono(2-ethylhexyl) ether/$C_{10}H_{22}$ at 25°C (From Ref. 17.)

$C_{12}H_{25}OC_2H_4OSO_4Ca_{1/2}$/glycerol mono(2-ethylhexyl) ether/$C_{10}H_{22}$ system. In this study, the HLB of the surfactant is modified by changing the composition of the mixture of the ionic surfactant with glycerol mono-(2-ethylhexyl) ether in place of medium-chain alcohol. At the composition at which the HLB of the mixed surfactants is approximately balanced, the solubilized oil and water in the O/W or W/O microemulsions are 10 times or more than the amount of total surfactants. Shinoda et al. [17] also reported the formation of microemulsions using sodium hexadecyl sulfate as the ionic surfactant. The solubilizing power of the system was very great, because 16 wt% liquid paraffin is solubilized in 5 wt% per system of the surfactant and 27 wt% hexadecane was solubilized in the same surfactant solution. The amount of solubilized oil decreases with increasing hydrocarbon chain length. The use of a surfactant with a longer hydrocarbon chain is effective in enhancing the solubilizing power. These results are in agreement with those for nonionic surfactant systems as shown in Figs. 2 and 3. It was also reported [22,23] that microemulsions are formed by employing a dialkyl-type ionic surfactant without using a cosurfactant, since the surfactant itself possesses a balanced hydrophile-lipophile property.

Furthermore, formulations of microemulsions using lecithin as the surfactant were studied. Lecithin is biocompatible and is a suitable surfactant for cosmetics and pharmaceutical products. Luisi et al. [24] reported that W/O microemulsion gels originate from a reverse micellar solution in organic solvent–lecithin systems. Gelation is induced by the addition of a small amount of water. Many organic solvents acceptable for cosmetics (e.g., isopropyl myristate) could be used. Shinoda et al. [25] reported that microemulsions with hexadecane can originate at a low lecithin concentration by adding a short-chain alcohol to water. The 2.3 wt% lecithin per system can mix 48.8 wt% aqueous solution containing 13 wt% propanol and 48.8 wt% hexadecane in a single phase. The propanol concentration in water can be reduced by incorporating a small amount of a hydrophilic nonionic surfactant (e.g., dodecyl oligoglucoside) with lecithin.

It is expected that safety problems with microemulsions can be eliminated by applying these results and that skin care products can be produced with these microemulsion formulations.

On the other hand, the preparation of unique microemulsions has been reported in patents and the microemulsions are used mainly for hair care products [26–28]. They contain an amino-functional polyorganosiloxane, a nonionic surfactant, and an acid and/or a metal salt. The preparation method involves completely admixing the silicones and the surfactants, adding water (and metal salt) to the mixture, and then adding an acid to the mixture to adjust the pH of the mixture to a suitable value (acid side). It is considered that microemulsions may be formed by changing the pH of the mixtures by the addition of acid, since the affinity for water of amino-functional polyorganosiloxanes depends on the pH of aqueous solutions. The phase transition of microemulsions is reversible with a change of temperature. Therefore, it is suggested that they are single-phase microemulsions.

C. Applications in Cosmetics

In spite of the active interests in microemulsions, a few cosmetic formulations have been developed. One important area of application of microemulsions is in the solubilization of fragance and flavor oils. But the usual cosmetic products utilize so-called solubilization systems but not microemulsions. The products, such as toilet waters and toning lotions, contain lower concentrations of odoriferous substances. Dartnell et al. [29] reported a microemulsion containing a perfuming concentrate, i.e., a nonalcoholic perfuming product. Current perfuming products with high concentrations of odoriferous substances are usually prepared by dissolution in alcohols such as ethanol and sometimes aqueous solutions with relatively high concentrations of ethanol. It is desirable, however, to prepare the products without using alcohol, because alcohol is known to be undesirable in terms of skin safety. This microemulsion contains approximately 5 to 50 wt% of the concentrate of odoriferous substances without using ethanol; polyethylene glycol derivatives are used as the primary surfactant and polyglycerol and ether phosphate derivatives are used as cosurfactants.

There have been studies of the utilization of functions of microemulsions. It was reported by Chiu and Yang [30] that a vitamin E microemulsion has high resistance to oxidation in air. The microemulsions are obtained by solubilizing vitamin E in aqueous solutions with pure nonionic surfactants, polyethylene glycol monoalkyl ethers. However, it is not clear why higher resistance to oxidation is achieved in microemulsions compared with macroemulsions with the same components, although kinetics and mechanisms involved in the dissolution of vitamin E are discussed. Parra et al. [31] reported microemulsions as vehicles for nucleophilic reagents in cosmetic formulations. The modifications of chemical reactivity induced in human hair during its treatment with a reductive agent ($NaHSO_3$) or oxidative agent (H_2O_2) with a micellar or a microemulsion system as the vehicle were investigated. The W/O type of microemulsion consists of sodium dodecyl sulfate, pentanol (heptane), and water. The treatment of human hair with $NaHSO_3$ carried in a micellar solution system favors reductive attack, and this effect is enhanced when the reagent is in a microemulsion vehicle. The breakage of the cysteine residues increases due to the decrease in water content. In oxidization of human hair with H_2O_2, the order of reactivity is the following: aqueous medium < micellar solution < microemulsion. The reactivity becomes higher as the water content decreases and the hydrocarbon content increases. Solans et al. [32] also reported the activity of thioglycolic acid, incorporated in microemulsions, toward cysteine residues. The influence of microemulsion structure on cysteine reactivity with keratin fibers was investigated using an appropriate model system. The realm of hydrocarbon-continuous microemulsion-type media was found to induce the highest activity. This can be interpreted as follows. The existence of water pools containing most of the reagent in inverse microemulsions appears to enhance cysteine reactivity. Another decisive factor may be that the reactive medium wets preferentially the hydrophobic fractions of keratin and spreads onto and penetrates into the individual fibers.

III. ULTRAFINE EMULSIONS (THERMODYNAMICALLY UNSTABLE MICROEMULSIONS)

As described at the beginning, the term microemulsion in a broad sense can include fairly stable but thermodynamically unstable transparent or translucent dispersions. Only a few studies of thermodynamically unstable microemulsions have been reported until recently. In patents in the cosmetic area, there are microemulsions of silicone oils [33,34] produced by emulsion polymerization that are transparent or translucent. It is difficult to regard them as single-phase microemulsions in spite of their long-term storage stability, since a large amount of silicone oil is not solubilized in these surfactant systems. Therefore, they may be regarded as thermodynamically unstable microemulsions, i.e., ordinary emulsions that have extremely small droplet sizes and are transparent or translucent. In these reports, however, this aspect is not discussed.

As described in the introduction, ultrafine emulsions have been studied that are regarded as thermodynamically unstable microemulsions. They are prepared by two methods: a condensation method and a dispersion method. For ultrafine emulsions, the preparation methods, characteristics, and some advantages for application are described below.

A. Ultrafine Emulsions Prepared by Condensation Method

1. Phase Behavior and Preparation Method

It was found from investigations of the phase behavior that ultrafine emulsions can be prepared by the condensation method [11] in systems of nonionic surfactant, oil, and water. Figure 5 [11] shows the effect of the chain length of alkane on the solubilization of oil in water/nonionic surfactant/alkane systems.

Decane, hexadecane, liquid paraffin, and squalane are used as hydrocarbon oils. All the systems contain 9.09 wt% of the pure nonionic surfactant, $C_{16}H_{33}O(C_2H_4O)_8H(C_{16}E_8)$. The homogeneous phase (W_m in Fig. 1), in which the oil-swollen micelles are dispersed in water, is bounded on one side by the cloud point curve and on the other side by the solubilization phase boundary. The single-phase microemulsions with large amounts of hexadecane appear within a narrow temperature range in which the hydrophile-lipophile properties of the surfactant are well balanced in the systems. The mixtures at temperatures below the solubilization phase boundary consist of two phases: the water phase with oil-swollen micelles and the excess oil phase. Accordingly, O/W emulsions are usually produced [35] by shak-

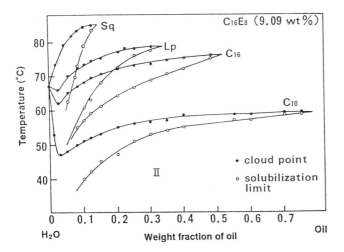

FIG. 5 Effect of carbon number on the solubilization of oil in $C_{16}E_8$/alkane/H_2O system. Sq, squalane; Lp, liquid paraffin.

ing the mixtures, which would separate into two phases at equilibrium conditions. We found that the ultrafine emulsions could be prepared by cooling the single-phase microemulsions to temperatures below the solubilization phase boundary without shaking. From the process of formation of ultrafine emulsions, this preparation method can be regarded as a condensation method. It is noteworthy that the turbidity of ultrafine emulsions immediately after cooling is similar to that of microemulsions and the ultrafine emulsions are transparent and translucent. The rate of change of the emulsion turbidity is dependent mainly on the chain length of oil. The stability of ultrafine emulsions will be described later in detail. It should be noted that the turbidity of the liquid paraffin and squalane ultrafine emulsions remains essentially unchanged for more than 1 year at 25°C. It is evident, however, that the emulsions consist of two phases and are thermodynamically unstable, since the precise solubilization phase boundaries can be determined in the liquid paraffin and squalane systems.

Ultrafine emulsions prepared by this method are similar to microemulsions in appearance. Accordingly, we can consider that they are O/W ultrafine emulsions with droplet size similar to that of microemulsions. The mean hydrodynamic droplet diameters of the ultrafine emulsions measured by the dynamic light scattering technique are approximately 10–100 nm. Therefore, the ultrafine emulsions can be regarded as thermodynamically unstable microemulsions.

2. Control of Droplet Size in Ultrafine Emulsions

Hexadecane and liquid paraffin ultrafine emulsions containing 5% $C_{16}E_8$ are prepared with various weight ratios (R) of the oil to $C_{16}E_8$. The mean droplet radius on the basis of volume versus R is shown in Fig. 6 [11]. The mean droplet radius increases with increasing R and is proportional to R. The desired droplet size can thus be obtained by choosing an appropriate value of R. To prepare ultrafine emulsions of controlled droplet size, however, it is necessary to restrict the composition to the range in which the oil can be completely solubilized at a temperature balancing the HLB of the surfactant as shown in Fig. 3 and Fig. 6.

Furthermore, if we assume that almost all the surfactant molecules in ultrafine emulsions are distributed at the oil-water interfaces, we can explain why the droplet size of the ultrafine emulsions is linearly related to R [12]. Figure 7 illustrates a droplet in an ultrafine emulsion. The oil droplet is covered with a saturated monolayer of $C_{16}E_8$; r is the radius of the droplet, r' is the radius of the core consisting of the oil and the alkyl chain of the surfactant, and d is the thickness of the layer of hydrated ethylene oxide chain. The total volume of the cores in a unit volume of an ultrafine emulsion, V, is given by

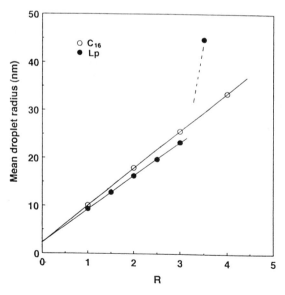

FIG. 6 Plot of mean droplet radius against the ratio (R) of oil to $C_{16}E_8$ in hexadecane and liquid paraffin ultrafine emulsions.

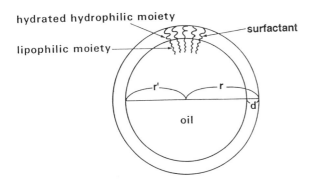

FIG. 7 Illustration of a droplet in an ultrafine emulsion.

$$V = \frac{W_a}{\rho_a} + \frac{aW_b}{\rho_c} = \frac{4}{3}(\pi \sum r_i'^3) = \tfrac{4}{3}(\pi r_v'^3 n) \tag{1}$$

where W_a and W_b are the weights of the oil and the surfactant in the ultrafine emulsion of unit volume, α is the weight fraction of the alkyl chain of the surfactant to the total weight of the surfactant, ρ_a and ρ_c are the densities of the oil and the alkyl chain of the surfactant, and r_v' is the mean core radius on the basis of volume. The total surface area of the cores, S, is given by

$$S = \frac{ANW_b}{M_b} = 4\pi \sum r_i'^2 = 4\pi r_A'^2 n \tag{2}$$

where r_A' is the mean core radius on the basis of area, A is the area occupied by a single surfactant molecule at the oil-water interface, N is Avogadro's number, and M_b is the molecular weight of the surfactant. Combining Eqs. (1) and (2), we obtain the relation between the mean core radius and the weight ratio (R) of the surfactant to the oil as

$$\frac{r_v'^3}{r_A'^2} = \left(\frac{3M_b}{AN\rho_a}\right) R + \left(\frac{3\alpha M_b}{AN\rho_c}\right) \tag{3}$$

Assuming that the droplet size is monodispersed, the relation between the droplet radius (r) and R can be expressed as

$$r = \left(\frac{3M_b}{AN\rho_a}\right) R + \left(\frac{3\alpha M_b}{AN\rho_c}\right) + d \tag{4}$$

where d is the thickness of the layer of hydrated hydrophilic moiety of surfactant.

If the size distribution is narrow, for the hexadecane and liquid paraffin emulsions, the A values of $C_{16}H_{33}O(C_2H_4O)_8H$ can be calculated from the experimental slopes and Eq. (4); both values are 0.52 nm². This seems reasonable compared with the A value of $C_{16}E_8$, 0.49 nm², estimated from the A values of $C_{16}E_7$ and $C_{16}E_9$ at the air–aqueous solution interface at 25°C by Elworthy andn McFarlane [36]. Therefore, in ultrafine emulsions it is considered that almost all the surfactant molecules are indeed distributed at the oil-water interface, and the total area of the oil-water interface in the ultrafine emulsions can reach its maximum. Accordingly, the minimum droplet size can be attained at given composition. This is a great advantage in cosmetic products, because it is often important to minimize the amount of surfactant for cost and safety reasons.

3. Stability of Ultrafine Emulsion

Figure 8 [11] shows the changes in droplet size of alkane ultrafine emulsions at 25°C. The emulsions contain 10 wt% $C_{16}E_8$, 80 wt% water, and 10 wt% dodecane, tetradecane, pentadecane, hexadecane, liquid paraffin, or squalane. Just after preparation all of the emulsions have droplet diameters of about 20 nm. For the alkane ultrafine emulsions, the change in the droplet size with time is clearly dependent on the carbon number of oil, and the stability increases dramatically with increasing the carbon number. Decane ultrafine emulsions cannot be obtained because they are extremely unstable. Dodecane ultrafine emulsions are fairly unstable and the turbidity changes occur within 1 h after preparation at 25°C. The tetrade-

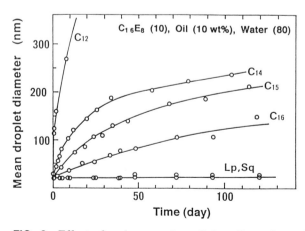

FIG. 8 Effect of carbon number of the oil on the stability of alkane ultrafine emulsions at 25°C.

cane emulsion turbidity increases several hours after preparation, and the pentadecane emulsion turbidity changes within 1 day after preparation. Hexadecane ultrafine emulsions are fairly stable and the turbidity remains practically unchanged for several days. On the other hand, liquid paraffin and squalane ultrafine emulsions are extremely stable, and the droplet size and the turbidity remain practically unchanged for more than 1 year at 25°C.

It may be presumed that the main cause of the instability of ultrafine emulsions is Ostwald ripening, which is a consequence of the Kelvin effect. Ostwald ripening in an emulsion has been discussed by Higuchi and Misra [37], Davis et al. [38,39], and other researchers [40,41]. Provided that the dispersed droplet size is small and the oil has finite solubility in the water phase, the emulsion can be coarsened by diffusion of oil molecules from the small droplets via the continuous phase into the large droplets, as the solubility of the oil in the small droplets is larger than that in the large droplets. Consequently, the small droplets decrease in size and finally can disappear (becoming oil-swollen micelles), and the large droplets increase in size.

Figure 9 [12] shows the stability of ultrafine emulsions with mixed oil in which a small amount of squalane is added to dodecane to provide unstable ultrafine emulsions (see Fig. 8). The vertical axis represents the ratio of the particle size after 18 days at 45°C to the initial particle size. The stability increases dramatically with increasing the concentration of squalane. Addition of more than 3% squalane stabilizes the emulsions, and the stability agrees with that of squalane alone. Such an effect is considered

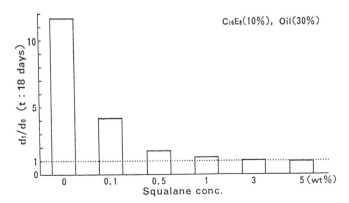

FIG. 9 Effect of small quantities of squalane on the stability of dodecane ultrafine emulsions at 45°C.

to be derived from the antagonism between Raoult's law and Kelvin's law [37,38]. Dodecane migrates from the small oil droplets to the large droplets according to Kelvin's law. But it is considered that squalane cannot migrate, because the droplet size remains practically unchanged for squalane ultrafine emulsions. Accordingly, squalane is diluted in the large oil droplets, whereas it is concentrated in the small oil droplets. This leads to the difference in concentration of squalane in the decane droplets. This difference in the chemical potential on the basis of Raoult's law can antagonize the difference in the chemical potential based on Kelvin's law.

4. Application

Ultrafine emulsions prepared by this method have some advantages in cosmetic and medical products. They have an appearance similar to that of single-phase microemulsions and they can be readily prepared. Furthermore, they have excellent stability and safety, and their droplet size can be readily controlled. Therefore, it is expected that they will be applied for cosmetic formulations utilizing their unique characteristics. Nakajima et al. [11] reported cosmetic formulations for skin care products using commercial nonionic surfactants and oils usually used in cosmetics.

B. Ultrafine Emulsions Prepared by Dispersion Method

Emulsions having a droplet size of 10–100 nm, ultrafine emulsions, are to be prepared by utilizing an emulsifying machine (a so-called dispersion method), if a machine with ideal power can be developed. However, it is difficult to obtain ultrafine emulsions with a droplet size below 50 nm using a Microfluidizer (Microdiscs Co.), which is a high-pressure homogenizer with very strong shear force. A new method for preparation of ultrafine emulsions has been developed [13]. The key process of the method involves homogenizing coarse emulsions with water phases containing large amounts of water-soluble solvents by using a high-pressure homogenizer.

1. Preparation Method

Figure 10 shows the effect of pressure on the mean droplet diameters of emulsions prepared with the Microfluidizer, which is operated at pressures from 3000 to 12,000 psi. The emulsions are recycled through the interaction chamber 10 times at about 30°C, at which the emulsions return to the reservoir. The emulsions contain 70 wt% water and 30 wt% liquid paraffin and nonionic surfactant. The surfactant used is the commercial products POE(30) oleyl ether ($C_{18}E_{30}$) and is safe for the skin. The weight ratios (R) of oil to $C_{18}E_{30}$ are 2, 4, and 10. The diameters are reduced with decreasing oil/surfactant ratio or with increasing pressure, but they are

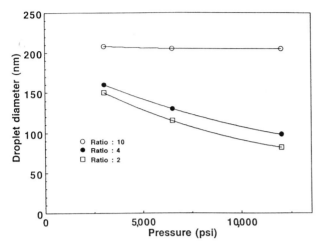

FIG. 10 Effect of the applied pressure on the droplet diameter of emulsions having weight ratios of oil to $C_{18}E_{30}$ of 2, 4, and 10.

almost constant at a ratio of 10. It is difficult to decrease the droplet diameters below 80 nm by this method.

Assuming the A value of $C_{18}E_{30}$ at the light liquid paraffin-water interface is about 1 nm^2 and d may be 3 nm in Eq. (4), the droplet diameters are calculated to be approximately 48, 88, and 208 nm at ratios of 2, 4, and 10, since ρ_a is 0.82, ρ_c is 0.8, M_b is 1590, and a is 0.16. At a ratio of 10, the diameters from calculations and experiments are approximately coincident. Emulsions for the ratios 2 and 4, however, cannot reach the calculated minimum size at the higher pressure of 12,000 psi. The ultrafine emulsions are prepared by homogenizing the coarse emulsions containing the water-soluble organic solvents with the Microfluidizer at pressures of 3000 psi and 12,000 psi. Figures 11 and 12 show the relation between the droplet diameters of emulsions and the concentrations of glycerol and 1,3-butanediol in the water phases. They contain 20 wt% oil, 10 wt% $C_{18}E_{30}$, and 70 wt% water or aqueous solution. The addition of these solvents is very effective in decreasing the droplet size except for 1,3-butanediol at over 70% at a pressure of 12,000 psi. The ultrafine emulsions were prepared with the Microfluidizer at 12,000 psi using a 50% glycerol aqueous solution, and the relation between the droplet diameters and the ratio of oil to $C_{18}E_{30}$ was investigated. The diameter is proportional to the ratio, and the droplet size can be controlled by changing the ratio. Accordingly, it is considered from Eq. (4) that the newly developed method can prepare

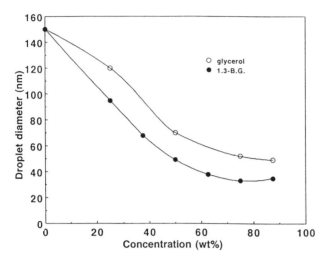

FIG. 11 Relation between the droplet diameter and the concentration of water-soluble solvent in the water phase at a pressure of 3000 psi.

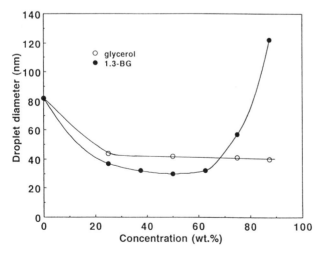

FIG. 12 Relation between the droplet diameter and the concentration of water-soluble solvent in the water phase at a pressure of 12,000 psi.

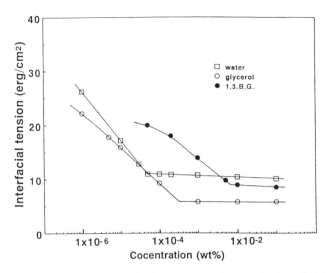

FIG. 13 Effect of the $C_{18}E_{30}$ concentration on the interfacial tension between liquid paraffin and 50 wt% aqueous solution of water-soluble solvent.

ultrafine emulsions with a droplet diameter of 20 nm in which almost all the surfactant molecules can be distributed at the oil-water interface.

2. Role of Water-Soluble Solvents

The droplet size of emulsions formed by applying the same mechanical energy can be determined mainly by the facility of droplet disruption and the stability at the time of emulsification. The facility is affected by interfacial tension. Therefore, the roles of water-soluble solvents in decreasing the droplet size have to be discussed in terms of the interfacial tension and the stability at the time of emulsification.

Figure 13 shows the effect of the concentrations of glycerol and 1,3-butanediol in water phases on the interfacial tensions between oil phases and water phases containing 1 wt% $C_{18}E_{30}$ at 30°C. The interfacial tension for both systems decreases with increasing concentration. Glycerol is more effective in decreasing it than 1,3-butanediol. This is because [42] the addition of glycerol [43] shifts the hydrophile-lipophile property of the surfactant toward lipophilic, that is, toward balance for this system, and 1,3-butanediol shifts it toward hydrophilic [44]. As shown in Figs. 11 and 12, however, the droplet size of glycerol is larger than that of 1,3-butanediol except for a concentration over 70% at 12,000 psi. Therefore, it can be considered that the decrease in the static interfacial tension does not always

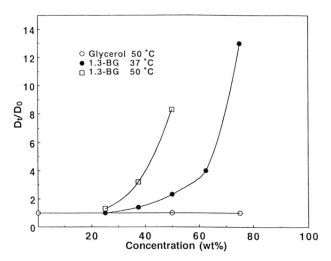

FIG. 14 Effect of the concentration of water-soluble solvent on the stability of emulsions. D_0, diameter immediately after preparation; D_t, diameter 24 h after preparation.

result in a reduction of the droplet size in the case of emulsification with the Microfluidizer.

Emulsions containine 20% oil, 10% $C_{18}E_{10}$, $C_{18}E_{15}$, and $C_{18}E_{30}$ as surfactant are homogenized at a pressure of 3000 psi. The interfacial tension between oil and aqueous solution of 1 wt% surfactant rises markedly with increasing ethylene oxide chain length. The values at 30°C are 1.0, 4.0, and 9.8 erg/cm^2. However, the droplet diameters are about 130 nm and are practically unchanged. From these results, it is confirmed that the interfacial tension between oil phase and water phase does not directly affect the droplet size when emulsions are prepared with a high-pressure homogenizer.

It should be noted that the interfacial tension is not dynamic but static. The droplet is disrupted by emulsification with a high-pressure homogenizer in a very short time (below 0.1 ms [45]) and the interface cannot be covered with a saturated monolayer of surfactant in the course of disruption even if the surfactant concentration is sufficiently higher than its critical micelle concentration (cmc). Figure 13 shows the interfacial tension between the liquid paraffin phase and water phase for water, 50 wt% glycerol aqueous solution, and 50 wt% 1,3-butanediol aqueous solution as a function of concentration of $C_{18}E_{30}$ in the water phase. It is noteworthy that the interfacial tension of the interface saturated with surfactant is increased drastically by desorbing a very small amount of surfactant molecule. The interfacial tension at the time of droplet disruption has to approach the surfactant-free interfacial tension. The surfactant-free interfacial tensions in these systems are 55 erg/cm^2 for water, 35 erg/cm^2 for glycerol, and 21 erg/cm^2 for 1,3-butanediol. This order agrees with that of the droplet size in the systems shown in Figs. 11 and 12. The interfacial tensions between liquid paraffin and the aqueous solutions of water-soluble solvents decrease with increasing solvent concentration, and 1,3-butanediol is more effective than glycerol [13].

It is clear that the droplet sizes shown in Figs. 11 and 12 are related to the interfacial tension of the surfactant-free interface. To reduce the droplet size of emulsions with a high-pressure homogenizer, it is important to decrease the interfacial tension of the interface without absorbed surfactant in the systems.

There is a relation between the initial droplet size and the stability in the course of emulsification. As shown in Fig. 12, the droplet diameters of 1,3-butanediol emulsions rise with increasing concentration when it exceeds 50 wt%. This is probably because the stability of ultrafine emulsions decreases with increasing concentration of 1,3-butanediol. Figure 14 shows the effect of the concentration of water-soluble solvent on the stability of ultrafine emulsions. The glycerol ultrafine emulsions are stable, but the stability of the 1,3-butanediol ultrafine emulsions decreases with increasing concentration and the ultrafine emulsion of 75 wt% 1,3-butanediol is extremely unstable at 50°C.

The temperature of homogenization, for a pressure of 3000 psi, is not much higher than 30°C, but the temperature for 12,000 psi must be much higher than 30°C due to a large amount of heat generation at homogenization. Therefore, it is considered that the increase in the droplet size in Fig. 12 is caused by the instability of 1,3-butanediol emulsions as shown in Fig. 14.

The stability of the ultrafine emulsions depends on the concentration of water-soluble solvents in the water phases as shown in Fig. 14. For liquid paraffin ultrafine emulsions, the droplet size is practically unchanged for more than 1 year at 25°C for 1,3-butanediol at a concentration below 15 wt% and for glycerol at a concentration below 50 wt%. In order to obtain stable ultrafine emulsions, the concentrated emulsions are first prepared by this method and then diluted with water.

3. Application

The dispersion method has an important advantage compared with the condensation method. Various kinds of surfactants and oils can be used, because this method does not require the formation of single-phase microemulsions. If the system is able to form O/W emulsions having usual stability except for creaming, ultrafine emulsions are to be prepared in the system. Actually, ultrafine emulsions using ionic surfactants [46] and silicone oil ultrafine emulsions [47] were prepared by this method. Skin care cosmetic products can be formulated in ultrafine emulsions, which provide excellent safety for the skin. Furthermore, it was reported [48,49] that ultrafine lipid emulsions for intravenous administration were prepared in which soy bean oil is employed and lecithin is used as the surfactant. Lipid emulsions having a droplet size below 50 nm may be used for drug delivery systems, since it is considered that these droplets could evade being taken into the reticuloendothelial system.

REFERENCES

1. L. M. Prince, in *Microemulsions* (L. M. Prince, ed.), Academic Press, New York, 1977, p. xi.
2. G. Gillberg, H. Lehitinen, and S. E. Ferberg, J. Colloid Interface Sci. *33*:40 (1970).
3. P. Ekwall, L. Mandell, and K. Fontell, J. Colloid Interface Sci. *33*:215 (1970).
4. K. Shinoda and H. Kunieda, J. Colloid Interface Sci. *42*:382 (1973).
5. K. Shinoda and S. Saito, J. Colloid Interface Sci. *26*:70 (1968).
6. I. Danielsson and B. Lindman, Colloids Surfaces *3*:391 (1981).
7. K. Shinoda and S. Friberg, *Emulsions & Solubilization*, Wiley-Interscience, New York, 1986.
8. K. Shinoda and B. Lindman, Langmuir *3*:135 (1987).
9. H. L. Rosano and A. Weiss, J. Phys. Chem. *85*:468 (1981).
10. S. Friberg, Colloids Surfaces *4*:201 (1982).
11. S. Tomomasa, M. Kochi, and H. Nakajima, Yukagaku *37*:1012 (1988).
12. H. Nakajima, S. Tomomasa, and M. Kochi, J. Soc. Cosmet. Chem. Jpn. *34*: 335 (1983).
13. H. Nakajima, S. Tomomasa, and M. Kochi, Proc. 1th. World Congress of Emulsion, Paris, 1993, vol. 1, p. 162.
14. H. Kunieda and K. Shinoda, J. Dispers. Sci. Technol. *3*:233 (1982).
15. K. Shinoda, Prog. Colloid Polym. Sci. *68*:1 (1983).
16. K. Shinoda and H. Kunieda, in *Microemulsions* (L. M. Prince, ed.), Academic Press, New York, 1977, p. 81.
17. K. Shinoda, H. Kunieda, T. Arai, and H. Saijo, J. Phys. Chem. *88*:5126 (1984).
18. G. Mathis, P. Leempoel, J. C. Ravey, C. Selve, and J. J. Delpuech, J. Am. Chem. Soc. *106*:6162 (1984).
19. A. Jayakrishinan, K. Kalalaiarasi, and D. O. Shah, J. Soc. Cosmet. Chem. *34*: 335 (1983).
20. H. Sagitani and E. J. Friberg, Dispers. Sci. Technol. *1*:151 (1980).
21. H. L. Rosano and J. L. Cavallo, J. Soc. Cosmet. Chem. *39*:201 (1988).

22. H. Kunieda and K. Shinoda, J. Colloid Interface Sci. *75*:601 (1980).
23. K. Shinoda and Y. Shibata, Colloids Surf. *19*:185 (1986).
24. P. L. Luisi, R. Scartazzini, G. Haering, and P. Schuetenberger, Colloid Polym. Sci. *268*:375 (1990).
25. K. Shinoda, Y. Shibata, and B. Lindman, Langmuir *9*:1254 (1993).
26. U.S. Patent 4,620,878 (1986).
27. European Patent 0 455 185 A1 (1991).
28. European Patent 0 532 256 A1 (1993).
29. U.S. Patent 5,252,555 (1993).
30. Y. C. Chiu and W. L. Yang, Colloids Surf. *63*:311 (1992).
31. J. L. Parra, J. J. Garcia Dominguez, J. Comelles, C. Sanchez, C. Solans, and F. Balaguer, Int. J. Soc. Cosmet. Sci. *7*:127 (1985).
32. C. Solans, J. L. Parra, P. Erra, N. Azemar, M. Clausse, and D. Touraud, Int. J. Soc. Cosmet. Sci. *9*:215 (1987).
33. European Patent 0 268 982 A2 (1988).
34. US. Patent 532,471 (1990).
35. F. Harusawa, T. Saito, H. Nakajima, and S. Fukushima, J. Colloid Interface Sci. *74*:435 (1980).
36. P. H. Elworthy and C. B. McFarlane, J. Pharm. Pharmacol. *21*:331 (1962).
37. W. I. Higuchi and J. Misra, J. Pharm. Sci. *51*:459 (1962).
38. S. S. Davis and A. Smith, *Theory Practice Emulsion Technol.*, Academic Press, London, 1976, p. 325.
39. S. S. Davis, H. P. Round, and T. S. Purewal, J. Colloid Interface Sci. *80*:508 (1981).
40. J. Ugelstad and P. C. Mrk, Adv. Colloid Interface Sci. *13*:101 (1980).
41. A. S. Kabal'nov, A. V. Pertzov, and E. D. Shchukin, Colloids Surfaces *24*:19 (1987).
42. K. Shinoda, Prog. Colloid Polym. Sci. *68*:1 (1983).
43. H. Sagitani, T. Hattori, and K. Nabeta, K. Nippon Kagaku Kaisi:1399 (1983).
44. L. Marszall, J. Colloid Interface Sci. *59*:376 (1977).
45. P. Walstra, in *Encyclopedia of Emulsion Technology*, vol. 1 (P. Becher, ed.), Marcel Dekker, New York, 1983, p. 116.
46. Japan Patent 63-126542A (1988).
47. Japan Patent 1-293131A (1989).
48. US. Patent 5,098,606 (1992).
49. European Patent 0 361 928 B1 (1994).

9
Microemulsions in Agrochemicals

Th. F. TADROS   Zeneca Agrochemicals, Bracknell, United Kingdom

I. INTRODUCTION

Most agrochemicals are water-insoluble compounds with various physical properties, which must be determined in order to decide on the type of formulation. Among the earliest types of formulations are wettable powders (WPs) that are suitable for formulation of solid water-insoluble compounds. The chemical (which may be micronized) is mixed with a filler such as china clay, and a solid surfactant such as sodium alkyl or alkyl aryl sulfate or sulfonate is added. When the powder is added to water, the particles are spontaneously wetted by the medium and, on agitation, dispersion of the particles takes place. The second and most familiar type of agrochemical formulation is the emulsifiable concentrate (EC). This is produced by mixing an agrochemical oil with another one such as xylene or trimethylbenzene or a mixture of various hydrocarbon solvents. Alternatively, a solid pesticide could be dissolved in a specific oil to produce a concentrated solution. In some cases, the pesticide oil may be used without any extra addition of oils. In all cases, a surfactant system (usually a mixture of two or three components) is added to enable self-emulsification on dilution of the concentrate with water on application. In recent years, there has been a great demand to replace ECs with concentrated aqueous oil-in-water (O/W) emulsions (EWs). This has the advantage of replacing the added oil with water, which is much cheaper and environmentally acceptable. In addition, removal of oil could help in reducing undesirable effects, such as phytotoxicity and skin irritation. By formulating the pesticide as an O/W emulsion, the droplet size can be kept to an optimum value, which may be crucial for biological efficacy. Water-soluble surfactants, which may be desirable for biological optimization, can be added to the aqueous continuous phase. A similar concept has been applied to replace WPs, namely with aqueous suspension concentrates (SCs). In this case, the agrochemical powder is first dispersed in an aqueous solution of a surfactant or macromolecule (usually referred to as the dispersing agent) using a high-speed mixer. The resulting suspension is then subjected to a wet milling process to break any remaining aggregates and agglomerates and reduce the particle size to smaller values (usually 1–2 μm).

A very attractive alternative for the formulation of agrochemicals is to use microemulsion systems. The latter are single optically isotropic and thermodynamically stable dispersions consisting of oil, water, and amphiphile (one or more surfactants) [1,2]. The thermodynamic stability arises from the low interfacial energy of the system, which is outweighed by the negative entropy of dispersion. These systems offer a number of advantages over O/W emulsions for the following reasons. Once the composition of the microemulsion is identified, the system is prepared by simply mixing all the components without the need for any appreciable shear. Because of their thermodynamicstability, these formulations undergo no separation or breakdown on storage (within a certain temperature range, depending on the system). The low viscosity of the microemulsion systems ensures pourability and dispersion on dilution, and they leave little residue in the container. Another main attraction of microemulsions is their possible enhancement of the biological efficacy of many agrochemicals. This, as we will see later, is due to the solubilization of the pesticide by the microemulsion droplets.

In this chapter, I summarize the basic principles involved in emulsifier selection for both O/W and W/O microemulsions and the possible enhancement of biological efficacy using microemulsions. The role of microemulsions in enhancing wetting, spreading, and penetration is discussed. Solubilization is another factor that may enhance the penetration and uptake of an insoluble agrochemical.

II. SELECTION OF SURFACTANTS FOR MICROEMULSION FORMULATION

The formulation of microemulsions is still an art, because understanding the interactions, at a molecular level, on the oil and water sides of the interface is far from being achieved. However, some rules may be applied for selection of emulsifiers for formulating O/W and W/O microemulsions. These rules are based on the principles applied for selection of emulsifiers for macroemulsions. Three main methods may be applied for such selection, using the hydrophilic-lipophilic balance (HLB), phase inversion temperature (PIT) and cohesive energy ratio (CER) concepts. The HLB concept is based on the relative percentages of hydrophilic and lipophilic (hydrophobic) groups in the surfactant molecule. Surfactants with a low HLB number, 3–6, normally form W/O emulsions, whereas those with a high HLB number, 8–18, form O/W emulsions. Given an oil to be microemulsified, the formulator should first determine its required HLB number. Several procedures may be applied for determining the HLB number, depending on the type of surfactant to be used, such as cloud point measurements and gas-liquid chromatography. However, Griffin [3] developed a simple equation that permits calculation of the HLB numbers of simple nonionic surfactants such as fatty acid esters and alcohol ethoxylates of the type $R(CH_2—CH_2—O)_n—OH$. For the polyhydroxy fatty acid esters, the HLB number is given by the expression

$$HLB = 20\left(1 - \frac{S}{A}\right) \tag{1}$$

where S is the saponification number of the ester and A is the acid number of the acid. Thus, a glyceryl monostearate with $S = 161$ and $A = 198$ will have an HLB number of 2.8; that is, it is suitable for a W/O emulsifier. For the simple ethoxylate alcohol surfactants, the HLB number can be simply calculated from the weight percent of oxyethylene E and polyhydric alcohol P:

$$HLB = \frac{E + P}{5} \tag{2}$$

If the surfactant contains polyethylene oxide as the only hydrophilic group, e.g., primary alcohol ethoxylates, the HLB number is simply $E/5$. Davies [4] derived a method for calculating the HLB number of surfactants directly from their chemical formulas, using an empirically determined group number. Thus, a group number is assigned to various emulsifier component groups and the HLB number is calculated from these numbers using the following empirical equation [5]:

$$\text{HLB} = 7 + \Sigma \text{ (hydrophilic group nos.)} - \Sigma \text{ (lipophilic group nos.)}$$

$$(3)$$

The group numbers for various component groups are given in Table 1. Davies and Rideal [5] have shown that the agreement between HLB numbers calculated using the empirical equation and those determined experimentally is quite satisfactory.

Once the HLB number of the oil is known, one must try to find the chemical type of emulsifier that best matches the oil. Hydrophobic portions of surfactants that are similar to the chemical structure of the oil should be looked at first.

The PIT system [6–8] provides information on the type of oil, phase volume relationships, and concentration of the emulsifier. The PIT system is established on the proposition that the HLB number of a surfactant changes with temperature and that an inversion of the emulsion type occurs when the hydrophile and lipophile tendencies of the emulsifier just balance. At this temperature no emulsion is produced. From a microemulsion viewpoint the PIT has an outstanding feature; it can throw some light on the

TABLE 1 HLB Group Numbers

Group	Group number
Hydrophilic	
$-SO_4-Na^+$	30.7
$-COO^-H^+$	21.2
$-COO^-Na^+$	19.1
N (tertiary amine)	9.4
Ester (sorbitan ring)	6.8
Ester (free)	2.4
$-COOH$	2.1
$-O-$	1.3
$-CH-$sorbitan ring	0.5
Lipophilic	
$-CH-$	0.475
$-CH_2-$	
$-CH_3$	
Derived	
$-CH_2-CH_2-O$	0.33
$-CH_2-CH_2-CH_2-O$	-0.15

chemical type of the emulsifier needed to match a given oil. Indeed, the required HLB values for various oils estimated from the PIT system compare very favorably with those prepared using the HLB system described above. This shows a direct correlation between the HLB number and the PIT of the emulsion.

The CER concept provides a more quantitative method for selection of emulsifiers. Beerbower and Hill [9] considered the dispersing tendencies of the oil and water interfaces of the surfactant or emulsion region in terms of the cohesive energies of the mixtures of oil with the lipophilic portion of the surfactant and the water with the hydrophilic portion. They used the Winsor [10] R, which is the ratio of the intermolecular attraction of the oil molecules and lipophilic portion of surfactant to that of the water and hydrophilic portion.

Several interaction parameters may be identified at the oil and water sides of the interface. Representing the lipophilic part by L, the oil by O, the hydrophilic portion by H, and the water by W, one can define at least nine interaction parameters: C_{LL}, C_{OO}, C_{LO} (at the oil side); C_{HH}, C_{WW}, C_{HW} (at the water side); and C_{LW}, C_{HO}, C_{LH} (at the interface). In the absence of the emulsifier, there will be only three interaction parameters, C_{OO}, C_{WW}, and C_{WO}. If $C_{OW} \ll C_{WW}$ the emulsion breaks. These interaction parameters may be related to the Hildebrand solubility parameter δ (at the oil side of the interface) [11] and the Hansen [12] hydrogen-bonding and polar contribution to δ at the water side of the interface. The δ of any component is related to the heat of vaporization, ΔH, by the expression

$$\delta^2 = \frac{\Delta H - RT}{V_M} \tag{4}$$

where V_M is the molar volume.

Hansen [12] considered δ to consist of three contributions: a dispersion contribution δ_d, a polar contribution δ_p, and a hydrogen-bonding contribution δ_h. These contributions have different weighting factors such that

$$\delta^2 = \delta_d^2 + 0.25\,\delta_p^2 + 0.25\,\delta_h^2 \tag{5}$$

Beerbower and Hills [9] used the following expression for the HLB number:

$$\text{HLB} = \frac{20M_H}{M_L + M_H} = \frac{20V_H\rho_H}{V_L\rho_L + V_H\rho_H} \tag{6}$$

where M_H and M_L are the molecular weights of the hydrophilic and lipophilic portions of the emulsifier, V_H and V_L are their corresponding molar volumes, and ρ_H and ρ_L are the densities, respectively.

The cohesive energy density was originally defined by Winsor [10] as

$$R_o = \frac{C_{LO}}{C_{HW}} \tag{7}$$

When $C_{LO} > C_{HW}$, $R_o > 1$ and a W/O emulsion forms. If $C_{LO} < C_{HW}$, $R_o < 1$ and an O/W emulsion forms, whereas if $C_{LO} = C_{HW}$, $R_o = 1$ and a planar system results. The last case denotes the inversion point. R_o can be related to V_L, ρ_L, V_H, and ρ_H by the expression

$$R_o = \frac{V_L \rho_L^2}{V_H \rho_H^2} = \frac{V_L(\delta_d^2 + 0.25\,\delta_p^2 + 0.25\,\delta_h^2)_L}{V_H(\delta^2 + 0.25\,\delta_p^2 + 0.25\,\delta_h^2)_H} \tag{8}$$

Combining Eqs. (6) and (8), one obtains the following expression for the cohesive energy ratio:

$$R_o = \left(\frac{20}{HLB} - 1\right) \frac{V_L(\delta_d^2 + 0.25\,\delta_p^2 + 0.25\,\delta_h^2)_L}{V_H(\delta_d^2 + 0.25\,\delta_p^2 + 0.25\,\delta_h^2)_H} \tag{9}$$

Thus, for an O/W microemulsion, HLB = 12 − 15 and R_o = 0.58 − 0.29 ($R < 1$). For a W/O system, HLB = 5 − 6 and R_o = 2.30 − 1.9 ($R > 1$), whereas for a planar system (bicontinuous microemulsion), HLB = 9 − 10 and R_o = 1.25 − 0.85 ($R \sim 1$). Thus, R_o combines both HLB and cohesive energy densities, giving a more quantitative estimate for emulsifier selection. The applicability of this concept relies on the availability of data for δ_d, δ_p, and δ_h for the various surfactant portions. Some values were tabulated by Beerbower and Hills [9] and later by Barton [13].

III. ROLE OF MICROEMULSIONS IN ENHANCEMENT OF BIOLOGICAL EFFICACY

The role of microemulsions in enhancement of biological efficiency can be described in terms of the interactions at various interfaces and their effect on transfer and performance of the agrochemical. The application of an agrochemical as a spray involves a number of interfaces, where interaction with the formulation plays a vital role [14]. The first interface during application is that between the spray solution and the atmosphere (air), which governs the droplet spectrum, rate of evaporation, drift, etc. In this respect the rate of adsorption of the surfactant at the air-liquid interface is of vital importance. Because microemulsions contain high concentrations of surfactant and usually more than one surfactant molecule is used for their formulation, on diluting a microemulsion on application, the surfactant concentration in the spray solution will be sufficiently high to cause efficient lowering of the surface tension γ. Two surfactant molecules are more efficient in lowering γ than either of the two components. Thus, the net effect will be production of small spray droplets, which, as we will see later, adhere better to the leaf surface. In addition, the presence of surfactants in sufficient amounts will ensure that the rate of adsorption (which is the situation under dynamic conditions) is high enough to ensure coverage of the freshly formed spray by surfactant molecules.

The second interaction is between the spray droplets and the leaf surface, whereby the droplets impinging on the surface undergo a number of processes that determine their adhesion and retention and further spreading on the target surface. The most important parameters that determine these processes are the volume of the droplets and their velocity—the difference between the surface energy of the droplets in flight, E_o, and their surface energy after impact, E_s. As mentioned above, microemulsions that are effective in lowering the surface tension of the spray solution ensure the formation of small droplets, which do not usually undergo reflection if they are able to reach the leaf surface. Clearly, the droplets need not to be too small, otherwise drift may occur. One usually aims at a droplet spectrum in the region of 100–400 μm. The adhesion of droplets is governed by the relative magnitude of the kinetic energy of the droplet in flight and its surface energy as it lands on the leaf surface. Because the kinetic energy is proportional to the third power of the radius (at constant droplet velocity), whereas the surface energy is proportional to the second power, one would expect that sufficiently small droplets will always adhere. For a droplet to adhere, the difference in surface energy between free and attached drop $(E_o - E_s)$ should exceed the kinetic energy of the drop, otherwise bouncing will occur. Since E_s depends on the contact angle, θ, of the drop on the leaf surface, it is clear that low values of θ are required to ensure adhesion, particularly with large drops that have high velocity. Dilution of microemulsions in the spray solution usually gives low contact angles of spray drops on the leaf surfaces as a result of lowering the surface tension and their interaction with the leaf surface.

Another factor that can affect the biological efficacy of foliar spray application of agrochemicals is the extent to which a liquid wets and covers the foliage surface. This, in turn, governs the final distribution of the agrochemical over the areas to be protected. Several indices may be used to describe the wetting of a surface by the spray liquid, of which the spread factor and spreading coefficient are probably the most useful. The spread factor is simply the ratio between the diameter of the area wetted on the leaf, D, and the diameter of the drop, d. This ratio is determined by the contact angle of the drop on the leaf surface. The lower the value of θ, the higher the spread factor. As already mentioned, microemulsions usually give low contact angles for the drops produced from the spray. The spreading coefficient is determined by the surface tension of the spray solution as well as the value of θ. Again, with microemulsions diluted in a spray both γ and θ are sufficiently reduced, and this results in a positive spreading coefficient. This ensures rapid spreading of the spray liquid on the leaf surface.

Another important factor for control of biological efficacy is the formation of deposits after evaporation of the spray droplets, which ensure the tenacity of the particles or droplets of the agrochemical. This will prevent removal of the agrochemical from the leaf surface by falling rain. Many microemulsion systems form liquid crystalline structures after evaporation, which have high viscosity (hexagonal or lamellar liquid crystalline phases). These structures incorporate the agrochemical particles or droplets and ensure their "stickiness" to the leaf surface.

One of the most important effects of microemulsions in enhancing biological efficacy is that on penetration of the agrochemical through the leaf. Two effects, which are complementary, may be considered. The first effect is due to enhanced penetration of the chemical as a result of the low surface tension. For penetration to occur through fine pores, a very low surface tension is required to overcome the capillary (surface) forces. These forces produce a high pressure gradient that is proportional to the surface tension of the liquid. The lower the surface tension, the lower the pressure gradient and the higher the rate of penetration. The second effect is due to solubilization of the agrochemical in the microemulsion droplet. Solubilization results in an increase in the concentration gradient, which enhances the flux due to diffusion. This can be understood from a consideration of Fick's first law:

$$J_D = D\left(\frac{\partial C}{\partial x}\right) \tag{10}$$

where J_D is the flux of the solute (amount of solute crossing a unit cross section in a unit time), D is the diffusion coefficient, and $(\partial C/\partial x)$ is the concentration gradient. The presence of the chemical in a swollen micellar system lowers the diffusion coefficient. However, the presence of the solubilizing agent (the microemulsion droplet) increases the concentration gradient in direct proportion to the increase in solubility. This is because Fick's law involves the absolute gradient of concentration, which is necessarily small as long as the solubility is small, but not its relative rate. If the saturation is noted by S, Fick's law may be written as

$$J_D = D\ 100\ S\left(\frac{\partial \%S}{\partial x}\right) \tag{11}$$

where $(\partial\%S/\partial)$ is the gradient in relative value of S. Equation (11) shows that for the same gradient of relative saturation, the flux caused by diffusion is directly proportional to saturation. Hence, solubilization in general increases transport by diffusion, as it can increase the saturation value by many orders of magnitude (which outweighs any reduction in D). In addition, the solubilization enhances the rate of dissolution of insoluble compounds and this has the effect of increasing the availability of the molecules for diffusion through membranes.

REFERENCES

1. I. Danielsson and B. Lindman, Colloids Surfaces *3*:391 (1981).
2. J. Th. G. Overbeek, P. L. de Bruyn, and F. Verhoecks, in *Surfactants* (Th. F. Tadros, ed.), Academic Press, London, 1984, p. 111.
3. W. C. Griffin, J. Cosmet. Chem. *5*:249 (1954).
4. J. T. Davies, *Proceedings, International Congress on Surface Activity*, 2nd ed., Vol. 1, Butterworth, London, 1968, p. 426.
5. J. T. Davies and E. K. Rideal, *Interfacial Phenomena*, Academic Press, New York, 1961.
6. K. Shinoda, J. Colloid Interface Sci. *25*:396 (1967).
7. K. Shinoda, *Proceedings, International Congress on Surface Activity*, 5th ed., Vol. 2, Butterworth, London, 1968, p. 295.
8. K. Shinoda, H. Saito, and H. Arai, J. Colloid Interface Sci. *35*:624 (1971).
9. A. Beerbower and M. W. Hill, Am. Cosmet. Perf. *87*(6):85 (1972).
10. P. Winsor, Ind. Eng. Chem. Prod. Res. Dev. *8*:2 (1969).
11. J. H. Hildebrand, *Solubility of Non-Electrolytes*, 2nd ed., Reinhold, New York, 1936.
12. C. M. Hansen, J. Paint. Technol. *39*:305 (1967).
13. A. F. M. Barton, *Handbook of Solubility Parameters and Other Cohesive Parameters*, CRC Press, New York, 1983.
14. Th. F. Tadros, Aspects Appl. Biol. *14*:1 (1987).

10

Microemulsions in Dyeing Processes

ERMANNO BARNI, PIERO SAVARINO, and GUIDO VISCARDI
Università degli Studi di Torino, Turin, Italy

I. INTRODUCTION

The titles of the chapters in this book show that the applications of microemulsions are both various and found in important areas. On the other hand, the applications of microemulsions to dyeing processes that have been reported in the literature are few, despite the fact that the embellishment of dye-substrate interactions in a homogeneous and isotropic medium with microemulsions appears to offer interesting technological and industrial prospects.

It is not difficult to find reasons for this lack. As in other industrial domains, processes based on microemulsions are used in the textile-dyeing field at an unconscious level. The applicability of the term microemulsion in its currently accepted meaning is not recognized. For example, in aspects of the widely used solvent-assisted dyeing process, microemulsions could play an important role. Furthermore, the use of microemulsions could arouse concern about the possible high concentrations of their components

209

and possible negative effects on costs and effluents (apart from the expected inhibition of the dyeing). On the other hand, it could be that we are on the brink of the wide appearance of microemulsions in dyeing processes.

We have recently run a bibliographic search using the keywords "microemulsion??+transparent(w)emulsion??*dye??" (an appendix lists the references obtained). All relate to recent years and, ignoring those of marginal relevance to the present subject, involve dye-microemulsion interactions (an important aspect, relevant to dyeing procedures) or the ternary system dye-microemulsion-substrate (in practice, a few items mainly concerning our investigations). In reviewing this work we find it convenient to divide the chapter into two headings, Dyes and Microemulsions and Dyeing Processes in the Presence of Microemulsions.

II. DYES AND MICROEMULSIONS

One important advantage offered by microemulsions is a reduction in the amount of potentially toxic solvents required to solubilize large molecules, such as dyes, upon dispersion of the solvents in water-rich microemulsions. Whereas the interactions of dyes with organized molecular assemblies, namely micelles or synthetic vesicles, has been extensively studied [1,2], little is known about the behavior of dyes in microemulsions. The use of fluorescent dyes to yield information on the microstructure of microemulsions has been reported [3,4] as well as studies on the location of dyes in the microenvironment [5,6] and on the photochemical behavior of dyes in microemulsions [7]. As far as the solubility of dyes in microemulsions is concerned, except for a study of the solubility of merocyanine 540 in a water-rich microemulsion [8], it was not until 1990 that the first systematic investigation of the solubility of dyes was reported. The dyes studied included azo (e.g., dye 1, Orange IV, C.I. Acid Orange 5, C.I. constitution number 13080) and triphenylmethane dyes (e.g., dye 2, Hidacid Wool Violet, C.I. Acid Violet 17, C.I. constitution number 42650) [9]. The main results of this study can be summarized as follows: (1) the solubility of

1

2

the dyes studied is much enhanced in microemulsions compared with that in pure solvents, (2) the primary variable affecting solubilization is dye amphiphilicity (i.e., the more surfactant-like the dye becomes, the greater its solubility in water-rich microemulsions), and (3) the ideal site for solubilization of the more surfactant-like dyes is the surfactant-rich interfacial region separating oil-rich and water-rich domains.

If the above work led to the conclusion that the solubility of the six dyes tested was much enhanced in microemulsions over pure solvents, more recent work [10] shows that this cannot be generalized. Studies of the solubility of dye 3 (Oil Red O, C.I. Solvent Red 27, C.I. constitution number 26125) and dye 4 [not shown] (Macrolex Blue RR, C.I. Solvent

3

Blue 97, Anthraquinone, C.I. constitution number not reported) in three different microemulsions, one water continuous, one oil continuous and one bicontinuous, showed that the dye solubilization depends on the structure of the dye. If the dye is only poorly soluble even in hydrocarbons, as is the case for Oil Red 0, then the microemulsion structure per se induces enhanced solubilization. If the dye shows good solubility in hydrocarbons with no or little solubility in long-chain alcohols, as in the case of Macrolex Blue RR, it does not enjoy superior solubilization in microemulsions.

The self-aggregation of dyes and the consequent attempts to obtain monomeric forms with the aid of amphiphilic auxiliaries are important for both fundamental and practical purposes. We have studied some heterocyclic hydrophobic dyes (e.g., dye 5) that are characterized by a marked tendency to form aggregates [11]. Whereas the addition of anionic, cationic, and

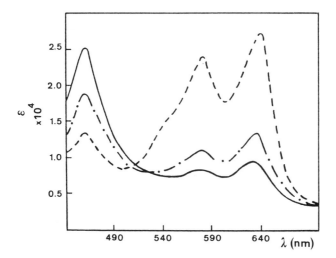

FIG. 1 Electronic absorption spectra: dye 5, 1.1×10^{-5} M (————) in the presence of 17.8% SDS + 18.2% butanol in water. (– – –) Microemulsion: 17.8% SDS + 18.2% butanol + 4.0% xylene in water. (—·—) Microemulsion: 17.8% SDS + 18.2% butanol + 4.0% hexadecane in water (From Ref. 11, p. 54.)

5

nonionic surfactants (below, at, and above their critical micelle concentration) did not promote satisfactory deaggregation, the use of a suitable microemulsion gave rise to an appreciable (even if not complete) deaggregation. This is shown in Fig. 1, where the dashed line shows the strengthening of the monomer peaks at about 590 and 640 nm and the weakening of the main aggregate peak at about 460 nm.

III. DYEING PROCESSES IN THE PRESENCE OF MICROEMULSIONS

The use of dyes and water alone for a dyeing process of industrial importance remains a dream for most dyers and colorists. Practically every colorant-substrate system requires the use of suitable auxiliaries to ensure the best performance of the final products. The primary functions of auxiliaries are (1) to prepare or improve the substrate in readiness for coloration, (2) to modify the sorption characteristics of colorants, (3) to stabilize the application media, (4) to protect or modify the substrate, (5) to improve the fastness of dyes, and (6) to enhance the properties of laundering formulations. Synthetic fibers are usually dyed with disperse dyes, which give rise, in the presence of dispersing agents, to thermodynamically unstable dispersions with respect to the tendency of fine particles to aggregate. Further steps involve the yield of the dyeing, i.e., the amount of dye uptake by the substrate, and the leveling, i.e., the uniformity of coloration on the substrate. A large number of auxiliaries are used, including surfactants that give rise to different organized assemblies. Among these, microemulsions show great potential for future developments. A recent book gives wide information on the function, chemistry, and classification of auxiliaries [12].

The present chapter will continue by treating the effect of microemulsions in dyeing processes, without considering microenvironments that are close to microemulsions. Examples of systems not covered are synthetic vesicles [13], liposomes [14], spontaneous and polyphase emulsions [15], atypical emulsions and liquid crystals [15,16], and solvent-assisted dyeing and coacervation [17].

Microemulsions have been successfully used in wool dyeing by Benisek et al. [18]. The microemulsion described in this note is now marketed with the label Lanalux (Precision Processes, Textiles, UK) and is based on modern silicone technology. It is a more sophisticated product than a conventional silicone softener, being a true microemulsion with particle sizes of the order of nanometers. Use of the microemulsion avoids the tendency of shrink-resist treated wools to harden during dyeing at high packing densities. Lanalux may be applied either in the last bowl of a continuous top shrink-proofing treatment, replacing the softener that is normally applied after the resin treatment, or as a pretreatment before dyeing. In both situations the product is applied at pH 5–6 and a temperature of 25°C. The auxiliary can be used in conjunction with all classes of dyes normally used to dye shrink-resist treated wool. These include reactive, chrome, 2:1 metal complex and acid milling dyes, no modifications of the normal dyeing procedures being necessary. The treatment is without effect on the shade or fastness of dyes. Unlike some silicone-based products, Lanalux does not require a high drying temperature to fix the product on the fiber, so they are permanent to washing and dry cleaning. There is no tendency to give a greasy handle when correctly applied, no adverse effect on the dyeing characteristics of wool, no unacceptable yellowing of pastels or bleached whites, no adverse effect on rubbing fastness even on heavy shades, and, finally, no adverse effect on the water vapor uptake, and hence comfort, of the fiber.

Over the past 4 years we have explored the use of microemulsion systems for the dyeing of various substrates [19–21]. Some preliminary considerations are appropriate.

As mentioned in the Introduction, an important objective has been the search for dyeing recipes based on additive concentrations as close as possible to those usually employed. We have therefore determined the monophasic domain of a microemulsion at 80°C (dyeing temperature), by titrating with the oil phase (n-octanol) solutions containing the chosen surfactant (Ethofor RO/40, ethoxylated castor oil and the cosurfactant (n-butanol) in a 1:1 weight ratio but at different concentrations (Fig. 2). Based on the shaded domain in Fig. 2, a typical recipe we are led to is surfactant, 1 g/L; cosurfactant, 1 g/L; n-octanol, 0.2 g/L (maximum). In subsequent tests we assumed the validity of the above findings and introduced more generic wording, e.g., oil additive or microemulsion precursor, for the mixtures we have used.

The subsequent dyeing tests were carried out at two distinct levels. The first is the thermodynamic level, according to the well-known procedures

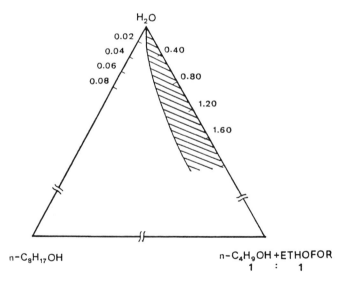

FIG. 2 The microemulsion domain (shaded area) in the pseudoternary system $H_2O/n\text{-}C_8H_{17}OH/n\text{-}C_4H_9OH\text{-}Ethofor$. (From Ref. 19, p. 264.)

of the physical chemistry of dyeing, dyeing isotherms and related parameters being obtained [22,23]. The second is the level of technological
laboratory tests on fabric specimens, performed with the aid of previously
obtained data [24].

Besides spectrophotometric dye titrations (performed with the aim of a
randomized check), the dye uptake by the fiber (and the related dye bath
exhaustion) and the dye leveling (uniformity of coloration) were evaluated
by a simplified colorimetric procedure. Measurements were made, with a
reflectance colorimeter, of the color difference (ΔE) [25] between the dyed
and the undyed sample in six (or nine) different positions. Calculation of
the mean value ($\overline{\Delta E}$) and its standard deviation ($\sigma \Delta E$) enables the conclusion that (1) the higher $\overline{\Delta E}$, the higher the dye uptake, and (2) the lower
$\sigma \Delta E$, the higher the leveling.

The model dye 6 (diethylaminoazobenzene) has been tested on polyamide 6.6. Linear partition isotherms of the Nernst type were obtained
[26], and the partition constants (K) showed that, at suitable n-octanol
concentration, the dye uptake is improved in both the presence of a single
surfactant (Ethofor RO/40) and the presence of a surfactant mixture (Ethofor RO/40 + DDDAB = didodecyldimethylammonium bromide) (Fig.
3). If the acid azo dye 7 is used, the interaction with the substrate follows
an acid-base mechanism and the isotherms are of the Langmuir type [26].

The microemulsion system lowers the affinity of the dye toward the fiber
and enhances the saturation, the latter effect being of importance for high-
intensity (depth) dyeings (Fig. 4). As a test of these indications, we studied
the commercial dye 8 (Nylosan Blue, C.I. Acid Blue 80, C.I. constitution
number 61585), well known for its performance but also for its tendency
to give nonuniform dyeings, in particular at low temperature and if the

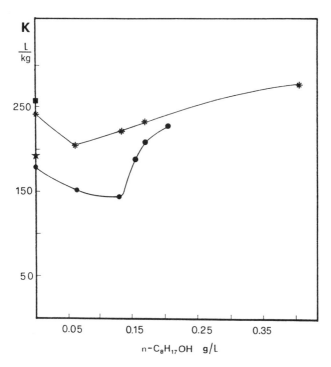

FIG. 3 Partition constant (K) values as a function of n-octanol concentration. ★ Ethofor alone. ● Ethofor + n-butanol + variable n-octanol. ■ Ethofor + DODAB alone. * Ethofor + DODAB + $n$butanol + variable n-octanol. (From Ref. 19, p. 266.)

8

temperature rise during the first step of the process is fast. For microemulsion systems, $\sigma_{\overline{\Delta E}}$ values are halved and $\overline{\Delta E}$ values increased by 10%.

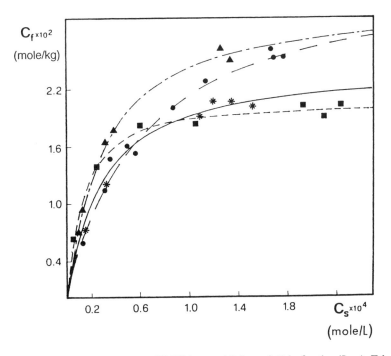

FIG. 4 Dyeing isotherms. ■ Without additives. * Ethofor 1 g/L. ▲ Ethofor 1 g/L + n-butanol 1 g/L. ● Ethofor 1 g/L + n-butanol 1 g/L + n-octanol 0.06 g/L. (From Ref. 19, p. 268.)

Finally, compared with other additives, microemulsions exhibit an interesting trend; better exhaustion and leveling are observed as the dyeing temperature decreases.

We have also used secondary cellulose-acetate rayon as a substrate in the presence of the model dye 6 and of dye 9, a disperse dye having

excellent fastnesses [27,28]. As part of the thermodynamic measurements with dye 6, K values have been determined as a function of surfactant-

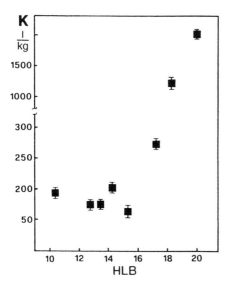

FIG. 5 Partition constant K versus cosurfactant HLB values in the presence of oil (*n*-octanol). (From Ref. 20, p. 29.)

cosurfactant mixtures, the latter being tuned to hydrophile-lipophile balance (HLB) values [sodium dodecyl sulfate (SDS) is the surfactant, poly(ethylene glycol methyl ether) is the cosurfactant]. Besides *n*-octanol as the oil phase, other alcohols with shorter and longer chains have been tested. From data in Fig. 5 and in Table 1, the following conclusions are

TABLE 1 Effect of the Oil Phase on Partition Constants

Surfactant SDS (g/L)	Cosurfactant $C_{12}E_{23}$ (g/L)	Oil phase	(g/L)	K (L/kg)
2.5	1	Hexanol	1.40	230
2.5	1	Octanol	0.60	270
2.5	1	Decanol	0.20	225
2.5	1	Dodecanol	0.10	265
2.5	1	Tetradecanol	0.08	240
2.5	1	C_8E_1	0.70	200

Source: Ref. 20, p. 29.

drawn: (1) as the cosurfactant HLB increases, the partition constant increases, and (2) n-octanol is by far the best oil additive.

Before transferring these conclusions to technical dyeings, it is appropriate to recognize their limitations:

Dye 9 must be tested for its performance.
If the cosurfactant HLB exceeds 19–20, K increases, but some precipitation occurs in the dye bath and the resulting dyeing has to be rejected.
For dye baths with a insoluble dye, pretreatment with ultrasound could be appropriate.
The possibility of reducing the total organic load is worthy of consideration.

Table 2 summarizes the results obtained in these studies. An arbitrary rating scale, whose scores are represented by asterisks, has been used. Sonication markedly enhances dye uptake and uniformity of coloration. Test 7 shows the advantage of cutting down the organic load but at the cost of a less than perfect dye uptake and very bad uniformity. Test 8 clearly shows that the presence of n-octanol leads to the best performance at the lowest organic load, thus saving both chemicals and environment.

Our most recent discovery related to the manyfold properties of microemulsions concerns the use of dye 8 for the dyeing of polyamide 6.6 fabrics. We have reduced the amount of water compared with that required

TABLE 2 Evaluation of Dyeings with Dye 9 in the Presence of Different Additive Systems

Test number	Total organic load (g/L)	Percentage composition of organic load				Sonication Y/N	$\overline{\Delta E}$	$\sigma_{\overline{\Delta E}}$
		SDS	$C_1E_{7.23}$	$C_{12}E_{23}$	Oct[a]			
1	3.5*	71.4	28.6	—	—	N	60*	5*
2	3.5*	71.4	28.6	—	—	Y	86***	1***
3	3.0**	100.0	—	—	—	N	52*	2**
4	3.0**	100.0	—	—	—	Y	77*	2**
5	3.0**	71.4	—	28.6	—	N	61*	3*
6	3.0**	71.4	—	28.6	—	Y	82**	1***
7	1.0***	71.4	—	28.6	—	Y	82**	4*
8	1.0***	61.0	—	24.4	14.6	Y	84***	1***

Asterisks denote arbitrary rating scale.
[a]Octanol.
Source: Ref. 20, p. 31.

TABLE 3 Evaluation of Dyeings of Nylon 6-6 Fabrics with Dye 8 in the Presence of Different Additive Systems

Composition of additive systems (g/L)			Liquor ratio 20:1	$\overline{\Delta E}$	$\sigma_{\overline{\Delta E}}$
SDS	Brij 35	Oct[a]			
—	—	—	20:1	71.2***	2.1*
1.0	—	—	20:1	62.9*	0.2***
0.55	0.45	—	20:1	67.4**	3.2*
0.50	0.40	0.15	20:1	71.8***	0.1***
—	—	—	10:1	65.7*	16.1
1.00	0.80	0.30	10:1	68.7***	0.2***
0.50	0.40	0.15	10:1	69.1***	0.4***

Asterisks denote arbitrary rating scale.
[a]Octanol.

by usual dyeing procedures. The results are reported in Table 3. When the liquor ratio (the ratio of the weight of dye bath to the weight of fabric treated) is reduced to 10:1 in the absence of additives, the quality of the dyeing is poor (it is also worth remarking the dramatic drop of uniformity to 16.1). In the presence of octanol the results are comparable with those obtained at a 20:1 liquor ratio. Industrially, this could be significant in terms of cost reduction: water saving, less energy needed for the heating of baths, and lower volume of effluent to be treated.

IV. CONCLUSIONS

The present chapter is probably the shortest of the volume. Some reasons for this have been given. We are now on the horns of a dilemma: is there some intrinsic or conceptual hindrance to the use of microemulsions in industrial dyeing processes, or will microemulsion systems be viable candidates as next-generation auxiliaries when their advantages, clear from the fragmentary but reliable tests available, become recognized? We hope that, with the aid of the synergic efforts of both fundamental and applied researchers, the next edition of the present volume will find it necessary to devote more space to this promising topic.

ACKNOWLEDGMENTS

Most of the work of the authors reported in the papers listed in the reference section was supported by a contribution from the Progetto Finalizzato

Chimica Fine of the Consiglio Nazionale delle Ricerche (CNR) and of the Ministero dell'Università e della Ricerca Scientifica (MURST) of Italy.

REFERENCES

1. M. E. Diaz Garcia and A. Sanz-Medel, Talanta *33*:255 (1986).
2. E. Barni, P. Savarino, and G. Viscardi, Accounts Chem. Res. *24*:98 (1991).
3. K. Kalyanasundaram, in *Photochemistry in Microheterogeneous Systems*, Academic Press, New York, 1987, Chapter 5.
4. D. S. Rushforth, M. Sanchez-Rubio, L. M. Santos-Vidals, K. R. Wormuth, E. W. Kaler, R. Cuevas, and J. E. Puig, J. Phys. Chem. *90*:6668 (1986).
5. K. S. Schanze and D. G. Whitten, J. Am. Chem. Soc. *105*:6734 (1983).
6. P. D. I. Fletcher, J. Chem. Soc. Faraday Trans. I *82*:2651 (1986).
7. C. A. Jones, L. E. Weaner, and R. A. Mackay, J. Phys. Chem. *84*:1495 (1980).
8. N. S. Dixit and R. A. Mackay, J. Am. Chem. Soc. *105*:2928 (1983).
9. K. R. Wormuth, L. A. Cadwell, and E. W. Kaler, Langmuir *6*:1035 (1990).
10. S. E. Friberg, T. Young, E. Barni, and M. Croucher, J. Dispers. Sci. Technol. *13*:611 (1993).
11. P. Savarino, G. Viscardi, and E. Barni, Colloids Surf. *48*:47 (1990).
12. T. M. Baldwinson, in *Colorants and Auxiliaries,* vol. 2. (J. Shore, ed.), Society of Dyers and Colourists, Bradford, England, 1990, chapters 8, 9, 10, 12, 13.
13. E. Barni, R. Carpignano, G. Di Modica, P. Savarino, and G. Viscardi, J. Dispers. Sci. Technol. *9*:75 (1988).
14. A. De la Maza, J. L. Parra, A. Manich, and L. Coderch, J. Soc. Dyers Colour. *108*:540 (1992).
15. *Ullmann's Encyclopedia of Industrial Chemistry*, 5th ed., VCH, Weinheim, 1987, pp. 298–299.
16. K. Knopf and E. Schollmeyer, Textil Praxis Int. *47*:474 (1992).
17. E. R. Trotman, *Dyeing and Chemical Technology of Textile Fibres*, 6th ed., Griffin, High Wycombe, England, 1984, p. 350.
18. L. Benisek, P. A. Duffield, R. R. D. Holt, and M. A. Rushforth, Melliand Textilber. *71*:604 (1990).
19. E. Barni, P. Savarino, G. Viscardi, R. Carpignano, and G. Di Modica, J. Dispers. Sci. Technol. *12*:257 (1991).
20. P. Savarino, G. Viscardi, P. Quagliotto, E. Barni, and S. E. Friberg, J. Dispers. Sci. Technol. *14*:17 (1993).
21. P. Savarino, G. Viscardi, P. Quagliotto, E. Montoneri, and E. Barni, J. Dispers. Sci. Technol. *16*:51 (1995).
22. H. H. Sumner, in *The Theory of Coloration of Textiles* (A. Johnson, ed.), Society of Dyers and Colourists, Bradford, 1989, chapter 4.
23. H. Zollinger, *Color Chemistry*, 2nd ed., VCH, Weinheim, 1991, chapter 11.
24. C. H. Giles, *A Laboratory Course in Dyeing*, Society of Dyers and Colourists, Bradford, 1983.
25. F. W. Billmeyer, Jr. and M. Saltzman, *Principles of Color Technology*, Wiley, New York, 1981, p. 62.

26. B. C. Burdett, in *The Theory of Coloration of Textiles* (A. Johnson, ed.), Society of Dyers and Colourists, Bradford, 1989, chapter 1.
27. R. Carpignano, P. Savarino, E. Barni, G. Di Modica, and S. S. Papa, J. Soc. Dyers Colour. *101*:270 (1985).
28. S. M. Burkinshaw, in *The Chemistry and Application of Dyes* (D. R. Warning and G. Hallas, eds.), Plenum, New York, 1990, chapter 7.

APPENDIX TO REFERENCES

The present appendix is the result of a bibliographic search using the keywords "microemulsion??+transparent(w)emulsion??*dye??" according to Host ESA-IRS (Frascati) and database CHEMABS file 2, and it covers volumes 66–120 (1966–1994). The Chemical Abstracts references are faithfully reported.

90: 164947y Reactions in microemulsions. 4. Kinetics of chlorophyll sensitized photoreduction of methyl red and crystal violet by ascorbate. Jones, C. E.; Jones, C. A.; Mackay, R. A. (Dep. Chem., Drexel Univ., Philadelphia, PA). J. Phys. Chem. 1979, 83(7), 805–10 (Eng).

90: 200565h Chlorophyll mediated photoreactions in microemulsion media. Jones, C. E.; Mackay, R. A. (Drexel Univ., Philadelphia, PA). Porphyrin Chem. Adv. [Pap. Porphyrin Symp.] 1977 (Pub. 1979), 71–88 (Eng).

92: 135481v Physicochemical studies of solutes in microemulsions. Mackay, Raymond A. (Dep. Chem., Drexel Univ., Philadelphia, PA (USA). Report 1979, ARO-13906. 7-CX; Order No. AD-AO73534, 33 pp. (Eng). Avail. NTIS. From Gov. Rep. Announce. Index (U.S.) 1979, 79(26), 148.

95: 176232c Picosecond fluorescence studies of xanthene dyes in anionic micelles in water and reverse micelles in heptane. Rodgers, Michael A. J. (Cent. Fast Kinetics Res., Univ. Texas, Austin, TX 78712 USA). J. Phys. Chem. 1981, 85(23), 3372–4 (Eng).

98: 162402h Absorption and emission characteristics of merocyanine 540 in microemulsions. Dixit, N. S.; Mackay, R. A. (Dep. Chem., Drexel Univ., Philadelphia, PA 19104 USA). J. Am. Chem. Soc. 1983, 105(9), 2928–9 (Eng).

104: 59257h Photooxidation of leuco dyes. 13. Photooxidation of thioindigo sol in microheterogeneous media. Vettermann, Stefan; Fassler, Dieter; Kirchhof, Barbara (Sekt. Chem., Friedrich-Schiller-Univ. Jena, Jena, Ger. Dem. Rep.). Z. Chem. 1985, 25(9), 333–4 (Ger).

104: 116579r Picosecond absorption spectroscopy of a dye probe in water in oil microemulsions. Pouligny, B.; Lalanne, J. R.; Ducasse, A. (Cent. Rech. Paul Pascal, Domaine Univ., 33405 Talence, Fr.). Chem. Phys. 1986, 102(1–2), 241–53 (Eng).

104: 159282f Microemulsions. Herrmann, Udo; Wagner, Joachim; Claussen, Uwe (Bayer A.-G.) Ger. Offen. DE 3,414,117 (Cl. B01F17/42), 24 Oct 1985, Appl. 14 Apr 1984; 14 pp.

109: 180195t Photochemical and spectral behavior of diolefinic laser dyes in microemulsion media. Ebeid, E. M.; Abdel-Kader, M. H.; Sabry, M. M. F.; Yousef, A. B. (Fac. Sci., Tanta Univ., Tanta, Egypt). J. Photochem. Photobiol., A 1988, 44(2), 153–9 (Eng).

109: 196900e Cosmetic compositions containing microemulsions of dimethylpolysiloxane. Harashima, Asao; Tanaka, Osamu, Maruyama, Tsuneo; Ohta, Yayoi (Toray Silicone Co., Ltd) Eur. Pat. Appl. EP 268,982 (Cl. A61K7/06), 01 June 1988, JP Appl. 86/274,799, 18 Nov 1986; 25 pp.

109: 237904e Dye interactions with surfactants in colloidal dispersions. Ortona, Ornella; Vitagliano, Vincenzo; Robinson, Brian H. (Dip. Chim., Univ. Napoli, 80134 Naples, Italy). J. Colloid Interface Sci. 1988, 125(1), 271–8 (Eng).

110: 78091q Solubilization by a mixture of nonionic surfactants and various surfactants. Ogino, Keizo; Uchiyama, Hirotaka; Abe, Masahiko (Coll. Sci. Technol., Tokyo Univ. Sci., Tokyo, Japan). Hyomen 1988, 26(9), 652–63 (Japan).

113: 12468y Solubilization of dyes in microemulsions. Wormuth, Klaus R.; Cadwell, Linda A.; Kaler, Eric W. (Exxon Res. and Eng. Co., Annandale, NJ 08801 USA). Langmuir 1990, 6(6), 1035–40 (Eng).

113: 193395r Low-temperature dyeing of synthetic fibers in emulsion. Carrion Fite, Francisco Javier (Universidad Politecnica de Cataluna) Span. ES 2,011,848 (Cl. D06P5/20), 16 Feb 1990, Appl. 8,802,603, 25 July 1988; 39 pp.

113: 210239n Determination of aniline in vegetable oils by diazotization and coupling in a microemulsion medium. Esteve Romero, J. S., Simo Alfonso, E. F.; Garcia Alvarez-Coque, M. C.; Ramis Ramos, G. (Fac. Quim., Univ. Valenica, Burjassot, Spain 46100). Anal. Chim. Acta 1990, 235(2), 317–22 (Eng).

114: 8252b Heterocyclic hydrophobic dyes and their interactions with surfactant and oil-in-water microemulsions. Savarino, P.; Viscardi, G.; Barni, E. (Dip. Chim. Gen. Org. Appl., Univ. Torino, 10125 Turin, Italy). Colloids Surf. 1990, 48(1–3), 47–56 (Eng).

114: 198756q Molecular fluorescence in micelles and microemulsions: micellar effects and analytical applications. Georges, Joseph (Lab. Sci. Anal., Univ. Claude Bernard-Lyon 1, 69622 Villeurbanne, Fr.). Spectrochim. Acta Rev. 1990, 13(1), 27–45 (Eng).

114: 249193p New developments in wool dyeing. Benisek, L.; Duffield, P. A.; Holt, R. R. D.; Rushforth, M. A. (IWS Dev. Cent., Ilkley/West Yorkshire, UK LS29 8PB). Schriftenr. Dtsch. Wollforschungsinst. (Tech. Hochsch. Aachen) 1990, 105(Aachener Textiltag., 1989), 263–85 (Eng).

115: 30940s New developments in wool dyeing. Benisek, Ladislav, Duffield, Peter Antony, Holt, Richard R. D.; Rushforth, Michael A., Ukley, UK) Melliand Textilber 1990, 71(8), E280 E283, 604 9 (Eng/Ger).

115: 258166e Microemulsions and their potential applications in dyeing processes. Barni, E.; Savarino P.; Viscardi, G.; Carpignano, R.; Di Modica, G. (Dip. Chim. Gen. Org. Appl., Univ. Torino, 10125 Turin, Italy). J. Dispersion Sci. Technol. 1991, 12(3–4), 257–71 (Eng).

115: 282282k Microemulsion thermal jet-printing inks. Miller, Robert J.; You, Young S. (Hewlett-Packard Co.) U.S. US 5,047,084 (Cl. 106-27; C09D11/00), 10 Sep 1991, Appl. 468,551, 23 Jan 1990; 4 pp.

116: 131663e Microemulsified silicones in liquid fabric care compositions containing dye. Coffindaffer, Timothy W.; Coffey, Geraldine M. (Procter and Gamble Co.) U.S. US 5,071,573 (Cl. 252-8.8; D06M13/34), 10 Dec 1991, Appl. 557,437, 23 Jul 1990; 8 pp.

117: 10019m Storage-stable jet printing inks for steady discharge. Suga, Yuko (Canon K. K.) Jpn. Kokai Tokkyo Koho JP 04 18,462 [92 18,462] (Cl. C09D11/00), 22 Jan 1992, Appl. 90/122,032, 10 May 1990; 12 pp.

118: 40795z Solubilization of organic dyes in microemulsions. Friberg, Stig E.; Young, Tim; Barni, Ermanno; Croucher, Mel (Cent. Adv. Mater. Process., Clarkson Univ., Potsdam, NY 13699-5810 USA). J. Dispersion Sci. Technol 1922, 13(6), 611–26 (Eng).

118: 90641s A hot melt ink for thermal jet printing. Sporer, A. H.; Kaler, E. W.; Murthy, A. K. (IBM Res. Div., San Jose, CA 95124 USA). J. Imaging Sci. Technol. 1992, 36(2), 176–9 (Eng).

118: 176484e Comparison between the kinetics of the alkaline fading of carbocation dyes in water/sodium bis(2-ethylhexyl) sulfosuccinate/isooctane microemulsions and in homogeneous media. Lesis, J. Ramon; Mejuto, Juan C.; Pena, M. Elena (Fac. Quim., Univ. Santiago, Santiago de Compostela, Spain 15706). Langmuir 1993, 9(4), 889–93 (Eng).

119: 205639n Solubilization of water-insoluble dyes via microemulsions for bleedless, non-threading, high print quality inks for thermal ink-jet printers. Wickramanayake, Palitha; Moffatt, John R. (Hewlett-Packard Co.)

U.S. US 5,226,957 (Cl. 106-25R; C09D11/14), 13 Jul 1993, Appl. 853,471, 17 Mar 1992; 6 pp.

120: 15710b Acid-base behavior of neutral red in compartmentalized liquids (micelles and microemulsions). Moulik, S. P.; Paul, B. K.; Mukherjee, D. C. (Dep. Chem., Jadavpur Univ., Calcutta, 700 032 India). J. Colloid Interface Sci. 1993, 161(1), 72–82 (Eng).

11

Microemulsions as Nanosize Reactors for the Synthesis of Nanoparticles of Advanced Materials

VINOD PILLAI and DINESH O. SHAH Center for Surface Science and Engineering, Departments of Chemical Engineering and Anesthesiology, University of Florida, Gainesville, Florida

I. INTRODUCTION

Particles with nanometer size in at least one dimension are of increasing scientific and technological interest [1]. There is clear evidence that small atomic clusters (10–1000 atoms) exhibit novel and hybrid properties between the molecular and bulk solid-state limits. These nanophase materials, which may contain crystalline, quasicrystalline, or amorphous phases, can be metals, ceramics, or composites with rather unique and improved mechanical, electronic, magnetic, and optical properties compared with normal, coarse-grained polycrystalline materials. Because of their ultrafine sizes and surface cleanliness, these particles can easily overcome conventional restrictions of phase equilibria and kinetics, leading to lowering of sintering and solid-state reaction temperatures and increase in sintering rate. The large fraction of atoms residing at the surfaces and grain boundaries of these materials also leads to materials with novel properties. Also, new microhomogeneous, multicomponent composites with nanometer-sized microstructures can be synthesized by reacting, coating, and mixing various types of nanophase materials. Such nanoparticles not only are of basic physical interest but also have resulted in important technological applications, such as (1) catalysis, (2) high-performance ceramic materials, (3) microelectronic devices, and (4) high-density magnetic recording [1–3].

Several techniques have been developed for the synthesis of various types of nanoparticles. These include gas-phase techniques such as gas evaporation, laser vaporization, and laser pyrolysis [4–8]; vacuum synthesis techniques like sputtering, laser ablation, and ionized beam deposition [9–12]; and liquid-phase techniques like precipitation from homogeneous solutions, sol-gel processing, and freeze drying [13–17]. Precipitation reactions in microemulsions offer a novel and versatile technique for the synthesis of a wide variety of nanophase materials with the ability to control precisely the size and shape of the particles formed [18–20].

II. REACTIONS IN MICROEMULSIONS

A microemulsion may be defined as a thermodynamically stable isotropic dispersion of two immiscible liquids consisting of microdomains of one or both liquids in the other, stabilized by an interfacial film of surface-active molecules [21]. In water-in-oil microemulsions, the aqueous phase is dispersed as nanosize droplets (typically 5 to 25 nm in diameter) surrounded by a monolayer of surfactant molecules in the continuous hydrocarbon phase. These aqueous droplets continuously collide, coalesce, and break apart, resulting in a continuous exchange of solute content [22,23]. The collision process depends on the diffusion of the aqueous droplets in the continuous medium (i.e., oil), and the exchange process depends on the attractive interactions between the surfactant tails and the rigidity of the interface, as the aqueous droplets approach close to each other [23].

For reactions in water-in-oil microemulsions involving reactant species totally confined within the dispersed water droplets, a necessary step prior to their chemical reaction is the exchange of reactants by the coalescence of two droplets. When chemical reaction is fast, the overall reaction rate is likely to be controlled by the rate of coalescence of droplets [23]. Therefore, properties of the interface such as interfacial rigidity are of major importance. A relatively rigid interface decreases the rate of coalescence and hence leads to a low precipitation rate. On the other hand, a substantially fluid interface in the microemulsion enhances the rate of precipitation. Thus, by controlling the structure of the interface, one can change the reaction kinetics in microemulsions by an order of magnitude [24]. It has been further shown that the structure of oil, the alcohol, and the ionic strength of the aqueous phase can significantly influence the rigidity of the interface and the reaction kinetics [25].

Conceptually, if one takes two identical water-in-oil microemulsions and dissolves reactants A and B respectively in the aqueous phases of these two microemulsions, upon mixing, due to collision and coalescence of the droplets, reactants A and B come in contact with each other and form precipitate AB. This precipitate is confined to the interior of the microemulsion droplets. This is the main principle utilized in producing nanoparticles with microemulsions. Figure 1 shows a schematic representation of this process. However, nanoparticles can also be produced in microemulsions by adding a reducing or precipitating agent, in the form of a liquid or a gas, to a microemulsion containing the primary reactant dissolved in its aqueous core.

Synthesis of ultrafine particles using reactions in microemulsions was first reported by Boutonnet et al. [19] when they obtained monodispersed metal particles (in the size range 3–5 nm) of Pt, Pd, Rh, and Ir by reducing corresponding salts in water pools of water-in-oil microemulsions with hydrazine or hydrogen gas. Since then, there have been several reports of the use of microemulsions for the synthesis of a variety of nanoparticles.

Metallic nanoparticles and metal clusters have a wide range of applications, including their use in catalysis, as biological stains, and as ferrofluids. Touroude et al. [26] synthesized bimetallic particles of platinum and palladium by reduction of H_2PtCl_6 and $PdCl_2$ with hydrazine in water/pentaethylene glycol dodecyl ether/hexadecane microemulsions. Similarly, platinum particles have also been synthesized in Aerosol OT (AOT) and cetyltrimethylammonium bromide (CTAB)–based microemulsions for applications in catalysis [27]. Colloidal gold and silver particles have potential applications as condensers for electron storage in artificial photosynthesis. This has led to studies by Kurihara et al. [28] on laser and pulse radiolytically induced colloidal gold formation in W/O microemulsions. Colloidal gold and silver particles have also been synthesized by reduction of corresponding metal salts in water-in-oil microemulsions using sodium borohydride [29,30].

Colloidal semiconductor particles, especially cadmium sulfide, have invoked a great deal of interest for their unique photochemical and photophysical properties. These properties are drastically dependent on the size of the semiconductor nanoparticles. Several researchers have therefore used

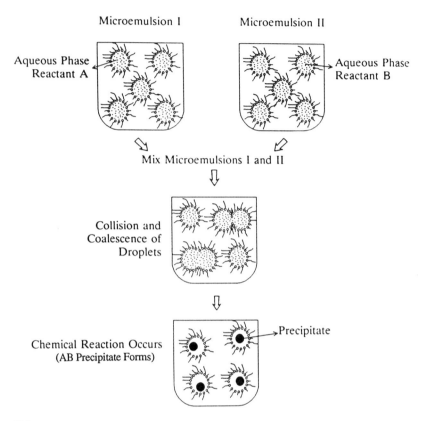

FIG. 1 Schematic representation of synthesis of nanoparticles in microemulsions.

water-in-oil microemulsions for the synthesis of such nanoparticles. Meyer et al. [31] have generated platinized colloidal cadmium sulfide in situ in AOT-based microemulsions. Several studies have been published on the synthesis, growth, and characterization of cadmium sulfide [32–38] and cadmium selenide [39] particles in microemulsions.

Colloidal particles of calcium carbonate, stabilized by surfactants, constitute an important class of additives for oils used in lubrication for internal combustion engines. Water-in-oil microemulsions thus provide an ideal medium for the synthesis of such particles. This has been exploited for the synthesis of calcium carbonate [40–43], barium carbonate [44], and strontium carbonate [45] particles by carbonation of their respective salts or hydroxides in the aqueous cores of water-in-oil microemulsions.

Microemulsions have also been used for the synthesis of monodisperse particles of nickel, iron, and cobalt boride particles, which have applications in heterogeneous catalysis, by the reduction of metal salts using sodium borohydride [46,47]. Other nanoparticles synthesized in water-in-oil microemulsions include monodisperse silica particles [48–51], nanosize molybdenum sulfide particles [52], and magnetic particles such as magnetite [53,54] and maghemite [55].

We have been successful in synthesizing a variety of nanoparticles using different microemulsion systems. These include in situ synthesis of nanoparticles of AgCl [56] and AgBr [25] and microemulsion-mediated synthesis of highly dense and microhomogeneous Y-Ba-Cu-O (123) and B-Pb-Sr-Ca-Cu-O (2223) high-temperature ceramic superconductors [57,58], high-coercivity barium ferrite for high-density perpendicular recording [59,60], and zinc oxide–based varistors [61]. The synthesis of these materials and their structure and properties are described in the following sections.

III. SYNTHESIS OF SILVER HALIDES

Nanoparticles of silver halides are extremely important for applications in photographic emulsions [62,63]. It is, however, difficult to obtain nanosize, monodispersed particles of silver halides by conventional methods. We have synthesized nanoparticles of AgCl [56] and AgBr [25] using water/AOT (sodium bis-2-ethylhexyl sulfosuccinate)/alkane microemulsion systems. Two microemulsions with the same composition but differing only in the nature of the aqueous phase were taken. Microemulsions were prepared by solubilizing the aqueous solutions into AOT/alkane mixtures under vigorous stirring. The concentration of AOT in alkane (for example, n-heptane) was 0.15 M. The aqueous phase in the first microemulsion was silver nitrate ($AgNO_3$, 0.4 M), whereas the aqueous phase in the second microemulsion was either NaCl or NaBr (0.4 M). The microemulsions had identical compositions and each solubilized 8 moles of aqueous phase per mole of surfactant (AOT). These two microemulsions were then mixed under constant stirring. This led to the precipitation of AgCl or AgBr particles within the aqueous droplets due to the reaction between the two reactants as the aqueous droplets collide, coalesce, and break again. The growth of the particles is restricted by the size of the aqueous droplets resulting in the formation of nanosize particles.

Transmission electron microscopy (TEM) was used to study the size and size distribution of these particles. Samples were prepared by dropping a small amount of particles dispersed in the microemulsion onto a TEM grid. Figure 2a and b show nanoparticles of AgCl precipitated in water/ AOT/heptane microemulsions, observed 1 day and 1 week, respectively, after the start of the reaction. These micrographs show that the particles formed in the microemulsions are in the size range 50–100 Å in diameter with a very uniform size distribution. These nanoparticles appear to have a spherical shape. It can also be seen that these dispersions are very stable with little or no aggregation of AgCl particles even after a week. This has

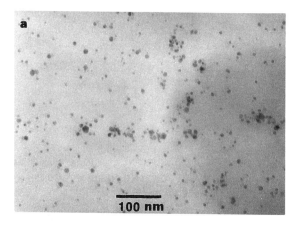

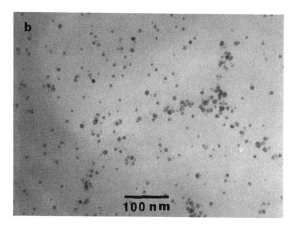

FIG. 2 Transmission electron micrographs of silver halide nanoparticles synthesized in water/AOT/heptane microemulsions (a) 1 day after start of reaction and (b) 1 week after start of reaction.

been explained on the basis of a low attractive interaction between the microemulsion droplets containing the AgCl nanoparticles [25,56].

Electron diffraction was used to identify the crystal planes of the AgCl nanocrystals. Four d planes of the face-centered cubic (fcc) lattice structure of AgCl, namely (111), (200), (220), and (420), were identified based on the analysis of the diffraction pattern.

Thus, water-in-oil microemulsions can be used for in situ synthesis of stable dispersions of nanoparticles of silver halides.

IV. SYNTHESIS OF BARIUM FERRITE

Barium ferrite ($BaFe_{12}O_{19}$) has been traditionally used in permanent magnets because of its high intrinsic coercivity and fairly large crystal anisotropy [64]. However, in the past decade, it has emerged as an important magnetic medium for high-density perpendicular recording [65]. Such technological applications require materials with strict control of homogeneity, particle size and shape, and magnetic characteristics [66].

The classical ceramic method for the preparation of barium ferrite consists of firing mixtures of iron oxide and barium carbonate at high temperatures ($\sim 1200°C$) [67,68]. Furthermore, the ferrite must then be ground to reduce the particle size from multidomain to single domain. This generally yields mixtures that are nonhomogeneous on a microscopic scale. Milling introduces lattice strains in the material, whereas high reaction temperatures induce sintering and coagulation of particles, both of which can be damaging if specific properties like uniform and reduced particle sizes and high coercivities are required in the final product [69,70].

In order to achieve homogeneity of ions at the atomic level and to overcome the effects of milling, various techniques such as chemical coprecipitation [71,72], glass crystallization [65,73], the organometallic precursor method [74,75], aerosol synthesis [76,77], and colloidal synthesis [78] have been developed to prepare ultrafine barium ferrite. We have used the aqueous cores of water-in-oil microemulsions as reaction media to produce uniform-sized, microhomogeneous nanoparticles of carbonate precursors for the synthesis of ultrafine barium ferrite [59,60].

We selected a microemulsion system with cetyltrimethylammonium bromide (CTAB) as the surfactant (12 wt%), n-butanol as the cosurfactant (10 wt%), n-octane (44 wt%) as the continuous oil phase, and a salt solution (34 wt%) as the dispersed aqueous phase. This system solubilizes a relatively large volume of aqueous phase in well-defined nanosize droplets of stable, single-phase water-in-oil microemulsions [57].

Microemulsions were prepared by solubilizing different salt solutions into CTAB/n-butanol/n-octane solutions. We took two microemulsions (microemulsion I and microemulsion II) with identical compositions but different aqueous phases. The aqueous phase in microemulsion I was a mixture of barium nitrate and ferric nitrate solutions in the molar ratio 1.1: 12. Even though, according to stoichiometry, a ratio of 1:12 should be sufficient, an excess of barium nitrate was used because it is necessary for the precipitation of Ba^{2+} and Fe^{3+} ions in the ratio 1:12 as barium carbonate is partially soluble in water. The aqueous phase in microemulsion II was the precipitating agent ammonium carbonate. These two microemulsions are then mixed under constant stirring. Because of the frequent collisions of the aqueous cores of water-in-oil microemulsions [22,23], the reacting species in microemulsions I and II come in contact with each other. This leads to the precipitation of barium-iron-carbonate within the nanosize aqueous droplets of the microemulsion. As the two microemulsions (I and II) are of identical composition, differing only in the nature of the aqueous phase, the microemulsion is not destabilized upon mixing. The aqueous droplets act as constrained nanosize reactors for the precipitation reaction, as the surfactant monolayer provides a barrier restricting the growth of the carbonate particles. This surfactant monolayer also hinders coagulation of particles.

The barium-iron-carbonate precipitate was separated in a centrifuge at 5000 rpm for 10 min. The precipitate was then washed with a 1:1 mixture of methanol and chloroform, followed by pure methanol to remove any oil and surfactant from the particles. The precipitate was then dried at 100°C. The dried precipitate was heated (calcined) at 950°C for 12 h to ensure complete conversion of the carbonates into the hexaferrite ($BaFe_{12}O_{19}$).

Transmission electron microscopy (TEM) was used to study the size and size distribution of the precursor particles (carbonates) as well as the calcined particles. Samples were prepared by ultrasonically dispersing the particles (precursors and calcined powder) in methanol prior to deposition onto a carbon-coated TEM grid. Figure 3 shows that the precursor particles formed within the aqueous droplets of the microemulsion were in the size range 5–15 nm. On the other hand, TEM of the calcined powder (Fig. 4) shows that these particles are in the size range 50–100 nm, implying that growth of particles has taken place during calcination.

The X-ray diffraction pattern of the calcined powder (Fig. 5) shows all the characteristic peaks for barium ferrite marked by their indices. No other phases are detectable. This confirms the complete conversion of the precursor carbonates into the hexagonal ferrite $BaFe_{12}O_{19}$.

Room temperature magnetic property measurements for the calcined powder were carried out on a vibrating-sample magnetometer using an

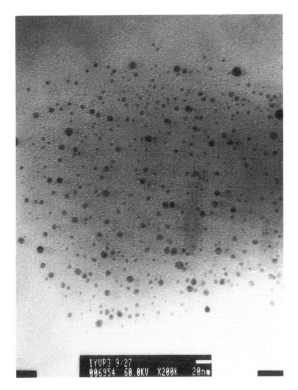

FIG. 3 Transmission electron micrograph of barium ferrite precursors synthesized in water/CTAB/1-butanol/octane microemulsions.

unoriented, random assembly of particles. The magnetization curve thus obtained yielded an intrinsic coercivity (H_c) of 5397 Oe, a saturation magnetization (M_s) of 61.2 emu/g, and a remanent magnetization (M_r) of 33.3 emu/g for the microemulsion synthesized barium ferrite particles. This high value of coercivity indicates that these barium ferrite particles are essentially single domain. This observation is also confirmed by the fact that the particles formed by us are in the size range 50–100 nm. Haneda and Morrish [79] have shown that using the theory of Kittel [80], the critical domain size for barium ferrite particles is about 1 μm. Also, direct observation of the domain structure of barium ferrite particles at room temperature by Goto et al. [81] confirms that the critical domain size is indeed in the region of 1 μm.

In summary, we have presented a new, microemulsion-mediated process for the synthesis of ultrafine particles of barium ferrite. This process

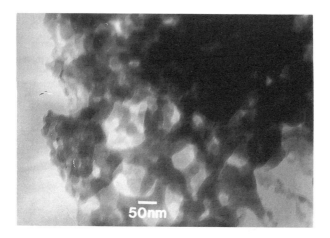

FIG. 4 Transmission electron micrograph of barium ferrite particles.

yielded high-coercivity, single-domain barium ferrite particles, which are
phase pure as confirmed by X-ray diffraction. These phase pure, single-
domain particles can be doped with Co and Ti to reduce the coercivity and
large anisotropy to make these particles suitable for high-density perpendicular recording [65].

V. SYNTHESIS OF SUPERCONDUCTORS

The discovery of a superconducting temperature (T_c) near 30 K in La-Ba-
Cu oxide by Bednorz and Muller [82] prompted many efforts to study
perovskite-like oxide superconductors. This led to the discovery of new
superconducting materials in the Y-Ba-Cu-O system with T_c above the
boiling point of liquid nitrogen [83,84]. Even higher T_c (>110 K) superconductors were discovered in the Bi-Sr-Ca-Cu-O [85] and Tl-Ba-Ca-Cu-
O [86] systems.

It is now well recognized that the properties of high-temperature oxide
superconductors are critically dependent on the microstructure of the sample. Control of particle size, size distribution, morphology of precursors,
and heat treatment conditions are critical to obtaining a desired microstructure. The conventional method for the preparation of oxide superconductors
by solid-state reaction has many inherent problems, such as poor homogeneity, large particle size, lack of reproducibility, and longer heat treatment times. In order to achieve a better level of homogeneity, wet chemical
techniques such as coprecipitation [87–90], freeze drying [91], amorphous

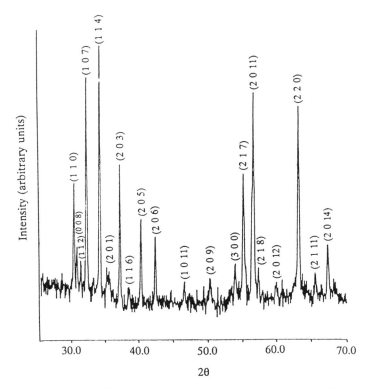

FIG. 5 X-ray diffractogram of barium ferrite particles synthesized using microemulsions.

citrate process [92,93], and chelation [94] have been used. One can achieve an even better level of homogeneity and smaller particles sizes if the wet chemical reactions are carried out in constrained nanosize reactors, like the aqueous droplets of water-in-oil microemulsions. We have used water-in-oil microemulsions for the synthesis of nanoparticles of oxalate precursors of both Y-Ba-Cu-O (123) and Bi-Pb-Sr-Ca-Cu-O (2223) superconductors.

A. Synthesis of Y-Ba-Cu-O (123) Superconductors

$YBa_2Cu_3O_{7-x}$ superconductors were synthesized by the coprecipitation of oxalates of yttrium, barium, and copper in the aqueous cores of water/cetyltrimethylammonium bromide/1-butanol/octane microemulsion systems [57]. Two microemulsions (each consisting of 29 wt% surfactant phase, 60 wt% hydrocarbon phase, and 11 wt% aqueous phase), one containing an aqueous solution of yttrium, barium, and copper nitrates in the molar ratio 1:2:3 and the other containing ammonium oxalate solution as the aqueous phase, were mixed. This led to the precipitation of the oxalate precursors within the aqueous cores of the microemulsion. Because the two microemulsions are identical in composition, there is no destabilization of the microemulsion phase on mixing. The barrier provided by the surfactant monolayer restricts the growth of the particles and hinders interparticle coagulation. Serial precipitation of the cations (which may occur in other wet chemical synthesis processes) is avoided in this approach due to the size of the "nanoreactors," which was about 130 Å in the present case. The particles are also expected to be microhomogeneous due to mixing of the cations at the nanoscale level in the microemulsion droplets. The oxalate precipitate was separated, washed, dried, and then calcined at 820°C for 2 h for complete conversion to the oxide. This powder was then pressed into pellets and sintered at 925°C for 12 h.

As a comparison, we also studied coprecipitation of yttrium, barium, copper oxalates in a bulk aqueous solution using ammonium oxalate. The oxalate precipitate thus obtained was calcined and then sintered under conditions similar to those for the microemulsion precipitate to produce the $YBa_2Cu_3O_{7-x}$ superconductor.

DC magnetic susceptibility measurements as a function of temperature were performed on a superconducting quantum interference device (SQUID) magnetometer. This revealed that the microemulsion-synthesized sample had a superconducting transition temperature (T_c) of 93 K and showed 90% of the Meissner shielding expected from an ideal sample ($-1/4\pi$). This sintered disk also had 98% of the theoretical single-crystal density. This high density is due to the ability of the nanoparticles to pack closely with very few voids, and upon sintering, these well-packed particles are able to form large grains (15–50 μm) with very low porosity. This can be seen from the scanning electron micrographs (SEMs) of the sintered disk shown in Fig. 6a and b. On the other hand, the bulk coprecipitated sample showed a Meissner shielding of 14% of that expected from an ideal sample. This can be explained by the fact that this sample was more porous (lower density) than the microemulsion-synthesized material due to the fact that the precursor particles were large and therefore did not pack well. This resulted in smaller grains (0.5–20 μm) in the sintered pellet with more pores (see Fig. 6c and d). A comparison of all properties of the microemulsion-synthesized and bulk coprecipitated samples are shown in Table 1. On the basis of the Meissner shielding results for the two samples, we can say that the microemulsion-derived 123 superconductor is seven times better than the one formed by bulk coprecipitation.

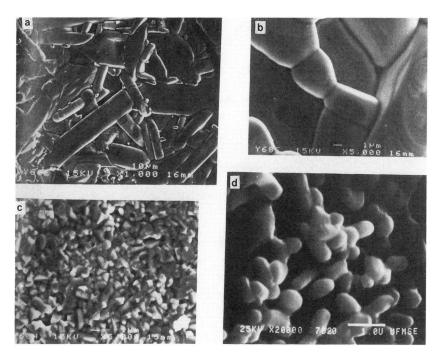

FIG. 6 Scanning electron micrographs of sintered $YBa_2Cu_3O_{7-x}$ pellet synthesized using microemulsions (a and b) and bulk coprecipitation (c and d). The magnifications used were (a) 1000, (b) 5000, (c) 5000, and (d) 20,000.

TABLE 1 Comparison of Selected Physical Properties of $YBa_2Cu_3O_{7-x}$ Synthesized Using Microemulsions and by Bulk Coprecipitation

Property	Microemulsion synthesis	Bulk synthesis
ESD[a] of Y-Ba-Cu oxalate precursor	47.4 nm	380.6 nm
ESD of Y-Ba-Cu-O powder	274.8 nm	626.6 nm
Calcination conditions	820°C, 2h	860°C, 6 h
Grain size of sintered pellet	15–50 μm	0.5–2.0 μm
Sintering conditions	925°C, 12 h	925°C, 12 h
Percent of theoretical density of sintered pellet	98% ($\pm$3)	90% ($\pm$2)
Fraction of ideal Meissner signal ($-1/4\pi$)	90.5%	14.4%
Superconducting T_c	93 K	91 K

[a]ESD, equivalent spherical diameter.

B. Synthesis of Bi-Pb-Sr-Ca-Cu-O (2223) Superconductors

We have also used water-in-oil microemulsions for the synthesis of Bi-Pb-Sr-Ca-Cu-O (2223) superconductors [87]. A nonionic surfactant Igepal CO-430 or nonylphenoxypoly(ethyleneoxy) ethanol was used for the formation of the microemulsion with cyclohexane as the external oil phase. The aqueous phase in the first microemulsion was a solution of salts of Bi, Pb, Sr, Ca, and Cu in the molar ratio 1.84:0.34:1.91:2.03:3.06 dissolved in a 50:50 (v/v) mixture of acetic acid and water. This aqueous solution was prepared by dissolving Bi_2O_3, $Pb(CH_3COO)_2$, $SrCO_3$, $CaCO_3$, and $Cu(CH_3COO)_2$ in the acetic acid/water mixture in the cationic ratio given above. The aqueous phase in the second microemulsion was a solution of oxalic acid in a 50:50 (v/v) acetic acid and water mixture. Both microemulsions consisted of 15 g of surfactant, 50 mL of oil, and 10 mL of aqueous phase. Mixing of these two microemulsions led to the formation of the oxalate precursor within the aqueous cores of the microemulsion. These precursors were in the size range 2–6 nm (Fig. 7). These oxalate particles were calcined at 800°C for 12 h and then pressed into a pellet and sintered in air for 96 h at 850°C.

DC magnetic susceptibility measurements as a function of temperature (shown in Fig. 8) indicated that this superconductor had a T_c of 112 K.

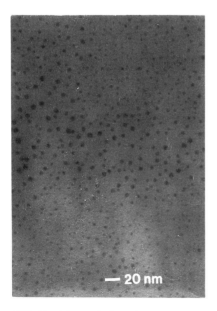

FIG. 7. Transmission electron micrograph of oxalate precursors for 2223 superconductors synthesized in water/Igepal CO-430/cyclohexane microemulsions.

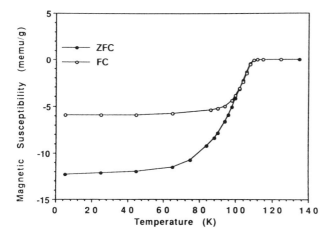

FIG. 8 Zero-field cooled (ZFC), diamagnetic shielding and field cooled (FC), flux expulsion signals for DC magnetic susceptibility measurements as a function of temperature for 2223 superconductor synthesized using microemulsions.

The zero-field cooled signal (diamagnetic shielding) corresponded to 93% of the Meissner shielding expected from an ideal sample ($-1/4\pi$). Phase analysis by X-ray diffraction showed characteristic peaks of the 110 K phase: (002) and (0010) at $2\theta = 4.8°$ and 23.9°, respectively. This confirmed that the material was phase pure 2223 superconducting oxide.

The sintered disk also had large, lamellar grains as can be seen from the scanning electron micrograph (Fig. 9). This disk has very low porosity and had a density corresponding to 97% of the theoretical value. As in the case of the 123 superconductor, the ability of the nanoparticles to pack closely, with very few voids, results in high-density disks with low porosity and high Meissner shielding.

Thus, we have shown that uniform-sized nanoparticles of precursors of superconductors precipitated in microemulsions, upon heat treatment, lead to the formation of microhomogeneous, high-density superconductors which show properties superior to those of superconductors synthesized by most wet chemical methods.

VI. SUMMARY

The importance of nanophase technology has resulted in tremendous research efforts to develop new techniques for the synthesis of nanosize particles. One such technique, developed a decade or so ago, was to use water-in-oil microemulsions as a reaction medium. The aqueous cores/water pools of these microemulsions can be used as nanosize reactors for

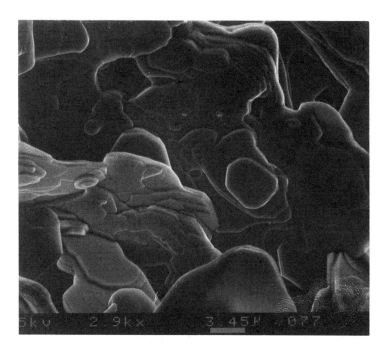

FIG. 9 Scanning electron micrograph of sintered Bi-Pb-Sr-Ca-Cu-O (2223) pellet synthesized using microemulsions.

the precipitation of a wide variety of nanoparticles. For this technique to be successful, the reactants should be confined to the aqueous cores of the microemulsions and should not interact with the other components of the microemulsion (namely oil and surfactant).

In this chapter, we have presented some of our research efforts in the field of synthesis of nanoparticles using water-in-oil microemulsions. We have described the synthesis of silver halides, barium ferrite, and Y-Ba-Cu-O and Bi-Pb-Sr-Ca-Cu-O superconductors using microemulsions. All these materials synthesized in microemulsions exhibit unique properties unlike those of materials synthesized using conventional techniques.

ACKNOWLEDGMENTS

The authors wish to thank the National Science Foundation (NSF), the Electric Power Research Institute (EPRI), and the Alcoa Foundation for supporting this research.

REFERENCES

1. R. P. Andres, R. S. Averback, W. L. Brown, L. E. Brus, W. A. Goddard III, A. Kaldor, S. G. Louie, M. Moscovits, P. S. Peercy, S. J. Riley, R. W. Siegel, F. Spaepen, and Y. Wang, J. Mater. Res. *4*:704 (1989).
2. R. W. Siegel, MRS Bull. *15*:60 (1990).
3. Y. T. Han, MRS Bull. *14*:13 (1989).
4. R. Uyeda, J. Cryst. Growth *24*:69 (1974).
5. Y. Liu, Q-L. Zhang, F. K. Tittel, R. F. Curl, and R. E. Smalley, J. Chem. Phys. *85*:7434 (1986).
6. M. D. Morse, Chem. Rev. *86*:1049 (1986).
7. K. LaiHing, R. G. Wheeler, W. L. Wilson, and M. A. Duncan, J. Chem. Phys. *87*:3401 (1987).
8. W. R. Cannon, S. C. Danforth, J. H. Flint, J. S. Haggerty, and R. A. Marra, J. Am. Ceram. Soc. *65*:324 (1982).
9. P. Fayet and L. Woste, Z. Phys. D *3*:177 (1986).
10. R. B. Wright, J. K. Bates, and D. M. Gruen, Inorg. Chem. *17*:2275 (1978).
11. M. L. Mandich, V. E. Bondybey, and W. D. Reents, J. Chem. Phys. *80*:4245 (1987).
12. T. Takagi, Z. Phys. D *3*:272 (1986).
13. E. Matijevic, Langmuir *2*:12 (1986).
14. T. Sugimoto, Adv. Colloid Interface Sci. *28*:65 (1987).
15. C. R. Veale, in *Fine Powders: Preparation, Properties and Uses,* Halsted Press Division, Wiley, New York, 1972.
16. A. Goldman and A. M. Lang, J. Phys. Colloq.:297 (1987).
17. M. Kagawa, M. Kikuchi, R. Ohno, and T. Nagae, J. Am. Ceram. Soc. *64*:C-7 (1981).
18. B. H. Robinson, Nature *320*:309 (1986).
19. M. Boutonnet, J. Kizling, P. Stenius, and G. Maire, Colloids Surfaces *5*:209 (1982).
20. M. A. Lopez-Quintela and J. Rivas, J. Colloid Interface Sci. *158*:446 (1993).
21. R. Leung, M. J. Hou, C. Manohar, D. O. Shah, and P. W. Chun, in *Macro- and Microemulsions* (D. O. Shah, ed.), American Chemical Society, Washington, DC, 1981, p. 325.
22. H. F. Eicke, J. C. W. Shepherd, and A. Steinemann, J. Colloid Interface Sci. *56*:168 (1976).
23. P. D. I. Fletcher, A. M. Howe, and B. H. Robinson, J. Chem. Soc. Faraday Trans. I *83*:985 (1987).
24. M. J. Hou, M. Kim, and D. O. Shah, J. Colloid Interface Sci. *123*:398 (1988).
25. C. H. Chew, L. M. Gan, and D. O. Shah, J. Dispers. Sci. Technol. *11*:593 (1990).
26. R. Touroude, P. Girard, G. Maire, J. Kizling, M. Boutonnet-Kizling, and P. Stenius, Colloids Surfaces *67*:19 (1992).
27. J. H. Clint, I. R. Collins, J. A. Williams, B. H. Robinson, T. F. Towey, P. Cajean, and A. K. Lodhi, Faraday Discuss. *95*:219 (1993).

28. K. Kurihara, J. Kizling, P. Stenius, and J. H. Fendler, J. Am. Chem. Soc. *105*: 2574 (1983).
29. P. Barnickel and A. Wokaun, Mol. Phys. *69*:1 (1990).
30. P. Barnickel, A. Wokaun, W. Sager, and H. F. Eicke, J. Colloid Interface Sci. *148*:80 (1992).
31. M. Meyer, C. Wallberg, K. Kurihara, and J. H. Fendler, J. Chem. Soc. Chem. Commun. *90* (1984).
32. P. Lianos and J. K. Thomas, Chem. Phys. Lett. *125*:299 (1986).
33. C. Petit and M. P. Pileni, J. Phys. Chem. *92*:2282 (1988).
34. S. Modes and P. Lianos, J. Phys. Chem. *93*:5854 (1989).
35. C. Petit, P. Lixon, and M. P. Pileni, J. Phys. Chem. *94*:1598 (1990).
36. T. F. Towey, A. K. Lodhi, and B. H. Robinson, J. Chem. Soc. Faraday Trans. *86*:3757 (1990).
37. B. H. Robinson, T. F. Towey, S. Zourab, A. J. W. G. Visser, and A. van Hoek, Colloids Surfaces *61*:175 (1991).
38. L. Motte, C. Petit, L. Boulanger, P. Lixon, and M. P. Pileni, Langmuir *8*:1049 (1992).
39. M. L. Steigerwald, A. P. Alivisatos, J. M. Gibson, T. D. Harris, R. Kortan, A. J. Muller, A. M. Thayer, T. M. Duncan, D. C. Douglass, and L. E. Brus, J. Am. Chem. Soc. *110*:3046 (1988).
40. K. Kandori, K. Kon-no, and A. Kitahara, J. Colloid Interface Sci. *115*:579 (1987).
41. K. Kandori, K. Kon-no, and A. Kitahara, J. Colloid Interface Sci. *122*:78 (1988).
42. J. P. Roman, P. Hoornaert, D. Faure, C. Biver, F. Jacquet, and J. M. Martin, J. Colloid Interface Sci. *144*:324 (1991).
43. K. Kandori, N. Shizuka, K. Kon-no, and A. Kitahara, J. Dispers. Sci. Technol. *9*:61 (1988).
44. K. Kon-no, M. Koide, and A. Kitahara, J. Chem. Soc. Jpn. *6*:815 (1984).
45. N. Shizuka, M. Yanagi, K. Kon-no, and A. Kitahara, in *Proceedings of the 37th Symposium on Colloid and Interface Chemistry,* Morioka, Japan, 1984, p. 360.
46. N. Lumfimpadio, J. B. Nagy, and E. G. Derouane, in *Surfactants in Solution,* Vol. 3 (K. L. Mittal and B. Lindman, eds.), Plenum, New York, 1986, p. 483.
47. J. B. Nagy, Colloids Surfaces *35*:201 (1989).
48. H. Yamauchi, T. Ishikawa, and S. Kondo, Colloids Surfaces *37*:71 (1989).
49. K. Osseo-Asare and F. J. Arriagada, Colloids and Surfaces *50*:321 (1990).
50. F. J. Arriagada and K. Osseo-Asare, Colloids Surfaces *69*:105 (1992).
51. W. Wang, X. Fu, J. Tang, and L. Jiang, Colloids Surfaces A: Physicochem. Eng. Aspects*81*:177 (1993).
52. E. Boakye, L. R. Radovic, and K. Osseo-Asare, J. Colloid Interface Sci. *163*: 120 (1994).
53. S. Bandow, K. Kimura, K. Kon-no, and A. Kitahara, Jpn. J. Appl. Phys. *26*: 713 (1987).

54. K. H. Lee, C. M. Sorensen, K. J. Klabunde, G. C. Hadjipanayis, IEEE Trans. Magn. *MAG-28*:3180 (1992).
55. P. Ayyub, M. S. Multani, M. Barma, V. R. Palkar, and R. Vijayaraghavan, J. Phys. C *21*:2229 (1988).
56. M. J. Hou and D. O. Shah, in *Interfacial Phenomena in Bio-technology and Materials Processing* (Y. A. Attia, B. M. Moudgil, and S. Chander, eds.), Elsevier, Amsterdam, 1988, p. 443.
57. P. Ayyub, A. N. Maitra, and D. O. Shah, Physica C *168*:571 (1990).
58. P. Kumar, V. Pillai, and D. O. Shah, Appl. Phys. Lett. *62*:675 (1993).
59. V. Pillai, P. Kumar, and D. O. Shah, J. Magn. Mag. Mater. *116*:L299 (1992).
60. V. Pillai, P. Kumar, M. S. Multani, and D. O. Shah, Colloids Surfaces A: Physicochem. Eng. Aspects *80*:69 (1993).
61. S. Hingorani, V. Pillai, P. Kumar, M. S. Multani, and D. O. Shah, Mater. Res. Bull. *28*:1303 (1993).
62. D. J. Locker, in *Encyclopedia of Chemical Technology*, Vol. 17 (R. E. Kirk, D. F. Othmer, M. Grayson, and D. Eckroth, eds.), Wiley, New York, 1982, p. 611.
63. B. H. Carroll, G. C. Higgins, and T. H. James, in *Introduction to Photographic Theory: The Silver Halide Process*, Wiley, New York, 1980, chapter 6.
64. B. D. Cullity, in *Introduction to Magnetic Materials*, Addison-Wesley, Reading, MA, 1972, p. 575.
65. O. Kubo, T. Ido, and H. Yokoyama, IEEE Trans. Magn. *MAG-18*:1122 (1982).
66. H. Sakai, K. Hanawa, and K. Aoyagi, IEEE Trans. Magn. *MAG-28*:3355 (1992).
67. H. Kojima, in *Ferromagnetic Materials*, Vol. 3 (E. P. Wohlfarth, ed.), North-Holland, Amsterdam, 1982, p. 305.
68. G. C. Bye and C. R. Howard, J. Appl. Chem. Biotechnol. *21*:319 (1971).
69. K. Haneda and H. Kojima, J. Am. Ceram. Soc. *57*:68 (1974).
70. M. Paulus, in *Preparative Methods in Solid State Chemistry*, Academic Press, New York, 1972, p. 488.
71. K. Haneda, C. Miyakama, and H. Kojima, J. Am. Ceram. Soc. *57*:354 (1974).
72. W. Roos, J. Am. Ceram. Soc. *63*:601 (1980).
73. B. T. Shirk and W. R. Buessem, J. Am. Ceram. Soc. *53*:192 (1970).
74. F. Licci and T. Besagni, IEEE Trans. Magn. *MAG-20*:1639 (1984).
75. M. Vallet, P. Rodriguez, X. Obradors, A. Isalgue, J. Rodriguez, and M. Pernet, J. Phys. *46*:C6-335 (1985).
76. Z. X. Tang, S. Nafis, C. M. Sorensen, G. C. Hadjipanayis, and K. J. Klahunda, IEEE Trans. Magn. *MAG-25*:4236 (1989).
77. W. A. Kaczmarek, B. W. Ninham, and A. Calka, J. Appl. Phys. *70*:5909 (1991).
78. E. Matijevic, J. Colloid Interface Sci. *117*:593 (1987).
79. K. Haneda and A. H. Morrish, IEEE Trans. Magn. *MAG-25*:2597 (1989).
80. C. Kittel, Rev. Mod. Phys. *40*:1294 (1969).
81. K. Goto, M. Ito, and T. Sakurai, Jpn. J. Appl. Phys. *19*:541 (1980).
82. J. Bednorz and A. Muller, Z. Phys. *B64*:189 (1986).

83. M. K. Wu, J. R. Ashburn, C. J. Torng, P. H. Hor, R. L. Meng, L. Gao, Z. J. Huang, Y. Q. Wang, and C. W. Chu, Phys. Rev. Lett. *58*:908 (1987).
84. H. Tabagi, S. Uchida, K. Kishio, K. Kitazawa, K. Fueki, and S. Tanaka, Jpn. J. Appl. Phys. *26*:L239 (1987).
85. H. Maeda, Y. Tanaka, M. Fukutonei, and T. Asano, Jpn. J. Appl. Phys. *27*: L209 (1988).
86. Z. Z. Sheng and A. M. Hermann, Nature *322*:138 (1988).
87. A. Manthiram and J. B. Goodenough, Nature *329*:701 (1987).
88. K. Kaneko, H. Ihara, M. Hirabayashi, N. Terada, and K. Senzaki, Jpn. J. Appl. Phys. *26*:L734 (1987).
89. J. D. Jorgensen, H. Schuttler, D. G. Hinks, D. W. Capone, K. Zhang, M. B. Brodsky, and D. J. Scalapino, Phys. Rev. Lett. *58*:1024 (1987).
90. P. Kumar, V. Pillai, and D. O. Shah, J. Mater. Sci. Lett. *12*:162 (1993).
91. S. M. Johnson, M. I. Gussman, D. J. Rowcliffe, T. H. Gebale, and J. Z. Sun, Adv. Ceram. Mater. *2*:337 (1987).
92. S. Vehida, H. Takagi, K. Kitazawa, and S. Tanaka, Jpn. J. Appl. Phys. *26*:L1 (1987).
93. C. Chu and B. Dunn, J. Am. Ceram. Soc. *70*:C375 (1987).
94. T. Fujisawa, A. Takagi, T. Honjoh, K. Okuyamia, S. Ohshima, K. Matsuki, and K. Muraishi, Jpn. J. Appl. Phys. *28*:1358 (1989).

12
Use of Microemulsions in the Production of Nanostructured Materials

M. ARTURO LÓPEZ-QUINTELA Department of Physical Chemistry, University of Santiago de Compostela, Santiago de Compostela, Spain

JOSÉ QUIBÉN-SOLLA Laboratory of R&D, La Artística de Vigo, Vigo, Spain

JOSÉ RIVAS Department of Applied Physics, University of Santiago de Compostela, Santiago de Compostela, Spain

I. INTRODUCTION

Production of nanoscale materials is one of the biggest challenges of science and technology today. Although the progress in this field has been very impressive in recent years, much has to be done in order to understand the properties of these systems and to achieve good control of the nanostructure of these materials.

Among the methods for producing nanomaterials, the technique of microemulsions has been developed rapidly in the past few years [1–5]. This technique offers different advantages than other techniques: it is a soft technique (i.e., it does not require extreme temperature or pressure conditions); it can be used, in principle, with almost all chemical reactions that have been developed to obtain particles in homogeneous solutions; and it does not require special equipment. Disadvantages include the difficulties in the control of the parameters that influence chemical reactions in microemulsions. For example, there are reactions that are slowed down greatly in microemulsions and others that do not proceed at all in these media [6]. Another disadvantage is that, at the moment and to our knowledge, this is only a laboratory technique, but different schemes have already been proposed to extend this technique to industrial scales [7].

The purpose of this chapter is to review the actual stage of our knowledge of this technique, focusing on general aspects of the mechanism involved in the technique and its possibilities rather than a detailed review of materials that have been synthesized by this technique, which can be found elsewhere (see, for example, Ref. 8 and Chapter 11).

In Sec. II we describe the foundations of the technique for producing nanoparticles in microemulsions. In Sec. III we discuss the mechanism of formation of particles. Section IV is devoted to examples of the application of this technique: the production of nanoscale mixed metals such as Fe-Cu and Co-Ag. Finally, the main conclusions on the use of microemulsions in the production of nanoscale materials are summarized.

II. DESCRIPTION OF THE METHOD

The microemulsion technique is based on the use of microemulsions as microreactors in order to control the growth of the particles obtained.

A microemulsion is a thermodynamically stable system with at least three components: two immiscible components (generally water and oil) and a surfactant. In general, such a system is a very complicated one in which one can find a great variety of complex structures, e.g., liquid crystals, gels, and vesicles [9]. For the purposes of the method described here for obtaining ultrafine particles, we are interested only in microemulsions with a simple structure, namely water-in-oil (W/O) microemulsions formed by nanodroplets of water dispersed in oil. The preparation of a W/O microemulsion for a particular system of oil, water, and surfactant is not an easy task, but although there is no general theory for predicting the structure of a microemulsion, there are at least several rules and phenomenological parameters that can help one to find particular microemulsions with the desired structure [10–14].

Once the right microemulsion system is obtained, the method consists of mixing two microemulsions carrying the appropriate reactants in order to obtain the desired particles [1,15]. A schematic picture of this process, based on several studies (see below), is represented in Fig. 1. It can be seen that after mixing both microemulsions containing the reactants, inter-

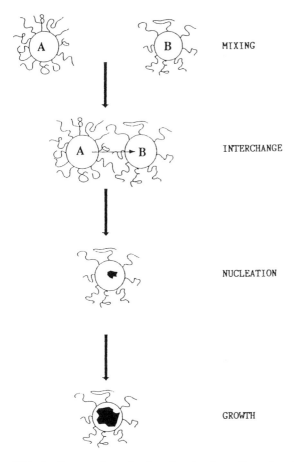

FIG. 1 Proposed mechanism for the formation of particles according to the experimental results obtained by different methods (see text).

change of the reactants (here referred to generally as A and B) takes place during the collisions of the water droplets in the microemulsions. The reaction then takes place inside the droplets (nucleation and growth), which control the final size of the particles.

III. MECHANISM OF THE FORMATION OF PARTICLES

After mixing a microemulsion, a uniform distribution of reactant species among the droplets of the microemulsion is attained. The rate of communication between droplets is very high, with $k = 10^6$–10^7 M^{-1} s^{-1} for the second-order communication-controlled rate constant, expressed in terms of the droplet concentration [16]. This implies that at a droplet concentration $> 10^{-3}$ M, exchange of reactants between droplets is complete within 1 ms. It is assumed that this exchange can take place only when an energetic collision between two droplets is able to establish a water channel between them [16]. This process is energetically unfavorable because it implies a change of the curvature of the surfactant film [17]. Therefore the rate constant of this process is not diffusion controlled ($k_D = 10^{10}$ M^{-1} s^{-1} in a solvent with a viscosity like that of n-heptane), and only approximately 1 in 10^3 collisions between droplets leads to reactant exchange [16].

Once the reactants are mixed inside the water pools, the chemical reaction to produce the particles can take place. This process can be divided into two steps. In the first one the chemical reaction gives rise to the formation of nuclei, and in the second nuclei grow to their final size. In order to understand how this process can occur, we can consider a particular example. Let us assume for that purpose the formation of Fe by chemical reduction of Fe^{2+} by $NaBH_4$ [18]. For a typical microemulsion, $\Phi = V_{droplet}/V_{total} = 0.1$ and $r_{droplet} = 3$ nm. Then the concentration of droplets is about 1.5×10^{-3} M. Taking into account the concentration of reactants usually employed (> 0.1–0.2 M), it can be deduced that there are approximately 100 reactant (ions or molecules) per water pool. It has also been observed that the final size of the particles coincides with the droplet size (see below). Therefore an iron particle with a radius of about 3 nm has approximately 6000 atoms. This implies that nuclei initially formed with the reactants inside the same droplet (assuming no loss of reactants) have to grow about 60 times to attain their final sizes. This growth can occur by interchange of nuclei between droplets or by interchange of reactants in the case in which chemical reaction is slow. For the first process, exchange of nuclei, the interchange becomes more and more difficult as these nuclei grow, because the channel between the droplets has to be large and this implies, as we have stated before, a great change in the curvature of the film, which is energetically very unfavorable, as shown in Fig. 2. Therefore, to control the final size of the particles by the microemulsion method it is important to have a rigid surfactant film. This can be achieved by choosing a microemulsion with a radius of curvature similar to the natural radius [19]. This is an important point because when the surfactant film is flexible collisions of droplets can form "transient droplet dimers" [17], leading to interchange of particles or nuclei and therefore making microemulsions not suitable for growth control. It should be pointed out that this may be why it has been found that in some cases the size of the

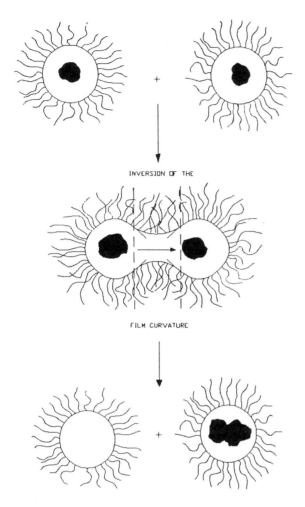

INVERSION OF THE

FILM CURVATURE

FIG. 2 Mechanism of particle growth by interchange of nuclei between droplets. The curvature of the surfactant film should be inverted during the interchange of nuclei.

droplets of the microemulsion does not control the final size of the particles [20–22].

When the chemical reaction is very slow, the growth of the nuclei can take place by having new reactants enter nanodroplets carrying nuclei. The reaction now proceeds over the nuclei already formed. Again the growth is controlled in this case by the surfactant film, which is not able to expand and thus hinders the further growth of the particles. The same rule, i.e., use of no flexible films, applies now to get real control of the particle size. It can then be said that the growth control by microemulsions can be achieved independently of the rate of the chemical reaction. This is one important difference from most of the methods developed so far for controlling the size of the particles. Those methods try to increase the rate of the reaction in order to get a burst of nuclei so that the particles can grow uniformly [23,24]. In the microemulsion technique this aspect is of minor importance and the control depends mostly on the flexibility of the surfactant film. If we take all these parameters into account, a good correlation between the droplet size of the microemulsions and the size of the particles can be achieved, as can be seen in Table 1.

It should be noted that the difference between a fast and a slow reaction is that in the first case a burst of nuclei occurs and then growth of the particles takes place by simple aggregation of nuclei. During this aggregation process the number of particles decreases. On the contrary, in a slow reaction nuclei are continuously formed during the whole process.

In studying the mechanism of formation of particles, different techniques have to be applied for some specific cases. As an example, in Fig. 3 we present results for Fe particles obtained in Aerosol OT (AOT) microemulsions (R = [H_2O]/[AOT] = 22; [AOT] = 0.05 M; reactant A:

TABLE 1 Radius of Particles Obtained by SAXS and DLS for Different AOT Microemulsions and Different Concentrations of Reactants

[AOT] (mol dm^{-3})	R	[$FeCl_2$] (10^4 mol dm^{-3})	[$NaBH_4$] (10^3 mol dm^{-3})	r (nm)	Technique
0.1	22	7.5	3.5	3.9	SAXS
0.1	22	3.8	1.75	3.9	SAXS
0.1	22	1.6	0.68	4.1	SAXS
0.1	22	0.38	0.35	3.9	SAXS
0.05	10	0.93	0.35	2.5	SAXS
0.025	4.44	48	24	1.9	DLS
0.025	8.88	48	24	2.5	DLS
0.025	13.3	48	24	3.3	DLS
0.025	17.8	48	24	4.3	DLS
0.025	22.2	48	24	4.7	DLS
0.025	26.6	48	24	5.4	DLS
0.025	31.1	48	24	5.8	DLS

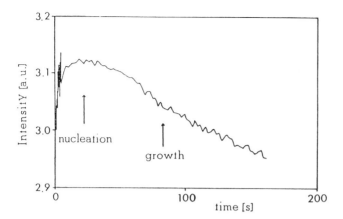

FIG. 3 Time course of the intensity obtained by SAXS during the formation of particles in microemulsions. Microemulsions: AOT/aqueous solution/heptane, $R = 22$, [AOT] = 0.05 M, [FeCl$_2$] = 1.9×10^{-4} M, [NaBH$_4$] = 8.8×10^{-4} M.

[FeCl$_2$] = 1.9×10^{-4} M; reactant B: [NaBH$_4$] = 8.8×10^{-4} M) with a stopped-flow technique and measurement of the small-angle X-ray scattering (SAXS) with synchrotron radiation (Hasy-Lab, Hamburg). Nucleation implies an increase in the number of scattering centers and therefore an increase in the scattering intensity. On the contrary, the growth of particles is associated with a decrease in the scattered intensity because the intensity was integrated (to decrease the noise) over a q range (0.01 Å^{-1} $\leq q = 2\pi/\lambda \sin \theta/2 \leq 0.015$ Å^{-1}) in the high-q part of the diffraction pattern; i.e., the observation window corresponds to the diffraction of the smallest particles which disappear during their growth. To check this interpretation we have also performed experiments by transmission electron microscopy (TEM) and ultraviolet-visible (UV-vis) spectroscopy. Figure 4a and b show two TEM photographs obtained by dropping a sample of the reaction mixture at different times onto a Cu grid and drying the grid to stop the reaction. Figure 4a corresponds to the beginning of the reaction (time = ~15 s) and Fig. 4b to the end of the reaction (time = 300 s). It can be observed that there is not much increase in the number of particles; there is an increase only in the size of the particles. This confirms the results obtained by time-resolved SAXS because due to the slow procedure used in the TEM experiments only the growth of the particles can be detected. We have also performed experiments by UV-vis spectroscopy. Small metallic particles, such as iron obtained in microemulsions, show a plasmon absorption band that in some cases can be coupled with electronic eigen-

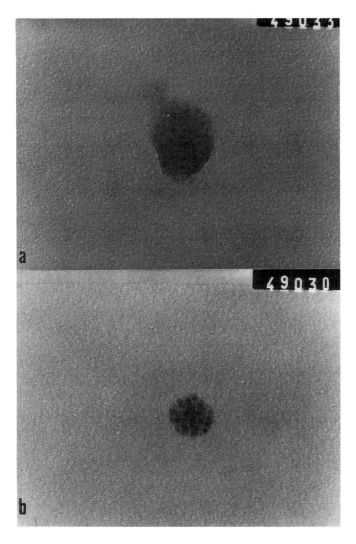

FIG. 4 Transmission electron microscopy (TEM) of particles formed in microemulsions of AOT/aqueous solution/heptane ($R = 22$, [AOT] = 0.05 M, [FeCl$_2$] = 1.9×10^{-4} M, [NaBH$_4$] = 8.8×10^{-4} M). Reaction time: 15 s (a) and 300 s (b). Magnification 382,000.

states of the same particles and/or energy levels of the surfactant attached to the surface of the particles [25]. Figure 5 shows the evolution of the plasmon bands for iron particles obtained in microemulsions. By a fitting procedure one can resolve the different bands, which are well described by a sum of Lorentzian terms [25]. It has been observed that for particles with radii in the range ~2–5 nm the position of the bands does not change and that only the bandwidth decreases as the particle size is reduced for particles of radius less than 4 nm [25]. By applying the same fitting procedure to the evolution of the bands observed in Fig. 5 we found that the observed changes can be explained by decreasing width of the absorption bands with time, which corresponds to an increase in the particle size. These results agree with those obtained by TEM and time-resolved SAXS.

Finally, it should be pointed out that the particles are usually very stable in the microemulsions in which they are formed. The particles are kept apart by the surfactant molecules attached to their surfaces. The presence of surfactant can be clearly seen in Fig. 6b, which represents chemical analysis by X-ray dispersive energy, performed for particles shown in Fig. 6a. This analysis shows, apart from the iron peak, the S and Na peaks corresponding to the surfactant AOT and NaCl from the reactant salts.

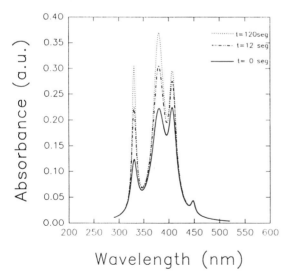

FIG. 5 Evolution of the UV-vis bands of the iron particles obtained in a microemulsion. The position of the bands does not change for particles with $r < 5$ nm.

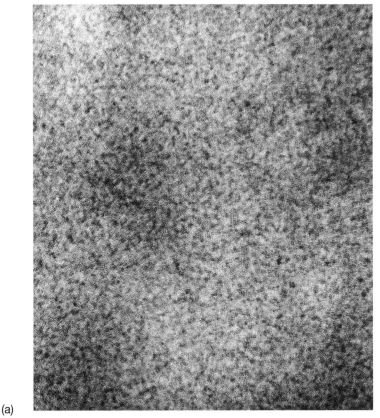

(a)

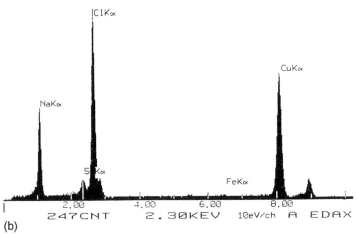

(b)

FIG. 6 (a) TEM photograph of iron particles obtained in a microemulsion of AOT/heptane/aqueous solution. Magnification 625,000. (b) X-ray dispersive energy (EDAX) analysis of the particles represented in (a). Note the presence of S and Na from the surfactant (AOT).

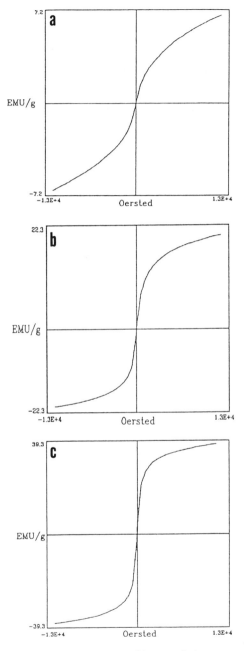

FIG. 7 Various shapes of hysteresis loops measured for magnetite particles at room temperature (see text). (a) Sample M1, coercive field $H_c = 0$ Oe, magnetization 7.2 emu/g at 13 kOe. (b) Sample M2, $H_c = 11$ Oe, $\sigma = 22$ emu/g. (c) Sample M3, $H_c = 38$ Oe, $\sigma = 39$ emu/g.

257

According to the mechanism discussed above, it could be thought that after the preparation of particles in microemulsions, introducing new reactants into the microemulsion could lead to particles of greater size because the reactants can enter a droplet carrying a particle, so that the reaction could proceed further over the previously formed particle, which can act as a nucleus for this second reaction. This speculation has proved to be right by the following experiment [18]: first, magnetite particles were obtained in Aerosol OT microemulsions of the composition described above for obtaining iron. The following reactant concentrations (hereafter the concentrations of reactants will be referred to the water content of the microemulsion) were employed: $[FeCl_2] = 0.1$ M, $[FeCl_3] = 0.2$ M (reactants A), and $[NH_4OH] = 2$ M (reactant B). Magnetite particles of about 4 nm were obtained (sample M1). Then new reactants (in the same concentrations as above) were added to aliquots of the final microemulsion containing the magnetite particles. New particles about 5 nm in size were obtained (sample M2). Again the same procedure was repeated with the last system, obtaining particles of about 5.7 nm (sample M3). Figure 7a–c show the magnetic hysteresis loop obtained at room temperature for these three samples. A change in the magnetic behavior is observed due to the increase in size of the particles. The behavior of smaller particles indicates that their blocking temperatures lie below room temperature.

It can be concluded that the size of the particles obtained by the microemulsion technique can be changed either by varying the droplet size of the microemulsion or by successive addition of reactants to previously formed particles in microemulsions.

IV. EXAMPLES OF APPLICATION

As examples of application of the method described above we describe the preparation of nanoscale multilayer particles of different metals in order to illustrate how powerful the method is (other examples can be found elsewhere [1,4]). The procedure for obtaining multilayer particles is based on the results obtained with magnetite particles by successive reactions. By using different kinds of reactants in the successive reactions, in principle, layers of different materials over the nuclei initially formed could be obtained.

A. Iron-Copper Composites

Fe nuclei were first produced in AOT/n-heptane/water microemulsions by reducing Fe^{2+} with $NaBH_4$ by a procedure similar to that described above (see Fig. 3, sample L1 [18]). With this microemulsion containing Fe par-

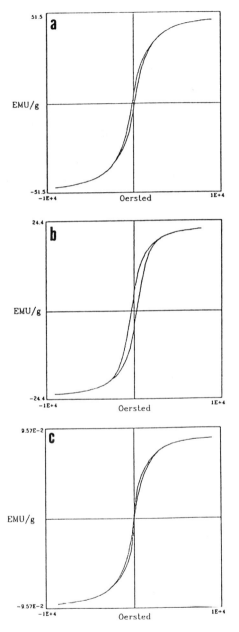

FIG. 8 Magnetic hysteresis loops at room temperature for iron-copper composites obtained by reaction in microemulsions of AOT/heptane/aqueous solution and reduced at 300 K in an H_2 atmosphere (see text). Coercive field, H_c, magnetization (at 10 kOe), σ, and squareness, SQ, defined as σ_r/σ_s, where σ_r is the remanence and σ_s is the magnetization at 10 kOe. (a) Sample L1, H_c = 105 Oe, σ = 51 emu/g, SQ (squared parameter) = 0.098. (b) Sample L2, H_c = 340 Oe, σ = 24 emu/g, SQ = 0.23. (c) Sample L3, H_c = 70 Oe, σ = 1.65 emu/g, SQ = 0.007.

ticles a second reaction was carried out, introducing in this case Cu^{2+} instead of Fe^{2+} (sample L2). A third reaction was then performed over the last system, again introducing Fe^{2+} and reducing it with $NaBH_4$ (sample L3). The concentrations of salts employed were the same in all cases: metallic salts 0.1 M and reducing agent 0.2 M. After the samples were prepared, they were separated by ultracentrifugation and washed several times with water and acetone. Finally, samples were reduced in an H_2 atmosphere at 300°C to eliminate any oxides that were present at the surface of the particles. Figure 8a–c show the hysteresis loops at room temperature for the samples obtained. It can be observed that the parameter SQ, which is related to the magnetic interactions between particles, increases from its initial value (sample L1) of 0.009 (great interactions between iron particles) to 0.23 (sample L2), showing that the presence of a layer of Cu over the Fe particles (see Fig. 9) gives rise to a drastic reduction of the magnetic interactions between the particles. In sample L3 the presence of a last layer of Fe over the Cu again increases the magnetic interactions, diminishing the value of SQ ($=0.007$).

B. Cobalt-Silver Composites

Interest in single-domain particles immersed in a nonmagnetic metallic matrix has increased [26–28]. Attention is paid to these granular systems, such as Fe-Ag and Co-Ag, because they show great magnetoresistances [28,29]. The process for obtaining Co-Ag particles, which was carried out in an inert globe box, can be divided into two steps. First, Co particles are synthesized in AOT microemulsions of the same composition as employed in the previous example. Co^{2+} (0.1 M) and $NaBH_4$ (0.2 M) were employed. After the preparation of Co particles (size ~7–10 nm), these particles were separated from the microemulsion by ultracentrifugation, washed several

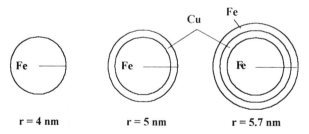

FIG. 9 Multilayer particles of Fe and Cu obtained by consecutive reactions in microemulsion of AOT ($R = 10$).

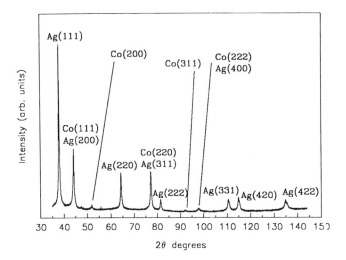

FIG. 10 X-ray diffraction pattern of particles of Co-Ag obtained by consecutive reactions in microemulsion.

times with *n*-heptane and ethanol, and finally dried with acetone. In a second step, the magnetic Co particles were redispersed with AOT in an aqueous solution containing $AgNO_3$ and EDTA (ethylenediaminetetraacetic acid). Silver ions are then adsorbed onto the particles. This solution was irradiated with UV light for 30 min to obtain a metallic cover on the magnetic particles [30]. The amount of EDTA and $AgNO_3$ used depends on the Ag/Co ratio to be prepared. Figure 10 shows the X-ray diffraction pattern of a typical Co/Ag sample. Long measurement times were needed in order to observe cobalt peaks. Only one Co peak is visible (200) because the others overlap with those due to the silver. Analysis of the width of the silver peaks (111) and (220) gives a size of about 30 nm for the coated Co particle. Figure 11a shows a TEM image of the same sample, and Fig. 11b shows the analysis of the particle by X-ray dispersive energy. It can be seen that the sizes agree with those obtained by X-ray diffraction (~30 nm). The chemical analysis clearly shows the presence of the cobalt and silver in the samples. Magnetic measurements performed on these granular systems after annealing at 500°C (Fig. 12) show coercive fields as high as $H_c \sim 600$ Oe at room temperature. These values are similar to those reported for granular materials which present giant magnet or resistance [28,29].

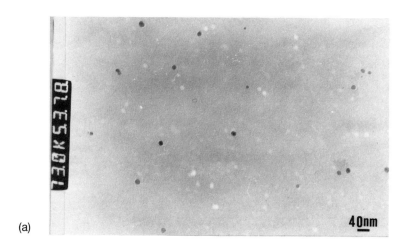

(a)

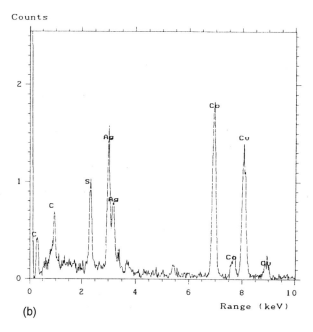

(b)

FIG. 11 (a) TEM picture of a sample of Co-Ag obtained by consecutive reactions in microemulsion. (b) EDAX analysis of the same sample performed over a particle of ~20 nm.

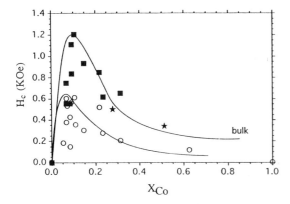

FIG. 12 Variation of the coercive field at 77 K (filled squares) and room temperature (open circles) with the composition for Co_x-Ag_{1-x} particles after annealing at 500°C. The filled stars are data for Co_x-Ag_{1-x} films [32]; solid lines are guides to the eye.

V. CONCLUSIONS

The use of microemulsions in the production of nanoscale complex materials is still at an early stage, but it is an area under active and growing development. Although the production of simple materials in microemulsions began approximately 15 years ago, this technique has evolved slowly, mainly because of our lack of knowledge of the structural and thermodynamic behavior of microemulsion systems. This shortcoming has probably discouraged use of the technique in the production of materials. Today, however, there is better knowledge of these systems and one can change the compositions of microemulsions without introducing major changes in their stability and structure [31].

We have briefly outlined in this chapter the use of microemulsions in the production of nanoscale materials and reviewed the mechanisms by which particles are produced inside the droplets of the microemulsion. It has been pointed out that the main parameters that control the growth of particles in microemulsions are the flexibility of the surfactant film and the concentration of reactants. By choosing the appropriate surfactants and concentrations, good control of the sizes of particles can be achieved. The technique can be further used to produce particles over cores of another kind of material previously formed, e.g., silver over cobalt or copper over iron, which indicates the power and richness of this promising technique for the production of nanomaterials.

ACKNOWLEDGMENTS

This work was partially supported by the "Fundación Ramón Areces," the "Dirección General Interministerial de Ciencia y Tecnología" (PB90-0934), "Xunta de Galicia" (XUGA 20612393), and NATO Project CRG 920255.

REFERENCES

1. M. A. López-Quintela and J. Rivas, J. Colloid Interface Sci. *158*:446 (1993).
2. M. P. Pileni, I. Lisiecki, L. Motte, C. Petit, J. Cizeron, N. Moumen, and P. Lixon, Prog. Colloid Polym. Sci. *93*:1 (1993).
3. J. P. Wilcoxon, U.S. Patent 5,147,841 to United States Department of Energy (1992).
4. J. B. Nagy and A. Claerbout, in *Surfactants in Solution 11* (K. L. Mittal and D. O. Shah, eds.), Plenum, New York, 1991, pp. 363–382.
5. V. Pillai, P. Kumar, and D. O. Shah, J. Mag. Mag. Mater. *116*:L299 (1992).
6. M. A. López-Quintela and C. Lalinde, to be published.
7. A. J. Russell and C. Komives, Chemtech. *JAN*:26 (1994).
8. M. P. Pileni, J. Phys. Chem. *97*:6961 (1993).
9. V. Degiorgio and M. Corti, eds. *Physics of Amphiphiles: Micelles, Vesicles and Microemulsions*, North Holland, Amsterdam, 1985.
10. Y. Talmon and S. Prager, J. Chem. Phys. *76*:1354 (1982).
11. T. Jouffrey, P. Levinson, and P. G. de Gennes, J. Phys. *43*1241 (1981).
12. B. Widom, J. Chem. Phys. *81*:1030 (1984).
13. S. A. Safran, D. Roux, M. E. Cates, and D. Andelman, Phys. Rev. Lett. *57*: 491 (1986).
14. M. Bourrel and R. S. Schechter, eds., *Microemulsions and Related Systems. Formulation, Solvency, and Physical Properties*, Marcel Dekker, New York, 1988.
15. M. A. López-Quintela, J. Rivas, and J. Quibén, U.S. Patent 4,983,217 (1991) and EC Patent 370,939 (1993).
16. P. D. I. Fletcher, B. H. Robinson, F. Bermejo-Barrera, and D. G. Oakenfull, in *Microemulsions* (I. D. Robb, ed.), Plenum, New York, 1982, pp. 221–231.
17. R. Zana and J. Lang, in *Microemulsions: Structure and Dynamics* (S. E. Friberg and P. Bothorel, eds.), CRC Press, Boca Raton, FL, 1987, p. 153.
18. L. Liz, Doctoral thesis, Univ. Santiago de Compostela, 1992.
19. P. D. I. Fletcher, A. Howe, N. Perrins, B. H. Robinson, C. Toprakcioglu, and J. Dore, in *Surfactants in Solution*, Vol. 3 (K. L. Mittal and B. Lindman, eds.), Plenum, New York, 1984, p. 1745.
20. K. Kurihara, J. Kizling, P. Stenius, and J. H. Fendler, J. Am. Chem. Soc. *105*: 2574 (1983).
21. J. B. Nagy, E. G. Derouane, A. Gourgue, N. Lufimpadio, I. Ravet, and J. P. Verfaillie, in Proceedings of Sixth Int. Symp. of Surfactants in Solution-Modern Aspects, New Delhi, 1986.

22. A. Khan-Lodhi, B. H. Robinson, T. Towey, C. Hermann, W. Knoche, and U. Thesing, in *The Structure, Dynamics and Equilibrium Properties of Colloidal Systems* (D. M. Bloor and E. Wyn-Jones, eds.), NATO ASI Ser., Ser. C, Vol. 324, Kluwer, Dordrecht, 1990, p. 373.
23. V. K. La Mer and R. H. Dinegar, J. Am. Chem. Soc. *72*:4847 (1950).
24. E. Matijevic, Chem. Mater. *5*:412 (1993).
25. M. C. Blanco, A. Meira, D. Baldomir, J. Rivas, and M. A. López-Quintela, IEEE Trans Magn. *30*:739 (1994).
26. J. R. Childress, C. L. Chien, and M. Nathan, Appl. Phys. Lett. *56*:95 (1990).
27. J. R. Childress and C. L. Chien, J. Apl. Phys. *70*:5885 (1991).
28. A. E. Berkowitz, J. R. Mitchell, M. J. Carey, A. P. Young, S. Zhang, F. E. Spada, F. T. Parker, A. Hutten, and G. Thomas, Phys. Rev. Lett. *68*:3745 (1992).
29. J. Xiao, J. S. Jiang, and C. L. Chien, Phys. Rev. Lett. *68*:3749 (1992).
30. J. Rivas, R. D. Sánchez, A. Fondado, C. Izco, A. J. García-Bastida, J. García-Otero, J. Mira, D. Baldomir, A. González, I. Lado, M. A. López-Quintela, and S. B. Oseroff, J. Appl. Phys. *76*:1 (1994).
31. S. H. Chen and R. Rajagopalan, eds., *Micellar Solutions and Microemulsions: Structure, Dynamics, and Statistical Thermodynamics*, Springer, New York, 1990.
32. S. H. Liou, S. Malhotra, Z. S. Shan, D. J. Sellmyer, S. Nafis, J. A. Woolam, C. P. Reed, R. J. de Angelis and G. M. Chow, J. Appl. Phys. *70*:5882 (1991).

13
The Microemulsion/Gel Method

STIG E. FRIBERG Center for Advanced Materials Processing and
Department of Chemistry, Clarkson University, Potsdam, New York

JOHAN SJÖBLOM Department of Chemistry, University of Bergen,
Bergen, Norway

I. INTRODUCTION

The sol/gel method [1] has been thoroughly investigated during past decades as evidenced by numerous conference proceedings and monographs concerned with different applications [2–5]. Especially noteworthy is the recent progress in the understanding of the kinetics of hydrolysis and condensation reactions. The early rate predictions by Kay and Assink [6] for the tetraethyoxysilane reactions were not in complete agreement with the experimental results on tetramethoxysilane of Yang et al. [7] and Pouxviel and Boilot [8]. The issue has been resolved by the contributions by Sanchez and McCormick [9], whose investigations revealed opposite effects of kinetics and thermodynamics.

The sol-gel method has also expanded into studies of reactions between hydrated metal salts and alkoxysilanes. This method, pioneered by Sjöblom et al. [10] in their studies of the structure and solubility of metal salt hydrates in alcohols [11,12], has already given rise to numerous investigations [13–15] of the reactions taking place. It should, however, be realized that these reactions take place in solutions and are outside the realm of the microemulsion/gel process. The reaction products are, however, of colloidal dimensions. The seemingly simple overall reaction

$$Si(OR)_4 + 2H_2O \overset{pH2}{\rightleftharpoons} SiO_2 + 4HOR \qquad (1)$$

forming an alcohol gel sustained by a complicated silica polymer in fact displays a plethora of hydrolysis and condensation species

$$Si_nO_{2n-m}(OH)_{2m-p}(OR)_p \qquad (2)$$

of colloidal size, as demonstrated by the results of light scattering determinations [16]. It should be emphasized that these colloidal fragments do not introduce well-defined hydrophilic-hydrophobic interfaces in the system; the solvent environment and the interior of the colloidal aggregate are both polar.

A colloidal system such as a microemulsion or a liquid crystalline phase contains such interfaces, and the effect of the interface of the reaction is one of the more important fundamental aspects of the microemulsion-gel process. The best-defined interface is the water-air one, and recent studies by Sjöblom et al. [17,18] are the first examples of well-defined studies of the alkoxysilane hydrolysis and condensation at a well-defined interface.

Using colloidal systems, microemulsions, as reaction media was independently pioneered by Guizard et al. [19] and Friberg and Yang [20] with different applications in mind. Guizard's aim [19] was to use the presence of water in microemulsion droplets to modify the reaction rates of metal organic compounds such as titanium alkoxides in order to prepare gels and, subsequently glasses of these oxides, whereas Friberg's group [20] emphasized the potential of microemulsions to serve as vehicles to include inorganic water-soluble salts in the glass. The investigations have so far clarified the colloidal issues involved and have also given rise to new applications.

II. COLLOIDAL STRUCTURES

The microemulsions consist of water dispersed in oil, oil dispersed in water, or bicontinuous structures of oil and water, all of colloidal dimensions [21–23], and the scientific problem to be solved is the influence of the colloidal structure on the sol-gel reactions per se and the effect on the colloidal structure of the reaction products. The two issues are, of course, not independent; a change in colloidal structure caused by the reaction products leads with necessity to a subsequent modification of the reaction rate and path.

Preservation of the colloidal structure can be achieved only by limiting the ratio of the ethanol formed to the colloidal polar solvent. Friberg and Ma [24] obtained this result by replacing water with glycerol in their investigation, and Dabodie et al. [25] limited the amount of tetraethoxysilane added to the system.

The results revealed that the colloidal structure had a most pronounced effect on the hydrolysis and condensation of tetraethoxysilane. Instead of the variety of molecules of the general formula $Si_nO_{2n-r}OH_{2r-x}OR_x$ that characterize the reactions in alcohol solution (Fig. 1A), ^{29}Si nuclear magnetic resonance (NMR) spectra of the microemulsion during reaction showed the presence of only two species, $Si(OEt)_4$ and SiO_2 (Fig. 1B), and the reaction in the liquid crystalline phase in addition showed the presence of $Si(OH)_4$ (Fig. 1C). The explanation for this difference between the out-

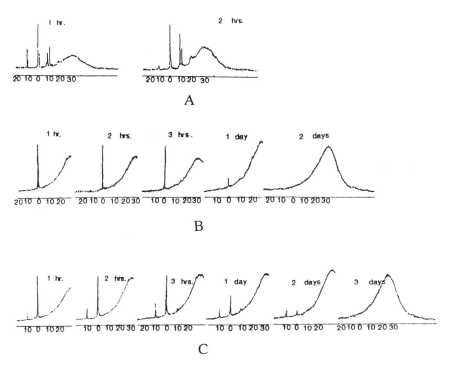

FIG. 1 The ^{29}Si NMR signal for different times after mixing of tetrathoxysilone and water for (A) a solution in ethylene glycol; (B) a microemulsion of ethylene glycol, sodium dodecyl sulfate, octanol, and tolene; and (C) a lamellar liquid crystal of lecithin and ethylene glycol.

comes of reactions in solutions and colloidal systems is due to the presence of an interface between water and nonpolar parts.

In an ethanol solution of water and triethoxysilane, the reaction between water and the ethoxy groups takes place statistically [6], although modified by the thermodynamics [9]. After an initial hydrolysis, the movement of the hydroxytriethoxysilane molecule is not restricted and the continued reaction depends on the probability of a collision with a water molecule. In a colloidal system, on the other hand, the tetraethoxysilane is located within a well-defined hydrophobic space and hydrolysis occurs only when the tetraethoxysilane reaches the interface toward the water region. Hence, the reaction takes place at the interface, changing the molecule from non-polar to polar. The movement of molecule now becomes restricted, and in fact it remains adsorbed at the interface and the hydrolysis continues without interruption to form tetrahydroxysilane [24]. Hence, the multitude of different molecules found in the solvent reaction do not appear. In the microemulsion, the water droplets frequently collide with each other and, in addition, disperse and reform within seconds, hence the condensation proceeds rapidly and $Si(OH)_4$ is not detected spectroscopically (Fig. 1B).

The lamellar liquid crystal is a more static system, with the diffusion of water-soluble compounds in practice limited to two dimensions in the aqueous lamellae. The condensation reaction is, hence, retarded and the $Si(OH)_4$ molecule is a significant part of the total number of compounds (Fig. 1C). The fact that the liquid crystals are relatively static structures has been utilized in an elegant manner by Guizard and his collaborators [25]. The SiO_2 formed retained the shape of the liquid crystal, which consequently served as a template to form materials with new properties.

The reaction rate is also different in a colloidal system, a fact that was used early by Guizard et al. [19] and later by Papoutsi et al. [26]. $Ti(OC_3H_1{}^i)_4$ is an extremely reactive compound in alcohol solutions and several restrictions apply in order to obtain acceptable gels [27]. The utilization of water-in-oil microemulsions alleviates these rate-related problems and excellent gels have been obtained.

It has to be realized that the preexistence of a microemulsion is not a requisite for forming an SiO_2-stabilized gel; the presence of huge amounts of alcohol formed during the hydrolysis causes distinctive changes in the colloidal structure. These changes are well illustrated by the reactions in the water, sodium dodecyl sulfate, pentanol system (Fig. 2). The microemulsion system contains two solutions, one with O/W droplets (A in Fig. 2) and one with W/O droplets (B in Fig. 2). In addition, two liquid crystals are found; one with a lamellar structure (C in Fig. 2) and one composed of micellar cylinders hexagonally close packed (D in Fig. 2).

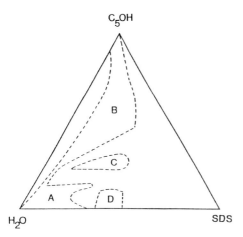

FIG. 2 Microemulsion base system water (H_2O) sodium dodecyl sulfate (SDS), and pentanol (C_5OH). The aqueous micellar solution (A) is connected with the pentanol inverse micellar solution (B) via a narrow channel. In addition, two liquid crystalline phases are found, one lamellar (C) and one (D) consisting of a hexagonal array of cylinders.

The formation of a transparent gel after addition of tetraethoxysilane was not limited to the original one-phase regions. Figure 3 shows that transparency was also found in areas that were of multiphase structure from the beginning. The results show beyond doubt that the reaction products of the hydrolysis and condensation broke the colloidal structures; the anisotropic liquid crystalline phases were changed to an isotropic gel without colloidal order. This is expected; addition of ethanol would lead to a breakdown of any colloidal order, but the potential additional influence of the polymeric silica formed deserves an evaluation. Figure 4 provides the basis for an answer to the question; it compares the region of transparent gel with the area of isotropic solution formed when ethanol was added to an amount equal to that liberated during the reaction. The difference in the amount of water corresponded to the quantity of water needed for the combined hydrolysis and condensation reactions, showing that the formed silica did not have a significant influence on the extent of the region for the transparent gel.

Such is, however, not the case for systems without a surfactant. The combination of pentanol and water (Fig. 5) shows the limit of the transparent gel area to coincide with the solubility region for the system

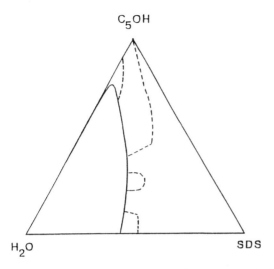

FIG. 3 Area for stable gels (---) in the system in Fig. 2 after addition of 60 wt% tetraethoxysilane based on the weight before addition. The phase diagram reflects the sum of water (H_2O), sodium dedecyl sulfate (SDS), and pentanol (C_5OH) only. The dashed curves show the structures in Fig. 2 and do not represent the phases after addition of tetraethoxysilane.

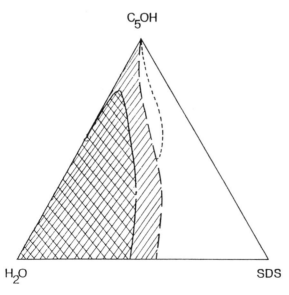

FIG. 4 Comparison of the regions for the isotropic transparent gel (※) in Fig. 3 and for the isotropic solutions formed when an alcohol content corresponding to that formed in Fig. 3 was added (/// + ※).

272

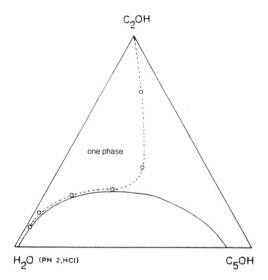

FIG. 5 Phase diagram for water at pH 2 (H$_2$O), pentanol (C$_5$OH), and ethanol (C$_2$OH) (———) and area of transparent gels when silicon tetraethoxide was added to a water/pentanol (C$_5$OH)/sodium dodecyl sulfate (SDS) mixture (- - -). The amount of ethanol in the latter case is that obtained from the hydrolysis of the silicon tetraethoxide; the water is what is left after completed hydrolysis and condensation of the silicon compound.

pentanol-ethanol-water with a high water content. At a low water content the gel was not stable when the water/pentanol ratio was less than a certain value. With too little water, the solvent was too nonpolar to maintain solubility of the silica. Obviously, the surfactant provided the needed polarity.

The influence of the ethanol liberated is also demonstrated in the formation of gels in emulsions. The results are similar (Fig. 6) [28], and the influence of the polarity is demonstrated by the fact that two ethanol molecules were replaced by one water molecule.

III. MICROEMULSION GELS

Microemulsion gels are important per se as an alternative to gelation of microemulsions using collagen [29,30] or lecithin [31]. Microemulsion gels were similar to those obtained by the traditional sol/gel method [1] with an increase in the viscous and elastic response by several orders of magnitude at gelation. The increase in viscous response took place before that in the elastic response, giving a maximum in the loss tangent

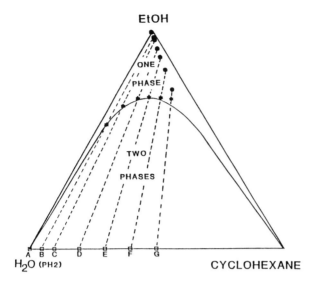

FIG. 6 Solubility region in the system water (H_2O) ethanol (EtOH), and cyclo-hexane (C_6H_{12}). The filled circles (●) and filled stars (★) represent the ethanol liberated according to experimental results for maximum and minimum solubility of Si(OEt)$_4$, respectively. The open squares (□) and dashed lines (- - -) represent the original composition of the water/cyclohexane emulsion and the gelation route of the system.

$$\tan \delta = \frac{G^{11}}{G^1} \tag{3}$$

in which $\tan \delta$ is the loss tangent and G^{11} and G^1 are the loss and elastic moduli. The lack of influence of the precursor structure on the rheological response was confirmed by the fact that applying the scaling theory for bond formation [32] gave results that agreed with random formation of bonds. According to Sakka and Kamiya [33],

$$\eta = C(t - t_g)^{-k} \tag{4}$$

in which η is viscosity, C a sealing coefficient, t time, t_g time for gelation, and k the critical exponent. At a shear rate of 50^{s-1} a value of $k = 1.0$ was obtained, strongly indicating random bond formation.

Nonaqueous microemulsions have also been used to prepare gels in order to evaluate the influence of the polarity of the medium on the prop-erties of the final gel [34]. A broad size distribution of clusters was formed as predicted by both percolation [35–38] and kinetic [39] theories. The actual influence of polarity was found to obey the following rules. For systems with a nonpolar solvent, the amount of smaller, linear clusters is substantial in the later stages of gelation and thus the spatial correlation length changes relatively late in the gelation process [40], whereas for polar solvent systems, the linear moieties are consumed gradually as the larger aggregates form [41]. The most important influence on the viscoelastic properties is the concentration of linear polymers at different stages during the gelation process. A polar solvent produces larger aggregates composed of smaller aggregates and relatively few interstitial linear polymers, which results in a low value of the loss tangent for the lowest frequencies of the shear. Obviously, the resultant large aggregates can rearrange themselves at these frequencies over long time scales, adjusting to the shear field without significant loss. This elastic relaxation is not possible for higher frequencies and shorter time scales, at which the large three-dimensional aggregates undergo loss.

IV. APPLICATIONS

The most elegant application is described in a recent contribution [25] demonstrating the use of liquid crystals as templates to form specific microgeometries of the silica.

The microemulsion method has also been applied in the preparation of titania using water-in-oil microemulsions [42]. Among recent results should be mentioned microemulsion/gel glasses doped with organic laser dyes [43,44], silica-supported spinel $LiMn_2O_4$ [13], combination materials with organic polymers [45], and bioactive materials [46].

It seems reasonable to forecast extensive applications utilizing the specific reactions and structures in microemulsions as well as the possibilities for preparing homogeneous materials with a large number of components.

ACKNOWLEDGMENT

This research was financed in part by the New York State Commission of Science and Technology at the Center for Advanced Materials Processing at Clarkson University, Potsdam, New York.

REFERENCES

1. C. J. Brinker and G. W. Scherer, *Sol/Gel Science: The Physics and Chemistry of Sol-Gel Processing*, Academic Press, New York, 1990.
2. L. C. Klein, *Sol-Gel Technology for Thin Films, Fibers, Preforms, Electronics and Specialty Shapes*, Noyes Publications, Park Ridge, NJ, 1987.

3. L. L. Hench and J. K. West, Chem. Rev. *90*:33 (1990).
4. J. Livage and C. Sanchez, J. Non-Cryst. Solids *145*:11 (1992).
5. R. Roy, Science *238*:1664 (1987).
6. B. D. Kay and R. A. Assink, Mater. Res. Symp. Proc. *107*:35 (1988).
7. H. Yang, Z. Ding, Z. Jiang, and X. Xu, J. Non-Cryst. Solids *112*:449 (1988).
8. J. C. Pouxviel and J. P. Boilot, J. Non-Cryst. Solids *94*:374 (1987).
9. J. Sanchez and A. McCormick, J. Phys. Chem. *96*:8973 (1992).
10. J. O. Saeten, H. Fördedal, T. Skodvin, J. Sjöblom, A. Amran, and S. E. Friberg, J. Colloid Interface Sci. *154*:167 (1992).
11. J. Sjöblom, R. Skurtveit, J. O. Saeten, and B. Gestblom, J. Colloid Interface Sci. *141*:329 (1991).
12. J. O. Saeten, J. Sjöblom, and B. Gestblom, J. Phys. Chem. *95*:1449 (1991).
13. B. Ammundsen, G. R. Burns, A. Amran, and S. E. Friberg, J. Sol-Gel Synth. Proc. *4*:23 (1995).
14. S. E. Friberg, A. Amran, and J. Sjöblom, J. Dispers. Sci. Technol., *16*:31 (1995).
15. S. E. Friberg, J. Yang, J. Sjöblom, and G. Farrington, J. Phys. Chem., *98*: 13528 (1994).
16. M. Dubois and B. C. Abane, Macromolecules *22*:2526 (1989).
17. J. Sjöblom, H. Ebeltoft, A. Bjorseth, S. E. Friberg, and C. Brancewicz, J. Dispers. Sci. Technol. *15*:21 (1994).
18. J. Sjöblom et al., Langmuir, in press.
19. C. Guizard, M. Stitou, A. Larbot, L. Cot, and J. Rouviere, in *Better Ceramics Through Chemistry III*, Mater. Res. Symp. Proceedings, Vol. 121, Material Research Society, Pittsburg, 1988, p. 115.
20. S. E. Friberg, C. C. Yang, in *Innovations in Materials Processing Using Aqueous, Colloid, and Surface Chemistry* (F. M. Doyle, S. Raghaven, P. Somasundaran, and G. W. Warrend, eds.), The Minerals, Metals and Materials Society, Wanendale, PA, 1988, p. 181.
21. S. Marteluccim and A. N. Chester, eds., *Progress in Microemulsion*, Plenum, New York, 1989.
22. D. Langevin, Accounts Chem. Res. *21*:255 (1988).
23. U. Pfüller, *Mizellen und Vesikeln*, Springer-Verlag, Berlin, 1986.
24. S. E. Friberg and Z. Ma, J. Non-Cryst. Solids *147* & *148*:30 (1992).
25. T. Dabadie, A. Ayral, C. Guizard, L. Cot, C. Lurin, W. Nie, and D. Rioult, J. Sol-Gel Technol., in press.
26. D. Papoutsi, P. Lianos, P. Yianoulis, and P. Koutsoukos, Langmuir *10*:1684 (1994).
27. D. Aunir, V. R. Kaufman, and R. Reisfeld, J. Non-Cryst. Solids, *74*:395 (1985).
28. S. E. Friberg and C. C. Yang, Particulate Sci. Technol. *11*:1 (1993).
29. G. Haering and P. L. Luisi, J. Phys. Chem. *90*:5892 (1986).
30. H. F. Eicke, C. Quellet, and G. Xu, J. Surf. Sci. Technol. *4*:2 (1988).
31. P. Schurtenberger, L. J. Magid, P. Lindner, and P. L. Luisi, Prog. Colloid Polym. Sci. *89*:274 (1992).

32. P. G. deGennes, in *Scaling Concepts in Polymer Physics*, Cornell University Press, Ithaca, NY, 1979.
33. S. Sakka and K. Kamiya, J. Non-Cryst. Solids 48:31 (1982).
34. S. E. Friberg, S. M. Jones, A. Motyka, and G. Broze, J. Mater. Sci. 29:1753 (1994).
35. D. W. Schaefer and K. D. Keefer, Phys. Rev. Lett. 53:1383 (1984).
36. W. Hess, T. A. Vilgis, and H. H. Winter, Macromolecules 21:2536 (1988).
37. T. Kojima, T. Kurotu, and T. Kawaguchi, Macromolecules 19:1284 (1986).
38. E. Bouchard, M. Delsanti, M. Adam, M. Daoud, and D. Durand, J. Phys. 47:1273 (1986).
39. J. E. Martin, J. Wilcox, and D. Adolf, Phys. Rev. A 36:1803 (1987).
40. M. Dubois and B. Cabane, Macromolecules 22:2526 (1989).
41. S. M. Jones and S. E. Friberg, J. Dispers. Sci. Technol. 13:6 (1992).
42. C. Guizard, A. Larbot, L. Cot, S. Peres, and J. Rouvière, J. Chim. Phys. 87:1901 (1990).
43. S. M. Jones and S. E. Friberg J. Non-Cryst. Solids 181:39 (1995).
44. S. E. Friberg, S. Jones, and J. Sjöblom, J. Mater. Syn. Proc. 2:29 (1994).
45. S. E. Friberg, C. C. Yang, M. B. Biscoglio, and H. F. Helbig, J. Mater. Sci. 11:1373 (1992).
46. S. M. Jones, S. E. Friberg, and J. Sjöblom, J. Mater. Sci. 29:4075 (1994).

14

Microemulsions in Enhanced Oil Recovery: Middle-Phase Microemulsion Formation with Some Typical Anionic Surfactants

MASAHIKO ABE   Faculty of Science and Technology, Science University of Tokyo, Chiba, Japan

I. INTRODUCTION

The application of microemulsion systems is of industrial and practical importance for enhanced oil recovery [1,2]. Microemulsions are homogeneous mixtures of hydrocarbons and water with large amounts of surfactants [3] and are thermodynamically stable isotropic single phases [4]. Qutubuddin et al. [5,6] have shown that the surfactant molecules are mostly located at the interface between the domains of polar and nonpolar fluids. In most surfactant formulations involving microemulsions, an alcohol and/or other amphiphilic cosurfactant is used in combination with the primary ionic surfactant [7–9]. The most fundamental role of alcohol is that of destroying liquid crystalline and/or gel structures that disturb the formation of a microemulsion. In fact, Winsor-type phase behavior [8,10,11] in ionic surfactant/oil/water systems could never be detected at a low temperature in the absence of alcohols.

The phase behavior of microemulsions clearly shows that a variety of phases can exist in equilibrium with one other. While in equilibrium, each phase will involve a different structure. The commonly observed Winsor-type systems indicate that the microemulsions can exist in equilibrium with excess oil, excess water, or both (Fig. 1). A Winsor-I-type system consists of a lower-phase microemulsion and excess oil; a Winsor-II-type system consists of an upper-phase microemulsion and excess water. The factors that affect the transitions between different types of systems include the temperature, the salinity, the molecular structure of surfactant and cosurfactant, the nature of the oil, and the oil-to-water ratio [12]. For example, the transition (lower → middle → upper) shown in Fig. 1 takes place with increasing salinity, cosurfactant concentration, surfactant concentration, molecular weight of surfactant, or brine/oil ratio but with decreasing tem-

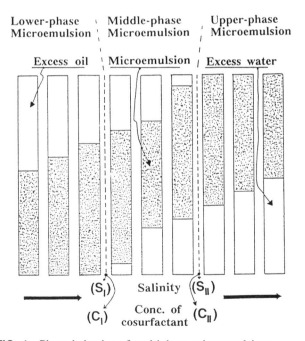

FIG. 1 Phase behavior of multiphase microemulsions.

perature or oil chain length. Under adequate conditions, the microemulsion system is miscible with both oil and brine. However, beyond its solubilization limit, the microemulsion system partitions into three phases, namely a surfactant-rich middle phase, a surfactant-poor brine phase, and an oil phase. This surfactant-rich middle phase (middle-phase microemulsion [13] is called a Winsor-II-type microemulsion. The systems of most interest are generally located in the neighborhood of this region. The lower-phase and the upper-phase microemulsions have structures in which droplets (dispersed phase) disperse in a continuous phase; the former microemulsion is an oil-in-water (O/W) type, and the latter is a water-in-oil (W/O) type. The structure of the middle-phase microemulsion, however, is presently a matter of controversy [14–18]. Clarifying the properties of the latter is more difficult than of the other microemulsions.

This chapter deals specifically with the preparation and properties of a middle-phase microemulsion formed with some typical anionic surfactants.

II. PHASE BEHAVIOR

Winsor [19] introduced a unifying concept that has proved useful in correlating many experimental observations. His primary concept is that the R ratio of cohesive energies, stemming from the interaction of the interfacial layer with oil, divided by the energies resulting from the interaction with water, determines the curvature of the interface. This concept (R theory), which is qualitative in nature, is based on the relative magnitude of the interaction energies (or cohesive energies in the terminology of Hildebrand and Scott [20]) of the amphiphile (C) with oil (O) and with water (W):

$$R = \frac{A_{CO}}{A_{CW}} \tag{1}$$

This concept has been advanced by Verzaro et al. [21], who developed the equation

$$R = \frac{A_{CO} - A_{OO} - A_{LL}}{A_{CW} - A_{WW} - A_{HH}} \tag{2}$$

Here R is the R ratio introduced by Winsor; A_{CO} and A_{CW} are the molecular interactions promoting miscibility between C and O or C and W, in other words, the cohesive energies; A_{OO}, A_{LL}, A_{WW}, and A_{HH} are the lipophilic- and hydrophilic-type contributions promoting segregation of the components as separate phases.

Most microemulsion systems include a high concentration of a cosurfactant, such as alcohol, to destroy any liquid crystalline materials [2]. Equation (2) can be rewritten in that case in the form

$$R = \frac{(A_{C1O} + A_{C2O}) - A_{OO} - (A_{L1L1} + A_{L1L2} + A_{L2L2})}{(A_{C1W} + A_{C2W}) - A_{WW} - (A_{H1H1} + A_{H1H2} + A_{H2H2})} \tag{3}$$

where C1 is the surfactant and C2 the cosurfactant, L1 and L2 are the hydrophobic portions, and H1 and H2 are the hydrophilic portions.

When the lipophilic-type contribution of the system increases (the numerator becomes larger than the denominator), a W/O-type microemulsion results. When the hydrophilic-type contribution of the system increases (the numerator becomes smaller than the denominator), an O/W-type microemulsion results. When the hydrophile and lipophile tendencies of the system are equilibrated, R is equal to 1. This case corresponds especially to the Winsor-III-type microemulsion optimal system, in which the volumes of oil and water solubilized in the middle phase, which exists in equilibrium with both oil and water, are equal; the so-called middle-phase microemulsion forms. Even at $R = 1$, a lamellar structure (liquid crystalline and/or gel) is formed by more or less regular arrangement of the surfactant molecules in parallel leaflets allowing alternate solubilization of oil and water [19]. In order to form a middle-phase microemulsion with a water-soluble surfactant (whose A_{CW} is quite large), the property of the system must become hydrophobic by changing the factors (salinity, temperature, etc.) that affect the R ratio of the system. For example, when the salinity increases, the ionic strength of aqueous solutions increases, this disturbs the dissociation of the counterions of an ionic surfactant and makes surfactant molecules more hydrophobic; as a result, the decrease in A_{CW} and the increase in A_{CO} occur simultaneously. The temperature rise results in an increase in the solubility of surfactant in water; the A_{CW} increases and the A_{CO} decreases simultaneously; therefore the amount of an inorganic electrolyte (salinity) must be increased to form a middle-phase microemulsion.

A. Effect of Salinity on the Phase Behavior

A Winsor-type phase transition takes place in a sodium alkyl sulfate/*n*-alkane/cosurfactant/brine system with increasing salinity [22]. When the alkyl chain length in surfactants (SCN) decreases, the salinity range in which the middle-phase microemulsion forms increases.

The optimum salinity for middle-phase microemulsion formation increases with increasing temperature and/or alkyl chain length in *n*-alkanes (alkane carbon number, ACN), but it decreases with increasing SCN and/

or alkyl chain length in the cosurfactant (cosurfactant carbon number, CCN).

The solubilization parameter at the optimum (equal volumes of oil and water in the middle-phase microemulsion) increases with an increase in SCN or CCN but decreases with an increase in ACN. Here, the solubilization parameter at the optimum is commonly defined as the volume of oil (or water) per unit volume of surfactant and is denoted as σ^*:

$$\sigma^* = 0.5\frac{V_m}{V_s} \tag{4}$$

where V_m is the volume of middle-phase microemulsion and V_s is the amount of surfactant in the middle-phase microemulsion.

B. Effects of SCN and ACN on the Phase Behavior

Table 1 shows the effects of SCN and of ACN on middle-phase microemulsion formation in a sodium alkyl sulfate/alkane/n-hexanol (4 wt%)/brine system. When ACN − SCN < 3, the middle-phase microemulsion is formed. For ACN-SCN > 4, however, it is not formed even at 17 wt% or greater salinity.

Israelachvili and Pashley [23] have proposed that the hydrophobic interaction free energy (G_H) between two hydrophobic particles is given by

$$G_H \simeq -20\frac{R_1 R_2}{R_1 + R_2}e^{-D/D_0} \quad (\text{kcal mol}^{-1}) \tag{5}$$

TABLE 1 Effect of SCN and ACN on Middle-Phase Microemulsion Formation in a Sodium Alkyl Sulfate/n-Alkane/n-Hexanol/Brine System at Various Temperatures (25–35°C)[a]

	SCN					
ACN	7	8	9	10	11	12
8	○	○	○	○	○	○
9	○	○	○	○	○	○
10	○	○	○	○	○	○
11	●	○	○	○	○	○
12	●	●	○	○	○	○
13	●	●	●	○	○	○
14	●	●	●	●	○	○

[a]○ denotes that a middle-phase microemulsion is formed; ● denotes that a middle-phase microemulsion is not formed.

where R_1 and R_2 are the radii (nm) of two unequal spheres, D is the distance (nm) between two particles, and D_0 is the distance at which G_H becomes equal to $1/e$.

If the hydrophobic portions of surfactant molecules and oil have radii R_1 and R_2, respectively, a decrease in ACN and SCN would lead to an increase in G_H. This means that the hydrophobic interaction between the hydrophobic portions of surfactant molecules and oil becomes small. Therefore, in the case ACN − SCN > 4, when using an alkane containing long alkyl chains and a surfactant having short alkyl chains, the interaction between oil and surfactant molecules (A_{CO}) becomes smaller than that between oil molecules (A_{OO}); the numerator of Eq. (3) becomes smaller than the denominator, so R cannot attain the value 1.

When SCN is rather large (for instance, C_{18}), a single-phase microemulsion without excess is observed. The R ratio of the system will be dependent not only on the balance of hydrophilic-hydrophobic interaction but also on the magnitudes of the respective strengths. In the case of an ionic surfactant containing the same hydrophilic groups, such as sodium alkyl sulfate, the magnitude of the respective strengths is larger for the surfactant that is able to form a middle-phase microemulsion at lower salinity than for that which can form it only at higher salinity. The effect of electrolyte on the dissociation of counterions is lower for long ACN than for short ACN, so the value of A_{CW} becomes large and the value of A_{CO} also becomes large.

III. PROPERTIES OF MULTIPHASE MICROEMULSIONS

The salinity range in which the middle-phase microemulsion forms becomes wider with decreasing SCN. Therefore, in order to investigate in detail the phase behavior with increasing salinity, we prepared a microemulsion with sodium octyl sulfate (SOS) [18].

Figure 2 depicts the changes in the volume fractions of the SOS/n-hexanol/n-decane/brine multiphase microemulsions with salinity. As can be seen in Fig. 2, Winsor type I, which is the two-phase region composed of a lower-phase microemulsion and excess oil, appears at salinity values lower than 10.3%. Winsor type II, which is two-phase region comprising an upper-phase microemulsion and an excess water phase, forms at salinity values higher than 15.7%. In the salinity region between 10.3% (S_I) and 15.7% (S_{II}), Winsor type III, which is the three-phase region containing a middle-phase microemulsion, is observed.

At the two characteristic salinities (S_I and S_{II}) where Winsor I ↔ Winsor III ↔ Winsor II phase transitions take place, the phase volume of the

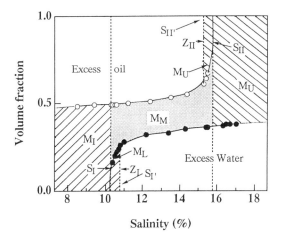

FIG. 2 Volume fractions of each phase in an SOS (2% by weight)/n-hexanol (4%)/n-decane (47%)/brine (47%) multiphase microemulsion system as a function of salinity at 35°C. M_L represents the lower-phase microemulsion, M_M the middle-phase microemulsion, and M_U the upper-phase microemulsion. S_I and S_{II} are the characteristic salinities where Winsor I ↔ Winsor III and Winsor III ↔ Winsor II phase transitions occur. Z_I and Z_{II} are the salinity zones where two different types of microemulsion coexist: from S_I to $S_{I'}$ and from $S_{II'}$ to S_{II}. (○) Microemulsion phase/excess oil phase interface; (●) microemulsion phase/excess water phase interface.

middle-phase microemulsion attains a maximum. At some intermediate salinity between S_I and S_{II} (optimum salinity for middle-phase microemulsion formation), however, the phase volume attains a minimum.

It is generally thought that a phase transition which is Winsor I type to Winsor III type and/or Winsor III type to Winsor II type occurs suddenly at the respective optimum salinity. In the system including an anionic surfactant having long alkyl chains (for example, sodium dodecyl sulfate), the region of the transition states of microemulsions with changing salinity is not enough wide to be detected.

One should note that two different types of microemulsion are in contact with each other in the vicinity of the characteristic salinities (S_I and S_{II}). A lower-phase and a middle-phase microemulsion (M_L and M_M) coexist at the salinity zone (Z_I) from 10.3% (S_I) to 10.7% ($S_{I'}$). A middle-phase and an upper-phase microemulsion (M_M and M_U) coexist at the salinity zone (Z_{II}) from 15.3% ($S_{II'}$) to 15.7% (S_{II}). Similar behavior is recognized in the report by Kunieda and Shinoda [24].

We measured the diffusion coefficient of each microemulsion with the salinity zones (Z_I and Z_{II}) where two different types of microemulsions coexist.

Figure 3 presents the reciprocal of the diffusion coefficient (D^{-1}) as a function of the magnitude of the scattering vector (q),

$$q = \left(\frac{4\pi n}{\pi\lambda}\right) \sin(\Theta/2) \tag{6}$$

where n is the refractive index of the solution, λ the wavelength of light, and Θ the scattering angle. The D^{-1} value of a spherical structure without interparticle interactions does not depend on q, but D^{-1} values of spherical structures with interparticle interactions and of nonspherical ones depend on q [25,26]. It can be seen in Fig. 3a that scattering from any type of multiphase microemulsion at which the salinity is very different from the two characteristic salinities (S_I and S_{II}) shows no dependence on the scattering vector. In Fig. 3b, however, as the characteristic salinities are approached, the D^{-1} values of Winsor-I- and Winsor-II-type systems come to depend on q. Even within the three-phase region, as the salinity moves away from the optimum salinity (13.0%) for middle-phase microemulsion

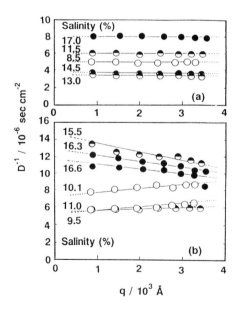

FIG. 3 Reciprocal of the diffusion coefficient of each microemulsion (D^{-1}) as a function of the scattering vector (q) at 35°C. ($\bigcirc$) M_L; ($\ominus$) M_M, ($\bullet$) M_U.

formation, a dependence of D^{-1} on q is found. This may be due to increasing interparticle interactions.

Figure 4 demonstrates the diffusion coefficients of each microemulsion versus salinity at 35°C. There exists a distribution of diffusion coefficients among the values obtained from dynamic light scattering (DLS) measurements, so that we adopted the z-average values as representative values of the distribution. As can be seen in Fig. 4, the D value initially decreases with increasing salinity until S_I, where the phase transition (type I ↔ type III) occurs; beyond this point, the value increases and attains a maximum through the optimum salinity, decreases again, and reaches a second minimum at S_{II}, where the phase transition (type III ↔ type II) takes place, and increases linearly with increasing salinity.

Figure 5 presents the hydrodynamic radii (R_H) of each microemulsion with changing salinity. Here, to obtain the hydrodynamic radii of middle-phase microemulsions, the viscosity value of the continuous phase in the microemulsions is taken to be 0.76 cp. This value is an average of water and n-decane viscosity values (0.72 and 0.80 cp, respectively). The apparent R_H value initially increases with increasing salinity until S_I; beyond this point, the value decreases and attains a minimum near the optimum salinity for middle-phase microemulsion formation again, reaches a maximum at S_{II}, and decreases with further increasing salinity. A similar tendency is evident for sodium decyl sulfate and sodium dodecyl sulfate results.

The hydrodynamic radii of the middle-phase microemulsion near the optimum salinity become greater with increasing ACN.

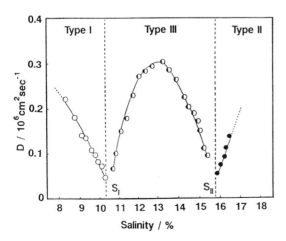

FIG. 4 Diffusion coefficients of each microemulsion as a function of salinity at 35°C. (○) M_L; (◑) M_M; (●) M_U.

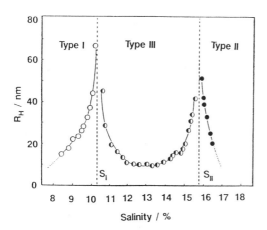

FIG. 5 Hydrodynamic radii of each microemulsion as a function of salinity at 35°C. (○) M_L; (◑) M_M; (●) M_U.

As mentioned earlier, there exist drastic changes in properties at the characteristic salinities (S_I and S_{II}); how these data change differs from one to the other depending on each Winsor type. It can be postulated that microemulsions having different structures would exist in the multiphase microemulsion system.

A large amount of water exists in the lower-phase microemulsion, while the amount contained in the upper-phase microemulsion is very small.

The polarity sensed by pyrene molecules (fluorescence probe) will be influenced by the moisture (polar solvent) content in the microemulsion. The fluorescence intensity ratio (I_1/I_3) of the pyrene dissolved in decane is large in the lower-phase but small in the upper-phase microemulsions. In addition, since the electrical conductivities in the lower-phase microemulsion increase with increasing salinity and those in the upper-phase microemulsion are very low and almost independent of salinity, the continuous phase in the lower- and/or upper-phase microemulsion will be water (brine) and/or oil. This suggests that the structures of the lower- and upper-phase microemulsions will be O/W type (oil droplets dispersing in water) and W/O type (water droplets dispersing in oil).

In the three-phase region, the moisture content decreases with increasing salinity and attains a 50% water content at the optimum salinity for middle-phase microemulsion formation [18]. However, the fluorescence intensity ratio (I_1/I_3) in it is not proportional to the amount of water and displays a minimum at the optimum salinity. This means that the structure of the middle-phase microemulsion will be different from that of other microemulsions. Even within the three-phase region, the moisture content and the fluorescence intensity ratio change with increasing salinity. With increasing salinity, the moisture content in the middle-phase microemulsion decreases, but the I_1/I_3 ratio in it is not proportional to the amount of water. Furthermore, the electrical conductivity of the microemulsion decreases with increasing salinity in the three-phase region. We suggest that the middle-phase microemulsion structure does not always remain constant; the middle-phase microemulsion formed at the optimum salinity will have a more regular structure than that at lower and/or higher salinities within the three-phase region. Furthermore, we have been able to measure the diffusion coefficient of the middle-phase microemulsion. Its particle size corresponds to the size due to apparent turbidity; the middle-phase microemulsion in the three-phase region will have the character of an assembly of droplets, in analogy with the lower- and/or upper-phase microemulsion.

One of the most characteristic properties of multiphase microemulsion systems is an ultralow interfacial tension between oil and water. Kunieda and Shinoda [27] have obtained approximate values of the interfacial tensions between oil and water (γ_{o-w}) by use of the values between the microemulsion phase and the excess oil phase (γ_{o-m}) [and/or between the microemulsion phase and the excess water phase (γ_{w-m})] in some multiphase microemulsion systems. In accordance with this approximation, we can regard γ_{o-w} in two-phase regions as γ_{o-m} and/or γ_{w-m}. Furthermore, γ_{o-w} in a three-phase region can be also taken as $\gamma_{o-m} + \gamma_{w-m}$. As a result, γ_{o-m} decreases with an increase in salinity and reaches zero at the characteristic salinity, S_{II}; γ_{w-m} decreases with decreasing salinity and attains zero at S_I. Consequently, γ_{o-w} becomes a minimum at the optimum salinity for middle-phase microemulsion formation.

The changes in the interfacial tensions observed in the three-phase region, γ_{o-m} and γ_{w-m}, would be mainly dependent on the interfacial tensions between the continuous phase of the middle-phase microemulsion and the excess phases. This is because the surfactant contents in the microemulsion hardly change. The decrease in γ_{o-m} with increasing salinity will be due to enhancement of cohesive forces between the middle-phase microemulsion and the excess oil, owing to an increase in the oil fraction in the middle-phase microemulsion with increasing salinity. On the other hand, the decrease in interfacial tension between water and middle-phase microemulsion with decreasing salinity will be caused by an increase in the water fraction in the middle-phase microemulsion with decreasing salinity.

In the salinity zones Z_I and Z_{II} where two microemulsion phases (a middle-phase microemulsion and a lower- (and/or an upper-) phase microemulsion) can exist, there will be positive interfacial tensions, although very low ones, between the two microemulsion phases. Thus, we consider that positive interfacial tensions mean the existence of different continuous phases between the middle-phase microemulsion and the lower- (and/or upper-) phase microemulsion. In other words, the medium phase of the middle-phase microemulsion will be neither water nor oil.

This suggestion is illustrated by the liquid-liquid dispersion model for these systems shown in Fig. 6. The figure shows the middle-phase microemulsion, which includes oil and/or water droplets as a dispersed phase as well as the lower and/or the upper phase one. The dispersion medium of the middle-phase microemulsion will be a mixed dispersion medium, which is a mixture of oil, water, and surfactant (cosurfactant). This dispersion medium may be formed due to a partial fracture of oil-water interfaces in the middle-phase microemulsion caused by stretching motions of the interfaces. Surfactant molecules in the middle-phase microemulsion may enhance the affinity between oil and water and thus facilitate their mixing. Furthermore, the fractions of oil and/or water in the middle-phase microemulsion would be factors that characterize its structure. The middle-phase microemulsion in the vicinity of the optimum salinity contains nearly the same amounts of oil and of water; that is, the oil/water ratio in the mixed dispersion medium would also be unity.

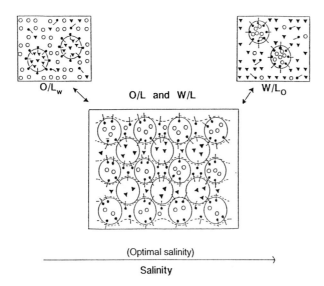

FIG. 6 Schematic illustration of models of microemulsion structures in a three-phase region. (○) Water; (▼) oil; (●) surfactant (cosurfactant); L, mixed dispersion medium; L_W, water-rich mixed dispersion medium; L_O, oil-rich mixed dispersion medium.

In this case, consequently, oil and/water droplets would disperse in the mixed dispersion medium (L). On the other hand, when the salinity is lower than the optimum salinity, the middle-phase microemulsion would have a water-rich mixed dispersion medium (L_W) and oil droplets as a dispersed phase. In contrast, at increased salinity, the middle-phase microemulsion would have an oil-rich dispersion medium (L_O) and water droplets.

IV. PARTITION OF SURFACTANTS IN MULTIPHASE MICROEMULSIONS AND PHASE EQUILIBRIUM RATE

Table 2 presents the effect of SCN on the partition of anionic surfactant in the excess water phase, middle-phase microemulsion, and excess oil phase [28]. It can be seen that most anionic surfactant exists in the middle-phase microemulsion, but a few surfactant molecules distribute into the excess water and oil phases. As SCN increases, the surfactant contents in

TABLE 2 Effect of SCN on Distribution Concentration and Partition Ratio of Anionic Surfactant at Optimum Salinity for Sodium Alkyl Sulfate (Surfactant)/n-Decane/n-Hexanol/Brine System at 35°C[a]

Phase	Lower	Middle	Upper
SCN = 8 system			
Phase volume (mL)	3.38	1.68	4.60
Concentration of surfactant (mol/L)	5.24×10^{-3}	3.96×10^{-1}	3.51×10^{-4}
Amount of surfactant (g)	4.12×10^{-3}	0.155	3.76×10^{-4}
Partition ratio (%)	2.58	97.2	0.236
SCN = 10 system			
Phase volume (mL)	2.00	4.05	3.25
Concentration of surfactant (mol/L)	1.00×10^{-3}	1.51×10^{-1}	5.59×10^{-4}
Amount of surfactant (g)	5.22×10^{-4}	0.159	4.75×10^{-4}
Partition ratio (%)	0.326	99.4	0.297
SCN = 12 system			
Phase volume (mL)	0.60	6.90	1.85
Concentration of surfactant (mol/L)	6.16×10^{-4}	8.02×10^{-2}	4.53×10^{-4}
Amount of surfactant (g)	1.07×10^{-4}	0.160	2.42×10^{-4}
Partition ratio (%)	0.0668	99.8	0.151

[a]Prepared amount of the anionic surfactant is 0.160 g.

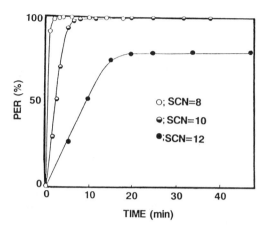

FIG. 7 Effect of SCN on the phase equilibrium rate of middle-phase microemulsion for the sodium alkyl sulfate/n-decane/1-hexanol/brine system at 35°C.

the middle-phase microemulsion increase. On the other hand, with increasing ACN, the sodium dodecyl sulfate (SDS) contents in the middle-phase microemulsion decrease slightly.

The effects of SCN and ACN on the phase equilibrium rate of a middle-phase microemulsion formation are shown in Figs. 7 and 8 [29]. Here, the phase equilibrium rate (PER) is given by

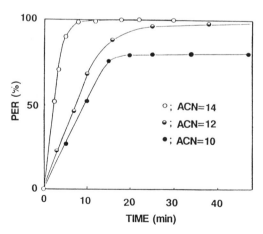

FIG. 8 Effect of ACN on the phase equilibrium rate of middle-phase microemulsion for the sodium alkyl sulfate/n-decane/1-hexanol/brine system at 35°C.

$$\text{PER} = \left(1 - \frac{V_t - V_e}{V_0 - V_e}\right) \times 100 \tag{7}$$

where V_0 is the volume of the system, V_t the volume of the middle-phase microemulsion at a given time (t), and V_e the volume of the middle-phase microemulsion at which equilibrium had been established. As can be seen in Figs. 7 and 8, the time at which equilibrium is attained increases with an increase in SCN or with a decrease in ACN. This may be attributed to the fact that the phase equilibration rate is closely related to the removal rate of excess solubilized water and/or oil in the middle phase; that is, the removal rate increases with a decrease in the magnitude of individual strengths of hydrophilic and hydrophobic interactions at the oil-water interface.

V. THERMODYNAMICS OF MICROEMULSION FORMATION

Important features of microemulsions include their thermodynamic stability and high solvent power [30]. Many thermodynamic studies have appeared which consider the stability of microemulsions. Rehbinder [31] was the first to recognize that a microemulsion can be stabilized by the small entropy increase resulting from the dispersion of one bulk phase into another, provided the interfacial tension is small enough that the net decrease in free energy due to dispersion overcomes the positive contribution resulting from the increase in the interfacial area. Ruckenstein and Krishnan [32] have extended the analysis to include the adsorption of both surfactant and alcohol, the change in chemical potential of the adsorbing species due to their changed solution concentrations, and the free energy of the drop dispersion. Lam et al. [33] have described the bending energy associated with the curvature of the interfacial zone separating oil from water regions in a micellar solution (microemulsion) as an essential feature governing the stability and phase behavior. However, these methods are still indirect.

We tried to measure directly the thermodynamic parameters of microemulsion formation with a calorimeter of the twin-conduction type [34]. Figure 9 demonstrates the changes in heat of microemulsion with salinity at 30°C. Here, open circles represent the enthalpy changes at the phase transition, which is macroemulsion to microemulsion in every system, and solid circles denote the heats of mixing of a surfactant-free system. It can be seen in Fig. 9 that the apparent heat of microemulsion formation initially increases with increasing salinity until the phase transition (10.3%) occurs; beyond this region, the value decreases and attains a minimum at the optimum salinity, increases again and reaches a second maximum at the sec-

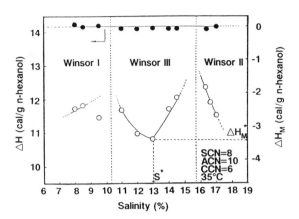

FIG. 9 Changes in the heat of microemulsion formation with salinity at 30°C.

ond phase transition (15.7%), and decreases with further increasing salinity. On the other hand, the heat of mixing of the system which is without surfactant is independent of salinity and remains at high values. We consider that the true heat of microemulsion formation (ΔH_M, right-hand side) is the difference between the enthalpy change of the phase transition, which is macroemulsion to microemulsion, and the heat of mixing. In other words, the true heat of microemulsion formation is exothermic, and the value (negative) is a maximum at the optimal salinity. A similar tendency is evident for other SCN and ACN results.

Figures 10 and 11 show the effect of SCN and/or ACN on ΔH_M^* (ΔH_M at the optimal salinity) at 30°C. The $-\Delta H_M^*$ value increases with increasing SCN and with decreasing ACN. As mentioned before, with increasing SCN, the solubilization parameter and particle size increase but the interfacial tension decreases; with increasing ACN, however, the solubilization parameter and particle size decrease but the interfacial tension increases. We consider that the value of ΔH_M^* (negative) is closely related to the values of the solubilization parameter, interfacial tension, and particle size. That is, surfactants having large ΔH_M^* (negative) values display large strengths at the oil-water interface, the interfacial tension becomes small, and the solubilization parameter is increased; as a result, the particle size becomes larger.

A Winsor-type phase transition will also take place with increasing amounts of cosurfactant. The details of the experiment are given in Ref. 32.

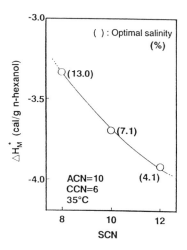

FIG. 10 Effect of SCN on the true heat of middle-phase microemulsion formation at the optimal salinity at 30°C.

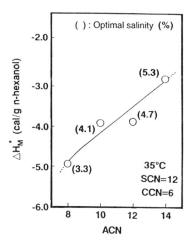

FIG. 11 Effect of ACN on the true heat of middle-phase microemulsion formation at the optimum salinity at 30°C.

Figure 12 exhibits the temperature difference between sample and reference cells with increasing cosurfactant concentration at 30°C. Here, ΔT denotes the temperature difference between the heat of microemulsion formation (heat of mixing with surfactant) and the heat of mixing without surfactant. It can be seen in Fig. 12 that ΔT becomes exothermic near the cosurfactant concentration at which the phase transition (Winsor II type to Winsor III type) occurs, that is, the concentration at which middle-phase microemulsion forms, and then becomes endothermic at the cosurfactant concentration at which the phase transition (Winsor III type to Winsor II type) takes place. A similar tendency is recognized in results for other SCN and ACN samples. The value of ΔT per time unit is closely related to ΔH per time unit and is given by

$$\frac{d(\Delta H)}{dt} = C_P\frac{d(\Delta T)}{dt} + K\,\Delta T \tag{8}$$

where $d(\Delta H)/dt$ is the change in enthalpy of microemulsion formation with time, C_P is the heat capacity, and K is a constant. In order to evaluate the middle-phase microemulsion formation due to different SCN thermodynamically, we adopt $d(\Delta H)^*/dt$, which is $d(\Delta H)/dt$ at the optimum cosurfactant concentration.

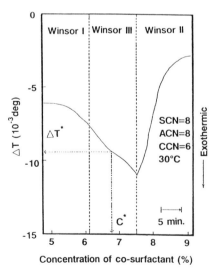

FIG. 12 Changes in temperature difference between sample and reference cells with increasing cosurfactant at 30°C.

The quantities $-\Delta T^*$ and $-d(\Delta H)^*/dt$ increase with SCN and CCN, but these decrease with increasing ACN. Typical data are shown in Fig. 13. Hence the heat of middle-phase microemulsion, $-d(\Delta H)^*/dt$, at the optimal cosurfactant concentration increases with increasing SCN and CCN but with decreasing ACN.

VI. COMPUTER SIMULATION FOR MULTIPHASE MICROEMULSION FORMATION

Surfactant molecules are amphiphilic substances consisting of hydrophilic (polar base) and hydrophobic (hydrocarbon chain) portions, and they are thermodynamically difficult to disperse in water against the cohesiveness of water molecules. The hydrophobic portion tends to coagulate due to its mutual effect of hydrophobicity. However, the hydrophilic base portion becomes thermodynamically stable because it makes contact with the water molecules.

A microemulsion belongs to the solubilization system, as aforementioned, and it is considered to be in a form in which swollen micelle is

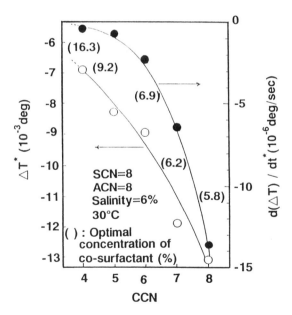

FIG. 13 Effect of CCN on ΔT^* and $d(\Delta H)^*/dt$ at 30°C.

dispersed in a thermodynamically stable condition. Therefore, the theory of Israelachivili et al. [35] regarding the size of water molecule aggregation can be applied. In other words, surfactant molecules in an equilibrium condition in the water have equal chemical potential regardless of the condition of aggregation. Hydrophobic bound free energy that acts between surfactant molecules is expressed as the product (γ_a) of interfacial free energy (γ) between water and liquid hydrocarbon and the cross-sectional area (a) occupied by surfactant molecules at the oil-water interface. The repulsion is expressed by C/a (where C is constant). Thus, the free energy $(\mu_N{}^0)$ per surfactant molecule at the oil/water interface is expressed in Eq. (9),

$$\mu_N{}^0 = \gamma_a + \frac{C}{a} \tag{9}$$

Generally, in the case of an ionic surfactant, its hydrophilic group is electrically charged because the counterion is dissociated. However, when the bulk ion concentration increases its central electric charge is weakened and the ionic atmosphere becomes thin; thus, the area per hydrophilic group decreases. Consequently, if the balance between the hydrophobic and hydrophilic interactions changes, as previously described, the type of microemulsion produced (O/W or W/O type) is assumed to be decided. If the curvature of a microemulsion can be ignored, the distance between the hydrophilic groups can be assumed to be equal to the condenser (near the flat plate) at the Debye length $(1/\kappa;$ thickness of dispersion electricity double layer), then Eq. (9) becomes Eq. (10).

$$\mu_N{}^0 = \gamma a + \frac{e}{Z\varepsilon_0\varepsilon_r a\kappa} \tag{10}$$

where κ denotes the Debye-Hückel parameter, which varies depending on the concentration of saline of the whole group. In a one-to-one electrolyte such as NaCl, for example, κ can be expressed as in Eq. (11).

$$\kappa^2 = \frac{2000N_A e^2 c}{\varepsilon_0\varepsilon_r kT} \tag{11}$$

where N_A denotes the Avogadro constant (6.02×10^{23}), e denotes the amount of an effluvium $(1.602 \times 10^{-19}$ C), ε_0 denotes the permittivity of a vacuum $(8.854 \times 10^{-12}$ F/m), ε_r denotes the relative permittivity of the medium $(25°C, H_2O = 78.54)$, k denotes the Boltzmann constant $(12.38 \times 10^{-23}$ J/K), T denotes absolute temperature (K), and c denotes the concentration of electrolyte (M). Equation (12) is obtained from Eq. (10) when both sides are differentiated by the cross-sectional area a.

$$\frac{d\mu_N^0}{da} = \gamma - \frac{e^2}{2\varepsilon_0\varepsilon_r}\frac{1}{a^2} = 0 \qquad (\therefore \mu_N^0 = \text{const.}) \tag{12}$$

a is given by Eq. (13).

$$a = \left(\frac{e^2}{2\varepsilon_0\varepsilon_r\kappa\gamma}\right)^{0.5} \tag{13}$$

Thus, when a satisfies Eq. (13), which is ($a = a_0$), μ_{N0} in Eq. (10) becomes minimal, and then the dispersion system will become stable.

Assuming the total volume V in a microemulsion using a geometric microemulsion model (Fig. 14) to simulate the multiphase microemulsion solubilized amount between water and oil based on salinity change, Eq. (14) is obtained.

$$V = MV_0 + NV_S = \frac{4\pi}{3}R_W^3 \tag{14}$$

The occupied area per surfactant molecule, which is attached to the surface of the microemulsion, is described in Eq. (15).

$$a = \frac{4\pi R_W^2}{N} \tag{15}$$

where V_0, M, N, and V_S denote the volume of oil, the number of oil molecules, the number of particles in the surfactant, and the occupied volume of the surfactant, respectively. As mentioned before, if $a = a_0$, then a stable

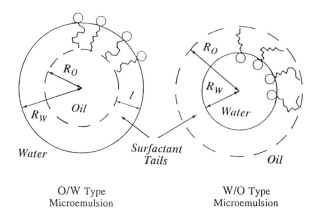

O/W Type
Microemulsion

W/O Type
Microemulsion

FIG. 14 Geometric model of microemulsion.

microemulsion in terms of energy is considered to be produced; thus, the next equation is acquired.

$$\frac{4\pi}{3}|R_0{}^3 - R_W{}^3| = NV_S \tag{16}$$

When N in Eq. (18) was substituted in Eq. (16) and the formulation was changed to $((R_0{}^3 - R_W{}^3) = (R_0 - R_W)(R_0{}^2 + R_0R_W + R_W{}^2))$ in order to use Eq. (17), we acquired Eq. (19) below.

$$|R_0 - R_W| = l \tag{17}$$

$$4\pi R_W{}^2 = Na_0 \tag{18}$$

$$\left(\frac{4\pi R_W{}^2}{a_0}\right)V_S = \frac{4\pi}{3}|(R_0 - R_W)(R_0{}^2 + R_0R_W + R_W{}^2)|$$

$$\frac{3V_S}{a_0 l} = \frac{R_0{}^2 + R_0R_W + R_W{}^2}{R_W{}^2} \equiv 1 + X + X^2 \left(X = \frac{R_0}{R_W}\right)$$

$$X = \frac{-1 + \sqrt{1 - 4\left(1 - \dfrac{3V_S}{a_0 l}\right)}}{2}$$

$$= -\frac{1}{2} + \sqrt{\frac{1 - 4\left(1 - \dfrac{3V_S}{a_0 l}\right)}{4}} = -\frac{1}{2} + \sqrt{\frac{3V_S}{a_0 l} - \frac{3}{4}}$$

$$X = \left(\frac{3V_S}{a_0 l} - \frac{3}{4}\right)^{0.5} - \frac{1}{2} \tag{19}$$

If the volume of oil is V_0, the volume of water is V_W, and total volume is V_S in an O/W type microemulsion and we disregard the size of the hydrophilic base, we have

$$V_0 = \frac{4}{3}\pi R_0{}^3$$

$$V_s = \frac{4}{3}\pi R_w{}^3 - \frac{4}{3}\pi R_0{}^3$$

$$V_w = \frac{4}{3}\pi R_w{}^3$$

Thus, the solubilizing ability of oil in a middle-phase microemulsion phase is expressed as follows.

$$\frac{V_0}{V_0 + V_s} = \frac{R_0{}^3}{R_w{}^3} = X^3 \tag{20}$$

Calculating from the reciprocal of Eq. (20), the solubilization ability of oil per surfactant unit is expressed in the following equation.

$$\frac{V_0}{V_s} = \frac{1}{X^{-3} - 1} \tag{21}$$

Likewise, the solubilization ability of water per surfactant unit is expressed in the following equation.

$$\frac{V_w}{V_s} = \frac{1}{X^3 - 1} \tag{22}$$

The solubilization parameter (σ) is expressed in Eq. (23) as the sum of Eqs. (21) and (22).

$$\sigma = \frac{1}{X^3 - 1} + \frac{1}{X^{-3} - 1} \tag{23}$$

An example of a multiphase microemulsion simulation for an octyl sodium sulfate/n-decane/1-hexanol/NaCl water solution system is as follows. Figure 15 shows the solubilized amount, which was derived from Eqs. (21), (22), and (23) using X, which was derived by substituting for volume V in Eq. (19) 233 Å^3. This value was calculated from the diameter of a microemulsion droplet, l, with 9.53 Å, which is the length of alkyl chain in octyl sodium sulfuric acid, and the value calculated from Eq. (13) with a [where ε_0, ε_r, and e are known, γ is the interfacial tension (measured value), and κ denotes the calculated values in each salinity derived from Eq. (11) in the above system].

Figure 15 clearly indicates a Winsor type of interaction (Winsor-type phase behavior) as the concentration of the saline increases, much like that

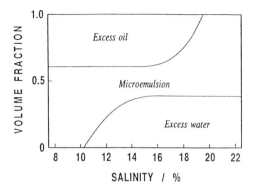

FIG. 15 Simulation for microemulsion in SOS/n-decane/1-hexanol/brine system.

shown in Fig. 1. When compared with the measured values of this system, the measured and simulated values coincided in the lower salinity area. However, with the concentration phase conversion from the middle-phase microemulsion to the W/O-type microemulsion in the higher salinity area, the measured values were more often found in the area toward the lower salinity side than the simulated value. This difference is considered to be caused by the addition of a cosurfactant (as a cosurfactant is not added in the simulation). In other words, the addition of 1-hexanol, which is a co-surfactant, changes the hydrophilic anionic surfactant to hydrophobic as in the effect of an inorganic electrolyte. Thus, it is assumed that the actual values were found more toward the lower salinity side than the simulation. It is also assumed that alcohol molecules adsorbed to form a mixed micelle and increased the solubilizing area, because the actual values of the solu-bilizing amount were larger than the calculated values.

REFERENCES

1. M. Abe, D. Schechter, R. S. Schechter, W. H. Wade, U. Weerasooriya, and S. Yiv, J. Colloid Interface Sci. *114*:342 (1986).
2. M. Abe, R. S. Schechter, R. D. Selliah, B. Sheikh, and W. H. Wade, J. Dispers. Sci. Technol. *8*:157 (1987).
3. J. H. Schulman, W. Stockenius, and L. M. Prince, J. Phys. Chem. *63*:1677 (1959).
4. K. Shinoda and K. Kunieda, J. Colloid Interface Sci. *42*:381 (1973).
5. E. Haque and S. Qutubuddin, J. Polym. Sci. *26*:429 (1988).
6. K. Chokshi, S. Qutubuddin, and A. Hassan, J. Colloid Interface Sci. *129*:315 (1989).

7. Y. Barakat, L. N. Fortney, R. S. Schechter, W. H. Wade, S. H. Yiv, and A. Graciaa, J. Colloid Interface Sci. *92*:561 (1983).

8. C. Lalanne-Cassou, I. Carmona, L. N. Fortney, A. Samii, R. S. Schechter, W. H. Wade, U. Weerasooriya, V. Weerasooriya, and S. Yiv, J. Dispers. Sci. Technol. *8*:137 (1987).

9. T. Yamazaki, M. Nakamae, M. Abe, and K. Ogino, Sekiyu Gakkaishi *33*:241 (1990).

10. B. M. Ninham, S. J. Chen, and D. F. Evans, J. Phys. Chem. *88*:5855 (1984).

11. F. D. Blum, S. Pickup, B. M. Ninham, S. J. Chen, and D. F. Evans, J. Phys. Chem. *89*:711 (1985).

12. D. O. Shah, ed., *Macro and Microemulsion*, American Chemical Society, Washington, DC, 1985.

13. R. N. Healy, R. L. Reed, and D. G. Stenmark, Soc. Pet. Eng. J. *16*:147 (1976).

14. L. E. Scriven, in *Micellization, Solubilization, and Microemulsion* (K. L. Mittal, ed.), Vol. 2, Plenum, New York, 1977, p. 877.

15. U. Olsson, K. Shinoda, and B. Lindman, J. Phys. Chem. *90*:4083 (1986).

16. S. J. Chen, D. F. Evans, B. W. Ninham, D. J. Mitchell, F. D. Blum, and S. Pickup, J. Phys. Chem. *90*:842 (1986).

17. Y. Tamlon and S. Prager, J. Phys. Chem. *69*:2984 (1978).

18. K. Ogino, M. Nakamae, and M. Abe, J. Phys. Chem. *93*:3704 (1989).

19. P. A. Winsor, Chem. Rev. *68*:1 (1968).

20. J. H. Hildebrand and R. L. Scott, *The Solubility of Nonelectrolytes*, 3rd ed., Reinhold, New York, 1950, p. 123.

21. F. Verzaro, M. Bourrel, and C. Chambu, in *Surfactants in Solution* (K. L. Mittal and P. Bothorel, eds.), Vol. 6, Plenum, New York, 1986, p. 1137.

22. M. Abe, M. Nakamae, and K. Ogino, Sekiyu Gakkaishi *31*:458 (1988).

23. J. N. Israelachvili and R. M. Pashley, J. Colloid Interface Sci. *98*:500 (1984).

24. H. Kunieda and K. Shinoda, J. Colloid Interface Sci. *75*:601 (1980).

25. D. S. Rushforth, M. Sanchez-Rubio, L. M. Santos-Vidals, K. R. Wormuth, E. W. Kaler, R. Cuevas, and J. E. Puig, J. Phys. Chem. *90*:6668 (1986).

26. H. M. Fijinaut, J. Chem. Phys. *74*:6857 (1981).

27. H. Kunieda and K. Shinoda, Bull. Chem. Soc. Jpn. *55*:1777 (1982).

28. M. Nakamae, M. Abe, and K. Ogino, J. Colloid Interface Sci. *135*:449 (1990).

29. M. Nakamae, M. Abe, and K. Ogino, Sekiyu Gakkaishi *33*:247 (1990).

30. K. Shinoda, H. Kunieda, T. Arai, and H. Saito, J. Phys. Chem. *88*:5126 (1984).

31. P. A. Rehbinder, *Proceedings of the 2nd International Congress on Surface Activity*, Vol. 1, Butterworths, London, 1957, p. 476.

32. E. Ruckenstein and R. Krishnan, J. Colloid Interface Sci. *76*:201 (1980).

33. A. C. Lam, N. A. Falk, and R. S. Schechter, J. Colloid Interface Sci. *120*:30 (1987).

34. K. Ogino and M. Abe, in *Surface and Colloid Science*, Vol. 15 (E. Matijevic, ed.), Plenum, New York, 1993, p. 85.

35. J. N. Israelachvili, D. J. Mitchell, and B. W. Ninham, Biochim. Biophys. Acta *470*:185 (1977).

15
Microemulsion and Optimal Formulation Occurrence in pH-Dependent Systems as Found in Alkaline-Enhanced Oil Recovery

HERCILIO RIVAS, XIOMARA GUTIÉRREZ, and JOSÉ LUIS ZIRITT
INTEVEP, S.A., Research and Technological Support Center of Petróleos de Venezuela, Caracas, Venezuela

RAQUEL E. ANTÓN and JEAN-LOUIS SALAGER   Laboratorio FIRP, Facultad de Ingeniería, Universidad de los Andes, Mérida, Venezuela

I. INTRODUCTION

Since the beginning of this century it has been well known that a relationship exists between the interfacial tension at the crude oil–water interface in a given reservoir and the percentage of oil recovered in a process of enhanced oil recovery (EOR). The efficiency of the process increases as the interfacial tension decreases. A system with an ultralow interfacial tension represents the ideal situation for achieving maximum oil recovery.

Several studies have shown that it is possible to obtain very low interfacial tensions in certain systems with crude oil/alkaline aqueous solutions [1–11]. The values of interfacial tension depend very much on the pH and ionic strength of the aqueous solution [12]. Riesberg and Doscher [7] suggested that the carboxylic acids present in the crude oil can be neutralized by the addition of caustic solution, producing carboxylate ions that have very high interfacial activity. This led to the use of caustic solutions in enhanced oil recovery in a process known as alkaline flooding.

In alkaline flooding, a caustic solution is injected into the reservoir to react with the naturally occurring acids in the crude oil to produce an ionized form of the acid, commonly called soap. Several mechanisms have been proposed to explain the process of alkaline flooding; however, all of them include emulsification at the reservoir and a change in the rock wettability.

In this work, we present the results of extensive research whose objectives are to gain an understanding of the mechanism of reduction of the interfacial tension in oil/acid/alkaline solution systems, as well as the properties of the emulsions formed. Therefore a series of phase behavior studies were conducted using model and real systems. In some cases the results obtained are compared with those from the literature.

II. ALKALINE FLOODING BASICS

Alkaline EOR processes are methods to be applied with crude oils containing acidic substances, often carboxylic acids. When an alkaline aqueous solution is injected into the reservoir, a surface-active mixture is produced in situ at the oil–water interface by the saponification of the acids that are contained in the oil [1–11].

There is actually an equilibrium between the nondissociated acid and its (sodium) salt, which will be referred to as a soap as a general term even though it is not necessarily a fatty acid salt. This equilibrium is affected by the pH, the mass inventory of acid in the different phases of the system, and the nature of the substances.

In this situation the surface-active composition is a mixture of lipophilic species, i.e., the nondissociated acids, and hydrophilic species, i.e., the ionized (sodium) soaps. If the linear mixing rule of the hydrophilic-lipophilic balance (HLB) is assumed to hold (at least qualitatively), the characteristics of the surfactant mixture depend not only on the acid/soap type but also on the relative amounts of acid and soap, which, in turn, are directly linked to the pH.

A variation of pH may thus produce a variation of the HLB of the interfacial surfactant mixture just as with the so-called formulation scans reported in the phase behavior studies of surfactant-oil-water systems. As mentioned in the literature [13] and recalled in other chapters of this book by Abe, Canselier, and Bavière, a formulation scan exhibits a special so-called optimum formulation, for which the affinity of the surfactant for the aqueous phase exactly equals its affinity for the oil phase. At the optimum formulation the interfacial tension goes though a very deep minimum, often an ultralow value in the millidyne/cm range, which is precisely the condition necessary to offset the capillary forces that trap the oil in the reservoir pores; in fact, the word "optimum" comes from the fact that it matches the best oil recovery [14].

However, it seems that the maximum oil recovery attained at the optimum formulation is due not only to the ultralow tension but also to other associated properties of the emulsified system. In effect, at the optimum formulation, the emulsion exhibits a very deep stability minimum [15–18] as well as a viscosity minimum [19]. As pointed out by some investigators [20–22], an ultralow interfacial tension would probably facilitate the formation of a fine emulsion with a very low stirring energy, such as the shear rate found in the flow through porous media. Because an emulsion is generally much more viscous than its external phase, such emulsified systems would quickly plug the porous medium or at least reduce the flow to a trickle. This does not happen at the optimum formulation because the emulsion droplets coalesce as soon as they are formed; thus there is really no emulsion flow in the porous medium at the optimum formulation. This is worth pointing out, because if an emulsion had an off-optimum formulation, it would be fairly stable and a completely different situation would arise that could end up with plugging of the porous medium instead of oil mobilization.

An optimum formulation can be attained by scanning practically any of the formulation variables, particularly the composition of a surfactant mixture made of a lipophilic surfactant and a hydrophilic one [15,23,24]. The best way to understand the features of alkaline flooding is to view the active surfactant as a mixture of two species: a lipophilic nondissociated acid and a hydrophilic soap. Variation of the pH can change the relative amounts of these two species and thus result in a formulation scan, with its optimum formulation (here optimum pH). This concept will be revisited later on.

Experimental evidence has also shown that other mechanisms can play a role in the mobilization and displacement of the oil globules in alkaline water flooding. Some of them are related to the transient attainment of ultralow interfacial tension [23–32] and/or the occurrence of a saturated zone by surfactant transfer in the so-called diffusion and standing [33] that result in spontaneous emulsification [34–38]. In these situations the emulsion formed corresponds to off-optimum formulation conditions and it can be stable, a source of trouble as far as the flow through porous media is concerned.

The alkaline soaps can absorb on the rock surface [4] and result in a wettability change [2], an effect that has been linked to a change in oil recovery, although not necessarily an increase. In the present case the soap might probably turn an oil-wet rock to a water-wet rock.

Interactions between the soaps and divalent cations can result in precipitation or gel formation, which may also impair the recovery process.

Finally, it must be remembered that soaps are anionic surfactants, which are thus sensitive to a change in salinity. As remarked by Antón and Salager [39], the introduction of NaOH or alkaline salts produces an increase in metallic ions that can affect the formation, whether or not the OH$^-$ ions play a role. On the other hand, Qutubuddin et al. [40] were the first to bring evidence that adding NaOH to a carboxylic acid/oil/water system first increased the hydrophilic tendency (because of an increase in the proportion of soap with the pH), then turned back the tendency to a less hydrophilic situation because of a dominant effect of the increase of ionic strength on the anionic surfactant (soap).

As pointed by Rudin and Wasan [22], the current impression is that many studies have been carried out by just adding some alkaline product, sometimes at a fixed concentration, to see if the soap produced would lower the tension or facilitate the emulsification, rather than exploring systematically the effect of the pH as a formulation variable until an optimum formulation is reached at which the concomitant occurrence of low tension plus rapid coalescence and low viscosity converge to produce the desirable effect.

In the following this approach will be stressed, because it is believed to be the key to the attainment of an efficient alkaline flooding process.

III. ALKALINE FLOODING STUDIES

Several researchers have shown that acidic crude oils contain mainly carboxylic acids in the molecular mass range 300–400 daltons, often referred to as naphthenic acids or resins [41–43]. These acids are readily activated when the crude oil is put in contact with an alkaline aqueous solution. The interfacial tension tends to decrease as the pH increases, until a minimum is reached [4]. This minimum has a very specific physicochemical meaning as will be discussed later on, and the corresponding pH can be regarded as a characteristic of the acid (Fig. 1).

Vega et al. [44,45] have studied systems in which a fatty acid is added to the oil phase to simulate an acidic crude; they found a relationship between the pH for minimum tension and acid carbon number. If these results are extrapolated, acidic crude oil can be considered to contain acids

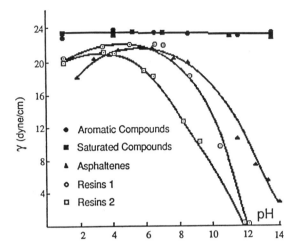

FIG. 1 Variation of interfacial tension versus pH for a system containing acid substances. (From Ref. 4.)

that are equivalent to $C_{20}-C_{24}$ fatty acids, in good agreement with the previously mentioned molecular mass.

Alkaline EOR studies have been carried out with a variety of basic substances, mainly sodium hydroxide, bicarbonate, carbonate, and silicate, as well as mixtures [1–7,22,25–32,35–38,46–58].

Some reports focused on the attainment of low interfacial tension, whereas other used the actual recovery as the goodness criterion. Some discussed different effects able to faciitate the mobilization of trapped oil, such as spontaneous emulsification or wettability change [2,36,37,59].

Because the change in pH is not in general a variable that can be manipulated up to an optimum formulation (which is not sought after anyway in many cases), some fine-tuning formulation adjustments can be carried out by adding pH-independent surfactants [60] or by other means.

IV. OPTIMAL FORMULATION AND MICROEMULSION FORMATION IN pH-DEPENDENT SYSTEMS

The concepts and reported experimental data deal with model systems that can be analyzed easily. The oil phase is a clean hydrocarbon phase, i.e., an alkane or a paraffinic/aromatic mixture; the acid is a well-defined fatty acid in the C_{12}–C_{18} range; and the water phase contains NaOH. To increase the accuracy of the results, the water phase also contains NaCl in such an amount that the total Na^+ content, i.e., the salinity, can be kept constant when the pH is changed by changing the added NaOH [44,45,61–64].

Because soaps are anionic surfactants, some alcohol is also added in systems with high soap concentrations so that a microemulsion forms instead of a solid mesophase, particularly in the high-solubilization cases near the optimum formulation. In low-concentration systems alcohol is not necessary, but there is not enough surfactant to form a three-phase system at the optimum formulation, the occurrence of which is detected through the interfacial tension minimum.

The fundamentals concepts that are outlined here are discussed in detail in a recent report [63], which makes use of the basic framework presented by Cratin [65] some time ago.

Let AcH be the nondissociated acid and Ac^-Na^+ the corresponding soap species. As may be remembered from the Griffin paper by Griffin [66] on HLB, carboxylic acids (respectively soaps) have very low (respectively high) HLB numbers. Thus their mixture can be hydrophilic or lipophilic (or indifferent, as at the optimum formulation) depending on the relative amounts of each of them. Actually, it is more correct to speak of surfactant affinity balance toward the oil and water, because the nature of these phases, as well as the temperature, plays a role in the balance. However, the discussion is based on the HLB for the sake of simplicity, and it is assumed to deal with a given oil-brine pair at constant temperature.

In water, the acid is dissociated according to

$$AcH \longleftrightarrow Ac^- + H^+ \tag{1}$$

with the equilibrium constant

$$K_a = \frac{[Ac_w^-]\,[H_w^+]}{[AcH_w]} \tag{2}$$

Keeping in mind that the relative amounts of acid and soap are linked with the formulation, it is seen that the fraction

$$\frac{[Ac_w^-]}{[AcH_w]} = \frac{K_a}{[H^+]} \tag{3}$$

depends on the pH. The higher the pH, the higher the ratio, and the more hydrophilic the acid/soap mixture, as seen from the HLB mixing rule written as a function of α, the dissociated fraction:

$$HLB_{mixture} = (1 - \alpha)HLB_{AcH} + \alpha HLB_{Ac^-} \tag{4}$$

Note that since K_a is typically 10^{-6}–10^{-7}, it is obvious that the soap exists in sizable quantity only at high pH. The HLB that corresponds to the optimum formulation of this particular system (oil/brine/temperature at unit water-to-oil ratio for the sake of simplicity) is attained for some value of α, which depends of course on the acid type, i.e., on HLB_{AcH} and HLB_{Ac^-}, that can be estimated for instance from an empirical relationship [63].

The previous reasoning is true only for the water phase; if there is an oil phase in equilibrium with the water phase, some of the surfactant species can partition into the oil phase. In the present case, it can be safely assumed that only the nondissociated species AcH is able to do so [63,67], as indicated in Fig. 2.

The partition coefficient of the acid is defined [for a water-to-oil (WOR) ratio = 1 system] by

$$P_a = \frac{[\text{AcH}_o]}{[\text{AcH}_w]} \tag{5}$$

As P_a is typically about 100, it is obvious that neglecting the nondissociated acid that has partitioned into the oil phase can be a gross error of judgment.

Anton [63] demonstrated that for a pH scan produced by adding more and more NaOH the relationship between the pH and the overall acid dissociated fraction f_i is

$$\text{pH} = -\log_{10}[\text{H}^+] = \log_{10}\left(\frac{f_i}{1 - f_i}\right) + \log_{10}\left(\frac{P_a}{K_a}\right) \tag{6}$$

where the dissociated fraction is

$$f_i = \frac{[\text{Ac}_w^-]}{[\text{AcH}_t]} \tag{7}$$

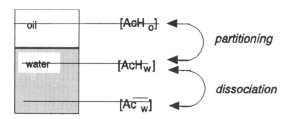

FIG. 2 Dual equilibria: partitioning of the nondissociated acid and acid dissociation in the aqueous phase.

[AcH$_t$] is the total acid originally in the system, supposed to be all in water or oil, which is the same because it was assumed that WOR $= 1$.

If f_n measures the amount of added NaOH with respect to the amount of acid originally present in the system, that is,

$$f_n = \frac{[Na^+]}{[AcH_t]} \tag{8}$$

Then the relationship between f_i and f_n is

$$f_i = f_n + \frac{[H^+] - [OH^-]}{[AcH_t]} = f_n + \frac{10^{-pH} - 10^{pH-14}}{[AcH_t]} \tag{9}$$

It is seen that they are essentially equal, except for very high or very low pH. A straightforward calculation [63] from Eqs. 4 and 6 leads to the relationship between the pH and the overall HLB of the system:

$$\frac{HLB - HLB_{AcH}}{HLB_{Ac^-} - HLB_{AcH}} = f_i = \frac{1}{1 + (P_a/K_a)[H^+]} = \frac{1}{1 + 10^{pH_{1/2}-pH}} \tag{10}$$

where pH$_{1/2}$ is the pH at which half the original acid has been dissociated ($f_i = 0.5$).

A pH-HLB plot of this relationship has a typical sigmoid shape with a buffered (almost pH-independent) variation near pH$_{1/2}$. Figure 3 shows that the intersection of the HLB-pH curve and an horizontal line at optimum HLB (which is always near HLB $= 10$–11) makes it possible to define the

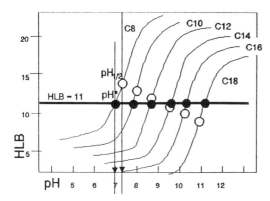

FIG. 3 HLB-pH relationship in oil/water systems for different fatty acids. (From Ref. 63.)

optimum pH, symbolized pH*, which occurs near $pH_{1/2}$, located at the inflection point.

This is true for different acids, each of them characterized by its pH* or $pH_{1/2}$. Note the linear variation of these characteristic pH values with the acid chain length and the agreement between the two series of data.

Acid	C_{08}	C_{10}	C_{12}	C_{14}	C_{16}	C_{18}
$pH_{1/2}$	7.2	8.0	8.9	9.5	10.3	11.0
pH*	7.0	7.9	8.7	9.6	10.35	11.15

System: xylene/heptane 25/75; water; NaOH; WOR = 1

When the system contains acid at a concentration exceeding the critical micelle concentration (CMC) of the acid/soap, a three-phase system is exhibited at the optimum formulation, provided that some alcohol, say 2% *sec*-butanol, is added to dissolve the mesophases. It may be thought that the microemulsion middle phase contains most of the surfactant species as in a typical anionic system [13]. This is not the case, probably because the nondissociated acid fully behaves as an anionic surfactant that is able to fractionate into the oil phase [68]. Anton [63] has proposed a very simple way to detect this partitioning without resorting to chemical analysis.

The volume of the microemulsion middle phase is plotted as a function of the total amount of acid originally poured in the system. Because the water phase concentration is roughly equal to the CMC, the surfactant inventory in the water phase (Q_{CMC}) is essentially negliglble. If it were the same with the oil phase the microemulsion volume would be proportional to the total acid. Figure 4 shows that the variation is linear but not proportional. However, it can be made proportional if a certain amount (calculated by extrapolation at zero microemulsion volume) is subtracted. This amount is AcH_o, i.e., the acid that has partitioned into the oil phase, which can be very important as in the data presented [67].

The influence of the formulation and the water/oil ratio on the phase behavior and emulsion type and properties has been studied by plotting the result on a bidimensional map, which has been found useful for comparison because it exhibits the typical patterns of a very general phenomenon [69–73] whatever the formulation variable [70,74]. Figure 5 shows the phase behavior and emulsion inversion line for both a conventional surfactant [69] and an acid/soap system [64]. It is clear that the phenomenology is identical, so that the pH can trigger exactly the same kind of change for any formulation variable [61,62]. The high-pH region is asso-

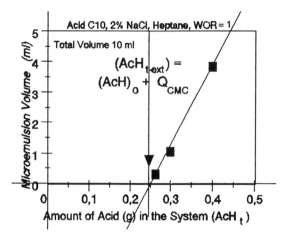

FIG. 4 Variation of the microemulsion middle-phase volume with the total amount of acid originally introduced into the system. (From Ref. 67.)

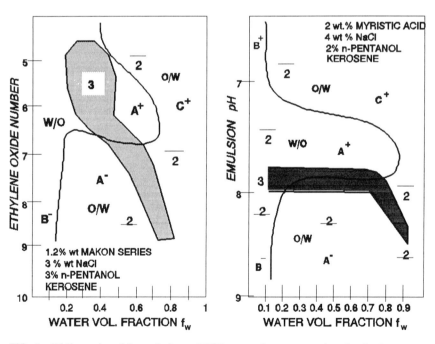

FIG. 5 Bidimensional formulation—WOR scans for a conventional anionic system (left) and an acid/soap system (right) [64,71]. The directions of variations of salinity and pH are selected so that they produce the same effects.

ciated with phenomena found with hydrophilic surfactants, whereas in the low-pH region the patterns exhibited by lipophilic surfactants occur. Note that the phenomenology concerns not only the phase behavior but also the emulsified system type and properties.

Acid/soap mixtures have also been associated with a conventional surfactant so that the formulation adjustment can be shared. Similar results are attained [22,45,60].

As a conclusion of this section, it must be stated that the general phenomenology of these acid/soap-oil-water systems is qualitatively similar to that exhibited by anionic surfactants, but with the extra fractionation feature found with low–ethylene oxide number (-EON) nonionic surfactant systems.

V. INTERFACIAL BEHAVIOR IN CRUDE OIL/ALKALINE SOLUTION SYSTEMS

Figure 6 shows a plot of the interfacial tension at the crude oil–water interface for two different Venezuelan heavy crude oils as a function of the aqueous phase pH, which was previously adjusted using NaOH. The interfacial tension decreases very rapidly at pH values above 8, reaching ultralow values in the range of pH between 11 and 12. A further increase

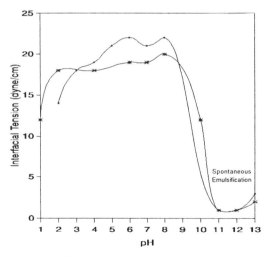

FIG. 6 Effect of pH on the interfacial tension for two Venezuelan heavy crude oils.

in the pH to 13 or higher promotes an increase in the interfacial tension, which is attributed to the formation of insoluble salts of sodium carboxylate [7]. It appears from these results that the amount of petroleum soap produced by the alkali-acid reaction was enough to lower the interfacial tension significantly. It should also be pointed out that in the range of pH between 11 and 12, the drop in interfacial tension was of such an extent that spontaneous emulsification occurred.

Phase behavior studies, carried out by mixing equal volumes of water and crude oil and varying the concentration of NaOH in the aqueous phase, showed that an optimum formulation is achieved at a pH value of 11.5 (optimum pH); at this point, the affinity of the surfactant (carboxylate ions) for the aqueous phase exactly equals its affinity for the oil phase. With reference to Fig. 3, it seems probable that the carboxylic acids present in this particular crude oil have an average carbon number higher than 18.

At the optimum formulation, the system showed a minimum in stability; however, at pH values slightly lower or higher than 11.5 (the optimum pH), the formation of rather stable W/O or O/W emulsions was observed. These emulsions have viscosities at least one order of magnitude higher than the viscosity at the optimum formulation. This aspect could be detrimental to an alkaline flooding process, because in order to keep the viscosity at the required low values, the pH should be controlled very precisely, which is quite difficult under reservoir conditions.

It is evident from the results discussed above that the pH is a very important parameter in controlling the variations in interfacial tension that take place in an oil/water system containing carboxylic acids. However, apart from the pH, the ionic strength should also play an interesting role in such processes. With this in mind, some experiments were conducted in a system characterized by a constant pH value and increasing electrolyte concentration in the aqueous phase.

The electrolyte selected in this case was sodium carbonate, which hydrolyzes according to the reaction:

$$CO_3^{2-} + H_2O \longrightarrow CO_3H^- + OH^- \cdot \qquad (11)$$

giving rise to a buffer solution whose pH remains constant at approximately 11.5, while the sodium carbonate concentration increased from 500 to 150,000 ppm.

Figure 7 shows the variations of interfacial tension for the crude oil/water system containing sodium carbonate in the aqueous phase. As the sodium carbonate concentration increases from 500 to 700 ppm, a sharp decrease in the interfacial tension is observed. For sodium carbonate concentrations in the range between 700 and 2000 ppm, spontaneous emulsification was observed and ultralow interfacial tension developed. A further

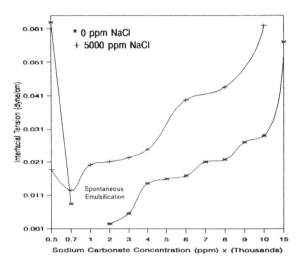

FIG. 7 Effect of sodium carbonate and sodium chloride concentrations on the interfacial tension for a heavy crude oil/water system.

increase in the carbonate concentration promoted a gradual increase in the interfacial tension.

When the pH remains constant, it is evident from the results discussed above that the effectiveness of the ionized carboxylic acids in reducing the interfacial tension depends very much on the concentration of Na^+. This behavior can be explained by the adsorption-desorption mechanism proposed in Fig. 8 [59].

At a sodium carbonate concentration of 500 ppm in the aqueous phase, the concentration of hydroxyl ions is high enough to promote the ionization of the carbolixylic acids (AH) present at the interface between crude oil and water (Fig. 8), forming the carboxylate ions (A^-), whose interfacial activity is well known [1–11]. Some of the carboxylate ions remain adsorbed at the interface, while the rest diffuse to the aqueous phase. This is due to the fact that the crude oil contains many types of acids, characterized by having different chain lengths and solubility that varies according to their chemical structure.

The presence of the carboxylate ions at the interface causes the interfacial tension to decrease initially to 6.3×10^{-3} dyne/cm (Fig. 7) for a concentration of sodium carbonate of 500 ppm. As the sodium carbonate concentration increases, the pH tends to remain constant; however, the concentration of Na^+ increases. This increase in the concentration of Na^+

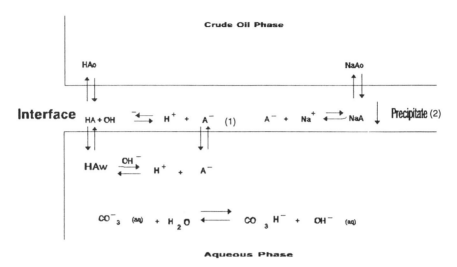

FIG. 8   Adsorption–desorption mechanism for carboxylate ions.

renders the carboxylate ions dissolved in the aqueous phase more hydrophobic, promoting their diffusion and adsorption at the interface, with a consequent decrease in the interfacial tension. A point is reached at which the concentration of sodium carbonate (between 700 and 2000 ppm) is enough to give the optimum formulation for ultralow interfacial tensions. Under these conditions, even when the pH remains constant, spontaneous emulsification is observed. An increase in the sodium carbonate concentration causes a decrease in number of carboxylate ions at the interface by shifting the equilibrium toward the formation of an insoluble sodium carboxylate (ANa), and hence the interfacial tension increases.

Figure 7 also shows the variation in the interfacial tension for the crude oil/water system as a function of sodium carbonate concentration in the presence of 5000 ppm NaCl. It can be seen that for any concentration of sodium carbonate, the interfacial tension is higher in the presence than in the absence of NaCl. It is also demonstrated that the region of spontaneous emulsification completely disappears in the presence of NaCl. The same is shown in Fig. 9, where the sodium carbonate concentration is kept constant while the NaCl concentration is increased. The increase in interfacial tension with the NaCl concentration is evident.

It is therefore concluded that the optimum formulation in acids crude oils/alkaline water systems, and in general the interfacial tension, is governed not only by the pH of the aqueous phase but also by the concentra-

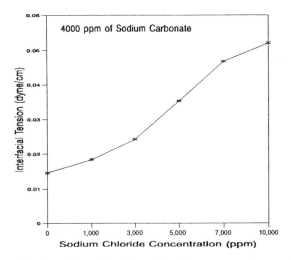

FIG. 9 Effect of sodium chloride concentration on the interfacial tension for the heavy crude oil/water system containing 4000 ppm of sodium carbonate.

tion of the cation of the alkalizing agent (Na^+ in this case), as both determine the fractions of ionized and unionized acids, which are responsible for the HLB of the system [Eq. (4)].

Phase behavior studies carried out by increasing the concentration of sodium carbonate in a system containing equal volumes of oil and water showed the formation of a third phase (microemulsion) in a range of sodium carbonate concentrations between 700 and 2000 ppm, which approximately coincide with the zone of ultralow interfacial tension in Fig. 7. An increase in the stability of the system and also in its viscosity is observed on either side of this concentration range.

There is no doubt that by controlling the pH by using a buffer solution, a wide range of electrolyte concentrations, for attaining ultralow interfacial tensions, can be achieved.

It is also important to consider the effect of the cation present in the alkalizing agent on the position of the optimum formulation. In Table 1 the ranges of concentrations required to obtain spontaneous emulsification for the crude oil/water system in the presence of different carbonates are shown. Lithium carbonate requires the lowest concentration, followed by sodium carbonate, and the highest one corresponds to potassium carbonate. This effect is associated with the cation ionic radius and the solubility of the carboxylate ion formed. The smaller the ionic radius, the lower the solubility of the carboxylate. Li^+, with the smallest ionic radius, forms the

TABLE 1 Range of Concentration of
Alkalies for Spontaneous Emulsification

Alkali	Range of concentration (ppm)
Li_2CO_3	500–600
Na_2CO_3	900–1500
K_2CO_3	2000–4000

most insoluble carboxylate, followed by Na^+ and K^+, respectively. This property should be taken into consideration when selecting the conditions for an alkaline flooding experiment.

It is evident from the results discussed and those in the literature [3,11,12] that the chemical composition of both phases (crude oil and aqueous) has great important in obtaining the low equilibrium interfacial tensions required for displacement in a given reservoir. It is therefore important to study the effect of the alkaline agents in the presence of different electrolyte solutions on the interfacial properties of crude oil/water systems. We believe that this information is important for establishing the correct conditions for the treatment of certain reservoirs by an alkaline flooding process. Several experiments were conducted using different reagents, such as sodium silicate, with results rather similar to those described above [75].

VI. FIELD APPLICATION

Microemulsions formed in situ during fluid flow in a porous medium, or formed externally and intentionally injected, can help (or hinder) oil recovery through two opposite mechanisms: liberation of oil trapped in pores and entrapment of oil droplets to force the displacing fluid to flow through unswept regions. Both mechanisms depend on the balance of forces on oil trapped in pores, which are determined by the capillary number, as illustrated in Fig. 10.

The capillary number is important in determining residual oil and the shape of the permeability saturation functions. As the capillary number increases, in general, residual oil and the curvatures of the permeability-saturation functions decrease, the curvature approaching a straight line at very high capillary numbers. During ordinary water flooding the order of magnitude of the capillary number is 10^{-6}, and because tensions are about 10 dyne/cm for typical crude-brine combinations, reducing the interfacial

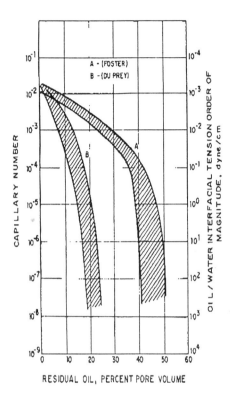

FIG. 10 Dependance of residual oil on capillary number (after Ref. 76).

tension on the order of 10^{-3} dyne/cm will yield capillary numbers in the required range to achieve residual oil saturations near zero.

Figure 11 shows an example of how the residual oil saturation in a Brea sandstone decreases when an alkaline solution of increased concentration is used to displace Mene Grande crude oil contained in the porous medium. The Mene Grande crude is a 16° API (American Petroleum Institute) oil with an acid number of 2.5 mg KOH/g. Through phase behavior studies and interfacial tension measurements it was determined that sodium silicate in this case promotes the ionization of the carboxylic acids present in the crude oil, reducing the interfacial tension to ultralow values.

The zone of spontaneous emulsification was found to be in the range of alkali concentration in the aqueous phase between 1300 and 9000 ppm, in a 1:1 oil/alkaline solution system containing equal volumes of water and oil. Considering the mixing, dispersion, and rock reaction effects in the

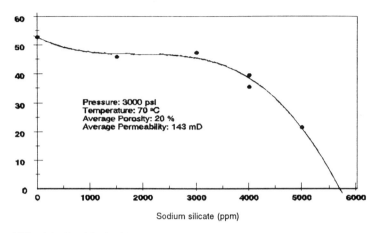

FIG. 11 Residual oil saturation after sodium silicate solution displacements at different concentration.

porous medium, the low residual oil saturations shown in Fig. 11 correlate with the attainment of spontaneous emulsification due to ultralow interfacial tension values.

As mentioned before, the permeability-saturation functions also vary with the change in the capillary number. Figure 12 shows the water imbibition relative permeability curves of a waterflood experiment with Mene

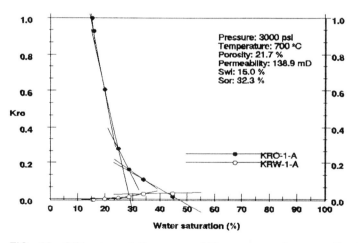

FIG. 12 Oil/water relative permeability curves of a waterflood in a Berea sandstone.

Grande crude oil in a Berea sandstone. The residual oil saturation end point is 32.3%. The same experiment was run using an 11,000 ppm sodium silicate solution as the imbition fluid (Fig. 13). The sodium silicate solution, injected at a concentration high enough to ensure interfacial tension values under 10^{-2} dyne/cm, changed the relative permeability curves, moving the residual oil saturation end point to 6.2%. The improvement in oil recovery or fluid displacement through porous media is caused not only by the changes in the capillary number, as demonstrated before, but also by the correction of heterogeneity of the porous medium.

Oil-in-water emulsions easily prepared from Midway Sunset crude oil and dilute solutions of sodium hydroxide were used by McAuliffe [78] to study the flow of emulsions through porous media. In experiments done with cores of different permeabilities mounted in parallel, oil-in-water emulsions decreased the flow in the higher permeability porous media and improved fluid distribution. Emulsion systems are formed and are stable if sufficient emulsifying agent is present to orient at the oil-water interface to form a film. During emulsion flow through porous media, oil droplets tend to coalesce and grow due to surfactant loss. When the droplets attain a size slightly larger than the pore throat, a higher pressure is required to force the droplets through, creating a flow restriction.

In waterflooding processes in which water displaces oil, fingering develops because of rock heterogeneity. When oil-in-water emulsions are

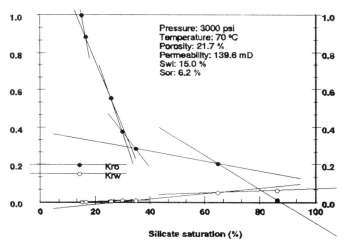

FIG. 13 Oil/water relative permeability curves of an 11,000-ppm sodium silicate solution displacement in a brea sandstone.

injected or in situ emulsification is induced or occurs naturally, a greater amount of emulsion enters the more permeable zones. As this occurs, flow becomes more restricted, so water begins to flow into less permeable zones, resulting in greater sweep efficiency. These results have been corroborated not only at the laboratory level but also through a field test as reported by McAuliffe [78].

In conclusion, microemulsions formed in situ, or formed externally and intentionally injected during waterflooding, can help oil recovery through two mechanisms: the reduction of residual oil saturation by reduction of the capillary forces and the decrease of reservoir heterogeneity (greater sweep efficiency) by oil droplet entrapment. In addition, to achieve the most effective alkaline recovery process it is necessary to select an adequate alkalinizing agent, considering the interfacial properties of the crude oil/water system. The type of electrolytes present in the rock reservoir should also be taken into consideration.

ACKNOWLEDGMENTS

The authors thank INTEVEP S.A. (Research and Technological Support, Center of Petróleos de Venezuela) for allowing us to publish the information and Mr. Félix Silva for his valuable collaboration in the preparation of the manuscript.

REFERENCES

1. H. Y. Jennings, C. E. Johnson, and C. D. McAuliffe, A caustic waterflooding process for heavy oils, J. Petrol. Technol. December:1344 (1974).
2. C. E. Cooke, R. E. Williams, and P. A. Kolodzie, Oil recovery by alkaline waterflooding, J. Petrol. Technol. December:1365 (1974).
3. V. K. Bansal, K. S. Chan, R. McCallough, and D. O. Shah, The effect of caustic concentration on interfacial charge, interfacial tension and droplet size: A simple test for optimum caustic concentration for crude oil, J. Can. Petrol. Technol. *17*:1 (1978).
4. L. W. Holm and S. D. Robertson, Improved micellar/polymer flooding with high-pH chemicals, J. Petrol. Technol. January:161 (1981).
5. N. Mungan, Enhanced oil recovery using water as a driving fluid. Part 4: Fundamentals of alkaline flooding, World Oil June:209 (1981).
6. M. Chan and T. F. Yen, A chemical equilibrium model for interfacial activity of crude oil in aqueous alkaline solution: The effects of pH, alkali and salt, Can. J. Chem. Eng. *60*:305 (1982).
7. J. Riesberg and T. Doscher, Interfacial phenomena in crude oil-water systems, Producers Monthly November:43 (1956).

8. M. M. Sherma, A thermodynamic model for low interfacial tension in alkaline flooding, Soc. Petrol. Eng. J. 125 (1983).

9. I. Layrisse, H. Rivas, and S. Acevedo, Isolation and characterization of natural surfactants present in extra heavy crude oils, J. Dispers. Sci. Technol. 5(1):1 (1984).

10. H. A. Nasreldin and K. C. Taylor, Dynamic interfacial tension of crude oil/alkali/surfactant systems, Colloids Surfaces 66(1):23 (1992).

11. J. Rudin and D. J. Wasan, Mechanisms for lowering of interfacial tension in alkali/acidic oil systems; effect of added surfactant, Ind. Eng. Chem. Res. 31(8):1899 (1992).

12. J. Rudin and D. T. Wasan, Mechanisms of lowering of interfacial tension in alkali/acidic oil systems. 1. Experimental studies, Colloids Surfaces 68:67 (1992).

13. M. Bourel and R. S. Schechter, *Microemulsions and Related Systems*, Marcel Dekker, New York, 1988.

14. D. O. Shah and R. S. Schechter, eds., *Improved Oil Recovery by Surfactant and Polymer Flooding*, Academic Press, New York, 1977.

15. M. Bourrel, A. Graciaa, R. S. Schechter, and W. H. Wade, The relation of emulsion stability to the phase behavior and interfacial tension of surfactant systems, J. Colloid INterface Sci. 72:161 (1979).

16. J. L. Salager, L. Quintero, E. Ramos, and J. Anderez, Properties of surfactant-oil-water emulsified systems in the neighborhood of three-phase transition, J. Colloid Interface Sci. 77:288 (1980).

17. J. E. Viniatieri, Correlation of emulsion stability with phase behavior in surfactant systems for tertiary oil recovery, Soc. Petrol. Eng. J. 20:402 (1980).

18. J. L. Salager, I. Loaiza-Maldonado, M. Miñana-Perez, and F. Silva, Surfactant-oil-water systems near the affinity inversion. Part I: Relationship between equilibrium phase behavior and emulsion type and stability, J. Dispers. Sci. Technol. 3:279 (1982).

19. J. L. Salager, M. Miñana-Perez, J. Anderez, J. Grosso, C. Rojas, and I. Layrisse, Surfactant-oil-water systems near the affinity inversion. Part II: Viscosity of emulsified systems, J. Dispers. Sci. Technol. 4:161 (1983).

20. J. L. Salager, J. L. Grosso, and M. A. Eslava, Flow properties of emulsified surfactant-oil-water systems near optimum formulation, Rev. Teen. INTEVEP 2:149 (1982).

21. R. E. Anton, J. M. Anderez, and J. L. Salager, Properties of three-phase surfactant-oil-water dispersed systems, presented at the First International Symposium on Enhanced Oil Recovery, SIREMCRU 1, Maracaibo, February 1985.

22. J. Rudin and D. T. Wasan, Surfactant-enhanced alkaline flooding: Buffering at intermediate alkaline pH, SPE Reservoir Eng. November:275 (1993).

23. J. L. Salager, J. Morgan, R. S. Schechter, W. H. Wade, and E. Vasquez, Optimum formulation of surfactant-oil-water systems for minimum tension and phase behavior, Soc. Petrol. Eng. J. 19:107 (1979).

24. M. Bourrel, J. L. Salager, R. S. Schechter, and W. H. Wade, A correlation for phase behavior of nonionic surfactants, J. Colloid Interface Sci. *75*:451 (1980).

25. F. G. McCafferty, Interfacial tensions and aging behavior of some crude oil against caustic solutions, J. Can. Petrol. Technol. *15*(3):71 (1976).

26. C. I. Chiwetelu, V. Hornof, and g. H. Neale, A dynamic model for the interaction of caustic reagents with acidic oils, AIChE J. *36*:233 (1990).

27. E. M. Trujillo, The static and dynamic interfacial tensions between crude oils and caustic solutions, Soc. Petrol. Eng. J. August:645 (1983).

28. E. Rubin and C. J. Radke, Dynamic interfacial tension minima in finite systems, Chem. Eng. Sci. *35*:1129 (1980).

29. R. P. Borwankar and D. T. Wasan, Dynamic interfacial tensions in acidic crude oil/caustic systems. Part 1: A chemical diffusion-kinetic model, AIChE J. *32*:455 (1986).

30. M. M. Sharma, L. K. Jang, and T. F. Yen, Transient interfacial tension behavior of crude-oil/caustic interfaces, SPE Reservoir Eng. May:228 (1989).

31. V. J. Cambridge, et al. An improved model for the interfacial behavior of caustic/crude oil systems, Chem. Eng. Commun. *46*:241 (1986).

32. E. F. De Zabala and C. J. Radke, A nonequilibrium description of alkaline waterflooding, SPE Reservoir Eng. January:29 (1986).

33. K. J. Rushak and C. A. Miller, Spontaneous emulsification in ternary systems with mass transfer, Ind. Eng. Chem. Fundam. *11*:534 (1972).

34. D. C. England and J. C. Berg, Transfer of surface active agents across liquid-liquid interface, AIChE J. *17*:313 (1971).

35. D. T Wasan, S. M. Shah, M. Chan, K. Sampath, and R. Shah, Spontaneous emulsification and the effects of interfacial fluid properties on coalescence and stability in causting flooding, in *Chemistry of Oil Recovery* (R. T. Johansen and R. L. Berg, eds.), ACS Symposium Series 91, American Chemical Society, Washington, DC, 1978, p. 115.

36. D. T. Wasan, F. S. Milos, and P. E. Di Nardo, Oil banking phenomena in surfactant/polymer and caustic flooding: Droplet coalescence and entrainment processes, AIChE Symposium Series 78 (Vol. 212), American Chemical Society, Washington, DC, 1982, p. 105.

37. T. P. Castor, W. H. Somerton, and J. F. Kelly, Recovery mechanism of alkaline flooding, in *Surface Phenomena in Enhanced Oil Recovery* (D. O. Shah, ed.), Plenum, New York, 1981, p. 249.

38. J. Rudin and D. T. Wasan, Interfacial turbulence and spontaneous emulsification in alkali/acidic oil systems, Chem. Eng. Sci. *48*:2225 (1993).

39. R. E. Antón and J. L. Salager, Effect of the electrolyte anion on the salinity contribution to optimum formulation of anionic surfactant microemulsions, J. Colloid Interface Sci. *140*:75 (1990).

40. S. Qutubuddin, C. A. Miller, and T. Fort, Phase behavior of Ph-dependent microemulsions, J. Colloid Interface Sci. *101*:46 (1984).

41. W. K. Seifert, and W. G. Howells, Interfacially active acids in a California crude oil. Isolation of carboxylic acids and phenols, Anal. Chem. *41*:554

(1969); W. K. Seifert, Effect of phenols on the interfacial activity of crude oil (California) carboxylic acids and the identification of carbazoles and indoles, Anal. Chem. 41:562 (1969).

42. G I. Jenkins, Occurrence and determination of carboxylic acids and esters in petroleum, J. Inst. Petrol. 51:313 (1965).

43. C. H. Pasquarelli, et al., The role of acidic, high-molecular weight crude components in enhanced oil recovery, paper SPE 8895, presented at the 1980 SPE California Regional Meeting, Los Angeles, April 9–11, 1980.

44. J. Vega, Sistemas ácido carboxílico-aqua-aceite—Influencia del pH y de los aditivos, Thesis, Informe Técnico FIRP No. 8306, Univ. de Los Andes, Mérida-Venezuela, 1983.

45. J. Villabona, L. Rodriguez, H. Delphin, J. y Vega, and I. Layrisse, Propiedades de los sistemas mezela acido-surfactante/salmuera/aceite vs pH, First International Symposium on Enhanced Oil Recovery, SIREMCRU I, Maracaibo, February 1985.

46. J. E. Strassner, Effect of pH on interfacial films and stability of crude oil–water emulsions, J. Petrol. Technol., March:45 (1968).

47. L. K. Jang, M. Sharma, Y. I. Chang, M. Chan, and T. F. Yen, Correlation of petroleum component properties for caustic flooding, in *Interfacial Phenomena in Enhanced Oil Recovery*, AIChE Symp. Ser. 78:97 (1982).

48. E. F. De Zabala, J. M. Vislocky, E. Rubin, and C. J. Radke, Theory for linear alkaline flooding, SPE 8997, presented at the 5th International Symposium on Oilfield and Geothermal Chemistry, Stanford, CA, May 1980.

49. M. Chang and D. T. Wasan, Emulsion characteristics associated with an alkaline waterflooding process, paper SPE 9001, 5th Symposium on Oilfield and Geothermal Chemistry, Stanford, CA, May 1980.

50. C. I. Chiwetelu, V. Hornof, and G. H. Neale, Mechanism for the interfacial reaction between acidic oils and alkaline reagents, Chem. Eng. Sci. 45:627 (1990).

51. L. Ghosh and P. Joos, Steady-state interfacial tensions during diffusion of organic acids from hexane to aqueous alkaline solutions, J. Colloid Interface Sci. 138:231 (1990).

52. T. S. Ramakrishnan and D. T. Wasan, A model for interfacial activity of acidic crude oil/caustic system for alkaline flooding, Soc. Petrol. Eng. J. August: 602 (1983).

53. T. C. Campbell, A comparison of sodium orthosilicate and sodium hydroxide for alkaline waterflooding, paper SPE 6514, presented at the 1977 SPE California Regional Metting, Bakerfield, April 1977.

54. H. Y. Jennings, A study of caustic solution–crude oil interfacial tensions, Soc. Petrol. Eng. J. June:197 (1975).

55. P. H. Krumrine, J. S. Falcone, and T. C. Campbell, Surfactant flooding. 1: The effect of alkaline additives on IFT, surfactant adsorption, and recovery efficiency, Soc. Petrol. Eng. J. August:503 (1982).

56. P. H. Krumrine, J. S. Falcone, and T. C. Campbell, Surfactant flooding. 2: The effect of alkaline additives on permeability and sweep efficiency, Soc. Petrol. Eng. J. December:983 (1982).

57. T. R. French and T. E. Burchfield, Deisng and optimization of alkaline flooding formulations, paper SPE 20238, presented at the 1990 SPE/DOE Symposium on Enhanced Oil Recovery, Tulsa, April 22–25, 1990.

58. J. S. Falcone, P. H. Krumrine, and G. C. Schweiker, The use of inorganic sacrificial agents in combination with surfactants in enhanced oil recovery, J. Oil Chem. Soc. *59*:826 (1982).

59. J. Rudin and D. T. Wasan, Mechanisms for lowering of interfacial tension in alkali/acidic oil systems. Part II: Theoretical studies, Colloids Surfaces *68*: 81 (1992).

60. L. Rodriguez, Influencia del pH sobre las propiedades de los sistemas surfactante aniónico-ácido-aceite-salmuera al equilbrio y emulsionados, Informe Técnico FIRP No. 8505, Univ. de Los Andes, Mérida-Venezuela, 1985.

61. R. Anton, I. Layrisse, L. Quintero, H. Rivas, and J. L. Salager, Optimum formulation of surfactant-oil-water systems containing carboxylic acids, presented at the 9th International Symposium Surfactants in Solution, Sofia, Bulgaria, June 1992.

62. J. M. Anderez, L. Quintero, H. Rivas, and J. L. Salager, Emulsion properties of surfactant-oil-water systems containing carboxylic acids, presented at the 9th International Symposium Surfactants in Solution, Sofia, Bulgaria, June 1992.

63. R. E. Anton, Contribution à l'étude du comportement de phase des systèmes: meelanges de surfactifs-eau-huile, Dr. thesis, Univ. de Pau P. A., Pau, France, 1992.

64. Z. Mendez, Influencia del pH sobre las Propiedades de los Sistemas Acido Carboxílico/Agua/aceite, MSc thesis, Informe Técnico FIRP No. 8606, Univ. de Los Andes, Mérida, Venezuela, 1986.

65. P. Cratin, A quantitative characterization of pH dependent systems, in *Chemistry and Physics of Interfaces*, S. Ross, Ed., Vol. 2, American Chemical Society, 1969, p. 37.

66. W. C. Griffin, Classification of surface-active agents by "HLB," J Soc. Cosmetic. Chem. *1*:311 (1949).

67. I. Avendaño, Reparto del ácido en sistemas acido carboxílico/agua/aceite a pH variable, MSc. thesis, Informe Técnico FIRP No. 9101, Univ. de Los Andes, Mérida, Venezuela, 1991.

68. A. Graciaa, J. Lachaise, J. G. Sayous, P. Grnier, S. Yiv, R. R. Schechter, and W. H. Wade. The partitioning of complex mixtures between oil/microemulsion/water phases at high surfactant concentration, J. Colloid Interface Sci. *93*:474 (1983).

69. J. L. Salager, M. Miñana-Perez, M. Perez-Sanchez, M. Ramirez-Gouveia, and C. Rojas, Surfactant-oil-water systems near the affinity inversion—Part III: The two kinds of emulsion inversion, J. Dispers. Sci. Technol. *4*:313 (1983).

70. R. E. Anton, P. Castillo, and J. L. Salager, Surfactant-oil-water systems near the affinity inversion—Part IV: Emulsion inversion temperature, J. Dispers. Sci. Technol. *7*:319 (1986).

71. M. Miñana, P. Jarry, M. Perez-Sanchez, M. Ramirez-Gouveia, and J. L. Salager, Surfactant-oil-water systems near the affinity inversion—Part V: Properties of emulsions, J. Dispers. Sci. Technol. *7*:331 (1986).

72. B. Brooks and H. Richmond, Dynamics of liquid-liquid phase inversion using non-ionic surfactants, Colloids Surfaces *58*:131 (1991).

73. H. T. Davis, Factors determining emulsion type: HLB and beyond, Proceedings, World Congress on Emulsion, Vol. 4, Paris, October 1993, p. 69.

74. K. Shinoda and H. Saito, The effect of temperature on the phase equilibria and the type of dispersions of ternary systems composed of water, cyclohexane and nonionic surfactants, J. Colloid Interface Sci. *26*:70 (1968).

75. Mendez Zuleica, personal communication.

76. W. R. Foster, A low-tension waterflooding process, J. Petrol. Technol. February:205–210 (1973).

77. S. Thomas and S. M. Farouq Ali, Flow of emulsions in porous media, and potential for enhanced oil recovery, J. Petrol. Sci. Eng. *3*:121 (1989).

78. C. D. McAuliffe, Oil-in-water emulsions and their flow properties in porous media, J. Pet. Tech. June:727–733 (1973).

16

Microemulsions in the Chemical EOR Process

MARC BAVIÈRE   Reservoir Engineering Department, Institut Français du Pétrole, Rueil-Malmaison, France

JEAN PAUL CANSELIER   Institut de Génie des Procédés, Ecole Nationale Supérieure d'Ingénieurs de Génie Chimique, Toulouse, France

I. INTRODUCTION: THE NEED FOR EOR

Oil is accumulated, together with water and gas, in reservoirs consisting of porous and permeable rocks. The amount of oil spontaneously produced due to the natural pressure that exists in the reservoir (primary oil recovery), and that produced by water or gas injection (secondary oil recovery) do not usually exceed 30–40% of the original oil in place (OOIP). This results from unfavorable reservoir characteristics. On the one hand, high oil viscosity and rock heterogeneity are responsible for areas unswept by the injected fluid and for delay in oil production. On the other hand, capillary forces, highly active in such porous media, result in poor microscopic displacement efficiency.

Oil remaining trapped after primary and secondary oil recovery represents considerable amounts: probably more than 100 billion tonnes (10^{11} tonnes), whereas the world consumption in 1992 was 3×10^9 tonnes.

Therefore, in order to improve both the oil production rate and recovery factor, various methods have been developed: miscible, thermal, and chemical processes, specifically addressed to different reservoir types [1]. The utilization of microemulsions for enhanced oil recovery (EOR) is performed within the framework of the chemical process, which concerns the use of polymers, surfactants, and, possibly, alkaline agents, generally in combination. More accurately, the process aims at making use of the properties of microemulsions that form when a surfactant solution injected into the reservoir mixes with oil in place. The capillary-trapped oil in the reservoir after waterflooding is the main target of the process summarized hereunder.

II. MAIN OBJECTIVES: MOBILITY CONTROL AND INTERFACIAL TENSION LOWERING

The proportion of unswept area greatly depends on the mobility ratio

$$M = \frac{\lambda_d}{\lambda_o} = \frac{k_d \mu_o}{k_o \mu_d} \qquad (1)$$

where μ_o and μ_d are the oil and displacing fluid viscosities and k_o and k_d the effective oil and displacing fluid permeabilities of the reservoir rock. The ratios $\lambda_o = k_o / \mu_o$ and $\lambda_d = k_d / \mu_d$ are called the mobilities of the oil and of the displacing fluid, respectively.

If $\lambda_d < \lambda_o$, the flow becomes less easy in the part of reservoir where oil has been replaced by the less mobile displacing fluid. Thus, fluid velocity will tend to decrease in this part, so that any initial irregularity in the displacement front will tend to decrease. This is the role granted to water-soluble polymers to decrease λ_d, not only by increasing μ_d but also by decreasing k_d (because of polymer adsorption on the rock).

Even in well-swept zones, capillary forces cause large amounts of oil to be left behind, as in any immiscible displacement. As a matter of fact, in these zones the oil phase becomes discontinuous (droplets). Let us assume, by oversimplification, that the rock is water wet; each droplet can then flow until it encounters a pore narrower than its own diameter, where it is trapped. This oil is called residual oil.

To enter such narrow pores, the surface curvature of the droplet should increase. According to Laplace's formula, the capillary pressure, P_c, is

$$P_c = P_o - P_w = \frac{2\sigma}{r} \tag{2}$$

where P_o and P_w are the pressures in the oil phase and in the water phase, respectively, σ is the oil/water interfacial tension, and r is the curvature radius of the supposedly spherical interface. So, a pressure gradient is required to exceed the capillary force retaining the droplet and to make it flow [2]. To keep this gradient within a range of values compatible with field situations, it appears that interfacial tension has to be reduced to about 10^{-2} mN/m or less.

Thus, the displacement of residual oil is related to the competition between viscous and capillary forces, usually expressed by the capillary number N_c (dimensionless)

$$N_c = \frac{(\Delta P/L)k_d}{\sigma \cos \theta} \tag{3}$$

where ΔP is the pressure drop over the distance L and θ the contact angle. N_c can also be expressed as

$$N_c = \frac{\mu v}{\sigma \cos \theta} \tag{4}$$

where μ and v are the displacing-fluid viscosity and Darcy velocity (flow rate per unit surface area), respectively. No residual oil can be displaced until a minimum critical value of N_c is reached [2]. For typical waterflood conditions

$$(N_c)_{\text{crit}} \approx 10^{-6}$$

This value needs to be increased by three or four orders of magnitude, even more if the oil is the wetting phase, to achieve near-zero oil satura-

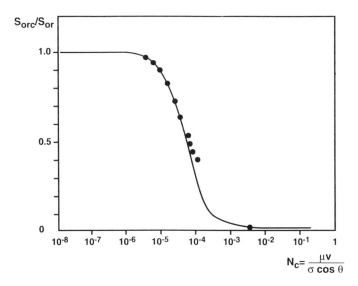

FIG. 1 Reduced oil saturation versus capillary number. S_{orc}, residual oil saturation; S_{or}, maximum trap saturation or residual saturation at an infinitely small N_c. (Filled circles) Taber experiments (Berea sandstone, oil/water/Triton X100 system) [2]; continuous curve was calculated [3]. (From Ref. 3.)

tion* (Fig. 1) [3]. This means that interfacial tension between oil and water has to be reduced by a factor 10^3 or 10^4. Values around 10^{-3} mN/m and less can be obtained with surfactants, usually under appropriate conditions that will be examined in the next section. This is the principle of surfactant flooding, also referred to as low-tension waterflooding, micellar flooding, or microemulsion flooding.

Usually, but not necessarily, applied after a water injection, the process consists of the following flooding sequence (Fig. 2):

Preflush: conditions the reservoir in order to protect surfactant against salt effect and retention on rock surfaces.

Micellar slug: mobilizes residual oil. The injected volume usually ranges from 5 to 50% of the pore volume (PV). The slug contains surfactant (0.1–10 wt%), cosurfactant, or cosolvent (0–5 wt%), brine and/or oil (0–90% wt%), and polymer (0.05–0.2 wt%).

Mobility buffer: protects the micellar slug against penetration by chase water due to an adverse mobility ratio (injected volume ranging from

---

*The saturation by a given fluid is the pore volume fraction filled with this fluid.

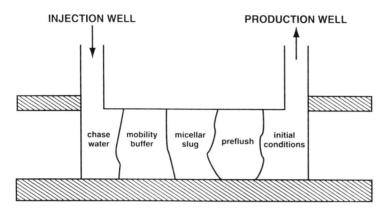

INJECTION WELL   PRODUCTION WELL

| chase water | mobility buffer | micellar slug | preflush | initial conditions |

FIG. 2 Schematic view of a microemulsion flood.

50 to 100% PV). It contains polymer (0.05–0.2 wt%), whose concentration ranges from a maximum at the front to zero at the back.
Chase water.

III.   SURFACTANT SELECTION

A.   Phase Behavior of Oil/Brine/Surfactant Mixtures

Surfactants are chosen in such a way that low interfacial tensions exist in multiphase systems that form as the injected fluid, usually an aqueous solution of surfactants, mixes with the crude oil. The degree of interfacial tension lowering depends on the phase behavior of the oil/brine/surfactant mixture. Such phase behavior has been extensively described [4–9]. The main features of interest for microemulsion flooding are briefly reviewed hereunder.

Surfactants are generally used at concentrations much higher than their critical micelle concentration (CMC). Phase behavior depends on the surfactant partition coefficient between oil and brine, resulting from the preferential surfactant solubility into one of the phases.

This is shown in Figure 3, where the phase behavior of such multicomponent systems is conventionally illustrated by (pseudo)ternary diagrams. The top apex more often represents a mixture of amphiphilic compounds.

With a hydrophilic surfactant, a mixture whose the overall composition representative point is located below the binodal curve will separate into two phases: an oil-in-water (O/W) microemulsion containing brine, surfactant, and oil solubilized in the micelles and an excess oil phase (essen-

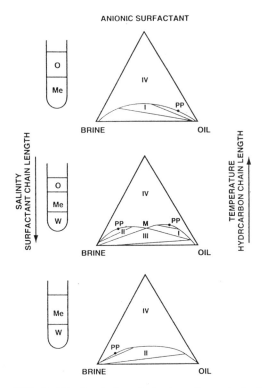

FIG. 3 Idealized phase behavior of a typical oil/brine/anionic surfactant system. O, excess oil phase; Me, microemulsion phase; W, excess water phase; PP, plait point; I and II, two-phase zones (with O and W as excess phases, respectively); III, three-phase zone; IV, single-phase zone; M, representative point of the microemulsion phase (middle phase) composition.

tially pure oil). This phase environment is called a type I system, according to Winsor's terminology [4]. With an oleophilic surfactant, phase splitting preferably yields a water-in-oil (W/O) microemulsion containing oil, surfactant, and water solubilized in inverted micelles and an excess brine phase. This phase environment is called a type II system. For intermediate surfactant solubility, a three-phase region forms, where a middle-phase microemulsion with a bicontinuous structure [10] is in equilibrium with excess oil and brine phases. This is the type III environment.

Any change of a variable that favors the surfactant partitioning in the oil tends to promote the I-III-II transition [also known as the II($-$)-III-II($+$) or $\underline{2}$-3-2 transition]. This variable could be an increase of salinity, a decrease of the oil molecular weight, or a lengthening of the lipophilic (nonpolar) part of the surfactant. The partitioning of anionic surfactants in the brine is favored by temperature rise, and that of nonionics is favored by temperature lowering. The liquid-liquid phase behavior is relatively insensitive to pressure.

With anionic surfactants, phase behavior may depend on surfactant concentration [11] through salt fractionation. NaCl is partly excluded from the surfactant-rich phase [12], whereas the opposite trend occurs with $CaCl_2$ due to the strong association of anionic surfactant with divalent cations [13]. This association makes the I-III-II transition easier. As a matter of fact, a given Ca^{2+} concentration has a greater effect on anionic surfactants than the equivalent Na^+ concentration [14]. That explains the strong effect of ion exchange phenomena with rock components (see Sec. IV).

The lowest interfacial tensions are reached when a type III environment occurs (Fig. 4) [15,16]. In such an environment, there are two types of

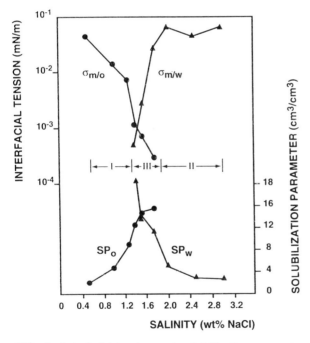

FIG. 4 Interfacial tension and solubilization parameter versus salinity. $\sigma_{m/o}$ and $\sigma_{m/w}$, interfacial tension between oil and microemulsion phase and between water (brine) and microemulsion phase, respectively; SP_o and SP_w, solubilization parameter of oil and water, respectively. (From Ref. 15.)

interfaces. Near a phase transition, the larger of the two interfacial tensions between microemulsion and excess phases is associated with a high surface pressure in the surfactant layer, whereas the lower tension is associated with the vicinity of critical end points in the phase diagram [17,18]. For evident reasons, the three-phase configuration is considered as an optimum, and any parameter value that leads to it is said to be optimal. Thus, the middle-phase microemulsion appears to be the central element of the chemical process.

B. Interfacial Tension and Solubilization

Interfacial tensions and amounts of oil and brine solubilized in the micellar phase are correlated: interfacial tension between the oil (or brine) excess phase and the microemulsion phase decreases as oil (or brine) solubilization increases [15]. In Fig. 4, solubilization is expressed in terms of the solubilization parameter SP, defined as the volume of oil or brine in the microemulsion phase per unit volume of surfactant.

According to Huh [19]:

$$\sigma = \frac{\alpha}{(SP)^2} \tag{5}$$

where α is a constant that depends on the surfactant.

Through solubilization, interfacial tension is related to molecular attractive interactions between surfactant, oil, and water molecules. Winsor expressed the overall effect resulting from these interactions in terms of the cohesive energy ratio R, defined as the ratio of solvent attraction between surfactant and oil to solvent attraction between surfactant and water [20]. The stronger the solvent attraction, the higher the solubilization.

If the surfactant is preferentially water soluble, R is less than 1 and the interfacial region will be convex toward water (Bancroft's rule), giving rise to a type I system. In the opposite case, the system will be of type II. When R is around 1, a type III system forms and leads to optimal conditions. However, optimized systems are not necessarily all equivalent: interfacial tensions are minimized when solubilization is maximized, i.e., when attractive interactions are as high as possible, for instance, at low salinity [9].

Thus, these considerations provide a fundamental basis for designing surfactant formulation.

C. Surfactants Used in Chemical EOR

In addition to being able to form microemulsions showing low interfacial tensions with oil and brine, surfactants for EOR must be available in large amounts and at a reasonable cost. They should be chemically stable (more than 1 year), brine soluble, and compatible with the other usual components (Table 1), especially polymers and alkaline agents.

Cationics and amphoterics are rejected because of their high adsorption on negatively charged clay surfaces. With anionics (usually sodium salts) and nonionics, conditions have to be established to minimize rock interactions (See Sec. IV).

Petroleum sulfonates, manufactured by sulfonating aromatic refinery products, and synthetic aromatic sulfonates, such as alkylxylene sulfonates, are most commonly used in EOR. However, their performances drop as salinity and divalent cation concentrations increase, resulting in low brine solubility and poor interfacial efficiency.

So, other surfactant families have been considered, which display much higher salt tolerance, namely sulfated or sulfonated ethoxylated alcohols or alkylphenols [21], α-olefin sulfonates [14,22], and nonionics [23]. The latter are often used as cosurfactants. All the surfactant families tested for EOR have been reviewed recently [24].

Cosolvents, generally C_3–C_5 alcohols, are sometimes combined with the primary surfactant to improve its solubility and to prevent the formation of highly viscous phases (lyotropic liquid crystals). But their effect on phase behavior has to be taken into account [25,26]. As far as possible, cosurfactants are preferred, in order to keep a higher level of interfacial activity.

TABLE 1 Components of Microemulsions or Micellar Solutions for EOR

Class	Composition
Oil	Crude oil (stock-tank oil) (but oil-free micellar solutions now preferred)
Brine	Formation water or water from sea, lakes, and rivers, containing from a few hundred ppm to more than 200 g/L salt (Na^+, Ca^{2+}, Mg^{2+}, Cl^-, SO_4^{2-}, etc.)
Chemicals	Primary surfactant
	Cosurfactant
	Cosolvent, e.g., C_3 to C_5 alcohols
	Water-soluble polymers
	Alkaline agents
	Bactericides, sacrificial agents, etc.

D. Phase Behavior Modeling

Compositional simulations of microemulsion flooding (see Sec. V) require an accurate knowledge of phase behavior. This can be established through long laboratory work. Another way is to build a model able to predict the stability of oil/brine/surfactant mixtures and the composition of phases that may form from these mixtures, and this with a moderate number of experiments.

A fruitful approach, developed in the 1980s, is based on the pseudophase model, which aims at representing the phase behavior of oil/brine/ surfactant/cosolvent multicomponent systems [27]. The cosolvent is assumed to be distributed among the other three components, thus forming three pseudophases. Therefore, a microemulsion phase consists of three microdomains: an aqueous pseudophase W', an oil pseudophase O', and an interface pseudophase ("membrane") M'. Oil and brine excess phases, if any, have the same composition as O' and W', respectively. Pseudophase composition can be calculated, without consideration of phase structure, from knowledge of the overall composition and the determination of several constants, among which is the partition coefficient of the cosolvent between pseudophases. Consequently, the multicomponent mixture is treated as a ternary one, in which the three pseudophases are the pseudocomponents.

In other respects, the calculation of the free enthalpy of mixing serves to determine whether excess phases exist. It takes into account the physicochemical characteristics of the system, namely the entropy of dispersion of oil and brine, molecular interactions, and the curvature energy of the interface.

IV. OIL DISPLACEMENT MECHANISMS AND ROCK-FLUID INTERACTIONS

A. Immiscible Displacement

Oil displacement by microemulsions mostly results from a flood with low-tension immiscible fluids. Going upstream toward the injection point, the following zones can be distinguished:

Zone of initial conditions and zone of preflush. Under tertiary conditions, i.e., after a continuous water injection, only brine flows, oil being at residual saturation (S_{or}).

Oil bank zone. Oil mobilized by surfactants accumulates in front of the micellar slug; then oil saturation increases and an oil bank forms. This is a region of high interfacial tension.

Between the oil bank and the micellar slug, a mixing zone develops, with two or three phases, depending on the surfactant phase behavior. This is a region of low interfacial tension. Some oil can be solubilized in the microemulsion phase and carried out as long as this phase is mobile.
Micellar zone,
Polymer zone,
Chase water zone.

The last three zones are characterized by a single-phase flow, at an oil saturation depending on interfacial tension lowering.

In these various zones, the flow is governed by the relative permeability values [28]. When two or three immiscible fluids flow simultaneously, the relative permeability of the porous medium toward a given fluid is

$$k_r = \frac{k_{eff}}{k} \tag{6}$$

where k_{eff} is the effective permeability of the medium toward the fluid considered, and k its absolute permeability.

For a given system of fluids and rock, and at given temperature and pressure, k_r is assumed to be dependent only on the local saturation of the fluid considered. Typical relative permeability curves are shown in Fig. 5 (top). Relative permeabilities depend on the capillary number through interfacial tensions and wettability. The end point relative permeability of each phase increases with N_c until it almost reaches unity at very high capillary numbers.

Relative permeabilities together with the corresponding viscosities are used to calculate the fractional flow of each phase as a function of the saturations [29,30] and to evaluate various design requirements, especially for mobility control.

Other displacement mechanisms can be involved, such as emulsification, wettability change, and mass transfer (e.g., oil solubilization with type I or type III phase behavior, or oil swelling when phase behavior corresponds to a type II system). But their effects are quite difficult to assess, and they are generally considered to be of less importance than those governing immiscible flow.

B. Mobility Control

It is crucial that the displacement of the oil bank by the micellar slug be stable (i.e., no fingering) in order to prevent the redispersion of mobilized oil. The same condition is required for the displacement of the micellar slug by the polymer slug in order to preserve surfactant slug properties. So, each bank must be displaced by a less mobile fluid.

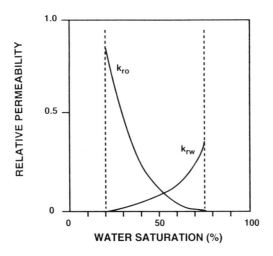

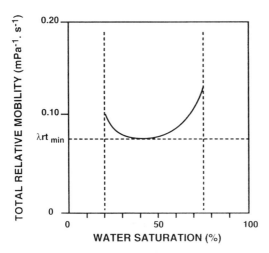

FIG. 5 Typical water-oil relative permeability (top) and total relative mobility (bottom) curves versus brine saturation. k_{ro} and k_{rw}, oil and water relative permeability, respectively. (From Ref. 31, © SPE 1970.)

The total relative mobility λ_{rT} of the oil bank is defined as the sum of the relative mobilities (relative permeability divided by viscosity) of the oil and water phases

$$\lambda_{rT} = \frac{k_{ro}}{\mu_o} + \frac{k_{rw}}{\mu_w} \tag{7}$$

The total relative mobility as a function of fluid saturation is deduced from the relative permeability curves (Fig. 5, bottom). It is advisable to take the minimum value of λ_{rT} for the design of the surfactant solution [31]. So, we need

$$\frac{k_{rs}}{\mu_s} \leq \lambda_{rTmin} \tag{8}$$

where k_{rs} and μ_s are the relative permeability and the viscosity of the surfactant solution, respectively. Therefore, μ_s must be equal to or higher than k_{rs}/λ_{rTmin}, or, assuming that k_{rs} is close to one,

$$\mu_s \geq \frac{1}{\lambda_{rTmin}} \tag{9}$$

Usually, the viscosity required for mobility control is obtained by adding water-soluble polymers, such as polyacrylamides and polysaccharides [32]. In the same way, the mobility of the polymer slug (mobility buffer) must be equal to or less than that of the surfactant slug.

C. Cation Exchange

During oil displacement, salinity may vary, especially because of cation exchange phenomena. This variation can affect phase behavior, as mentioned above, and then interfacial properties and surfactant retention.

Almost all oil reservoirs contain clays with a more or less significant cation exchange capacity (CEC). Based on the mass action law, the ionic equilibrium between clay and brine can be represented by

$$CaM_2 + 2Na^+ \rightleftharpoons 2NaM + Ca^{2+}$$

where M is a negatively charged site on clay surface. By simplification, only sodium and calcium ions are mentioned here. The equilibrium constant is called the selectivity coefficient

$$K_{Ca-Na} = \frac{X_{Na}^2 a_{Ca}}{X_{Ca} a_{Na}^2} \tag{10}$$

where X_i is the mole fraction of ion i ($i \equiv Na^+$ or Ca^{2+}) adsorbed on clay and a_i the ion activity in the brine.

Cation exchange occurs if the calcium/sodium ratio in the injected fluid differs from that of the initial fluid in place [33]. This is always the case with anionic surfactants. Moreover, cation exchange can be amplified by the preferential association of Ca^{2+} with micelles [34,35]. Generally, an increase of the divalent cation concentration in the micellar slug is observed.

If optimal salinity is high or if the exchange is low with respect to this salinity, cation exchange may not be important for surfactant design, whereas at low salinity great changes in interfacial properties can be expected from ion exchange. Anyway, it is possible either to take advantage of the cation exchange or to prevent it be removing calcium ions prior to surfactant injection, for instance, by a preflush, preferentially implemented in alkaline conditions (see Sec. V).

D. Surfactant Retention

The commercial success of a microemulsion flood depends mainly on the degree of surfactant retention. Several mechanisms, directly or indirectly related to rock-fluid interactions, are responsible for the loss of surfactants in porous media: precipitation, phase trapping, and adsorption.

Precipitation may occur when the surfactant slug comes in contact with brines of higher salinity, especially with higher divalent cation content, or because of cation exchange phenomena. Usually, in the presence of oil, the surfactant partitions into the oil rather than precipitates.

Due to surfactant transfer into the oil phase, leading to a type II environment, interfacial tensions increase, and consequently the W/O microemulsion risks being capillary trapped. The resulting surfactant loss may be very high [36]. This trapping does not occur in a type I environment because O/W microemulsions are miscibly displaced by the aqueous polymer slug. However, with a highly viscous microemulsion phase, the loss of surfactant can be significant.

Adsorption results from attraction between surfactant molecules (monomers) and mineral surfaces (mainly clays), due to van der Waals and Coulombic forces as well as hydrogen bonding. It increases reversibly with surfactant concentration until a plateau value is reached at the CMC.

The S shape of the adsorption isotherm of an alkylbenzenesulfonate on various minerals (Fig. 6) [37] points out the likelihood of a capillary condensation process. However, this isotherm can be conveniently represented as a Langmuir-type isotherm by the following equation

$$\frac{q}{q_{max}} = \frac{bC_s}{1 + bC_s} \tag{11}$$

where q is the amount of surfactant adsorbed on the rock, q_{max} the plateau adsorption value, C_s the bulk surfactant concentration at equilibrium, and b a parameter depending, as well as q_{max}, on the surface charge of the clay.

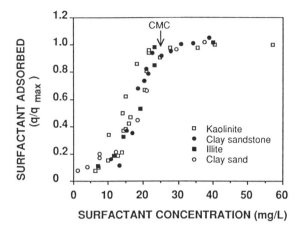

FIG. 6 Adsorption of sodium p-(1-propyl)nonylbenzenesulfonate in 10 g/L NaCl brine on kaolinite, illite, clay sandstone, and clay sand as a function of bulk surfactant concentration at equilibrium. (From Ref. 37.)

Adsorption depends on a great number of parameters, summarized in Table 2.

With sulfonates, the use of alkaline agents (sodium hydroxide, sodium carbonates, etc.) to reduce adsorption is highly efficient. The mechanism involved is a lowering of the number of positively charged sites (Fig. 7). Other ways to reduce adsorption are to use cosolvents (alcohols) in order to improve surfactant solubility in brine, or to inject sacrificial products, such as lignosulfonates and their derivatives, which are preferentially fixed on adsorption sites.

TABLE 2 Main Parameters Affecting Surfactant Adsorption

Parameter increased	Effect on adsorption
Temperature	↓
Specific surface area of rock	↑
Brine salinity	↑
pH	↓ (if anionic)
Surfactant chain length	↑

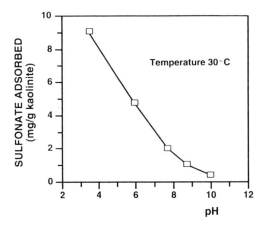

FIG. 7 Adsorption of sodium p-(1-propyl)nonylbenzenesulfonate in 10 g/L NaCl brine at the CMC (0.115 mg/L) as a function of pH.

The amount of surfactant injected needs to be higher than that required for adsorption. Practically, surfactant loss due to adsorption must be less than about 0.5 mg of surfactant per g of rock.

V. IMPLEMENTATION OF THE PROCESS

Microemulsion flooding is often referred to as tertiary recovery process. However, its use is not restricted to a particular phase in the production life of a reservoir. On the pilot scale, the process must be tested under tertiary conditions to make an evaluation of the incremental oil recoverable by microemulsion, but, for field-scale operations, it is recommended to start surfactant injection at an early stage of production.

A. Alkaline-Surfactant-Polymer Process

The first step in the process design is the selection of a surfactant able to mobilize the residual oil by lowering the interfacial tension. However, as mentioned earlier, controlling the propagation of the micellar slug across the entire oil reservoir first requires that fluid mobilities be properly controlled. This necessitates the use of water-soluble polymers. Second, the surfactant slug must be protected against excessive mass transfer and adsorption. This can be done by adding alkaline agents to the injected fluids.

For acidic crude oils, some authors recommend injecting alkalis in order to generate surface-active agents in situ by reaction with the acidic components of the crude, such as carboxylic acids (naphthenic acids). This method generally gives poor performance, unless it is applied with both surfactants (injected from the surface and now called cosurfactants) and polymers [38–40].

Thus, the microemulsion process finally consists of implementing surfactants, polymers, and alkalis in combination in order to make the best use of the properties of each additive. This serves to minimize the amount of surfactant injected and to maximize oil recovery.

The efficiency of the selected surfactants in displacing the residual oil should be checked by flood experiments with reservoir cores. Oil recovery must be higher than 90% of residual oil, under reservoir conditions. The criteria for selecting an oil field as a candidate for microemulsion flooding are shown in Table 3 (values are indicative only of orders of magnitude).

B. Salinity Gradient Technique

The main idea in the salinity gradient concept [41] is that the highest displacement efficiencies are required both at the front of the surfactant slug, to displace residual oil, and at the rear, to prevent microemulsion trapping.

At the front, the lowest interfacial tensions between surfactant and oil will be obtained if the phase behavior is near the type II–type III transition, whereas at the rear the lowest tensions between surfactant and polymer slug will be obtained near the type III–type I transition. As shown in Figs.

TABLE 3 Screening Criteria for Microemulsion Flooding

Field parameter	Requirement
Rock	Sandstones, sands, carbonates (with moderate clay content)
	Thickness <25 m
	Permeability >20 mD[a]
Crude oil	Viscosity < 60 mPa s
Brine salinity	Above 20–30 g/L salts, surfactants other than aromatic sulfonates are required (nonionics, ethoxysulfates or sulfonates, etc.)
Temperature	<80°C

[a]Permeabilities (dimension [L^2]) are expressed here in millidarcies (mD) with 1 mD $\sim 10^{-3}$ μm^2.

3 and 4, this evolution of the phase behavior can be provided by decreasing salinity.

A certain salinity of the formation water is desirable, so that a lower salinity can be applied in the surfactant slug, then a still lower salinity in the polymer slug.

At the rear of the surfactant slug, this technique also provides lower interfacial tensions (see Sec. III.B) and reduced adsorption (Table 2), thanks to lower salinity.

High displacement efficiencies both at the front and at the rear of the slug can also be provided, at constant salinity, by using multiple surfactant slugs with different compositions. The slug is split into two or three parts: at the front, the phase behavior is close to type II, while at the rear it is close to type I. This evolution of the phase behavior can be obtained, for instance, either by increasing the surfactant hydrophilicity [42] or by increasing the water/oil ratio of the slug composition [43].

C. Oil Displacement Modeling

Simulation is a necessary step for the process design and interpretation at field scale. For this purpose, multicomponent, multiphase simulators have been developed [44], especially compositional models used to treat each phase as a mixture of several components [45,46].

These simulators are based on material balance equations describing the flow processes, using Darcy's law with appropriate boundary and initial conditions. In addition to these equations, equations describing the main physical properties, such as phase behavior, interfacial tension, ion exchange, and adsorption, are needed. The main assumption is local thermodynamic equilibrium.

Simulations are used to calculate pressure, saturations, and phase composition, in order to obtain oil recovery and process efficiency. Generalized models take into account the chemical reactions associated with the use of alkaline agents [47,48]. Despite numerous simplifications and approximations, reasonably good agreement with experiments can be obtained (Fig. 8).

VI. DEVELOPMENT: FIELD OPERATIONS

The microemulsion process has developed slowly so far because of high risks and complex technology, associated with high costs. However, numerous field pilot tests, especially in the United States, have been reported, most of them concerning previously waterflooded reservoirs [44,49]. A

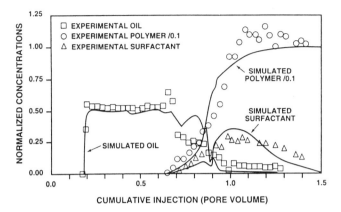

FIG. 8 Simulated and experimental total oil, surfactant, and polymer production histories. (From Ref. 46, © SPE 1987.)

strong decline in the number of projects occurred after 1986 with the drop in crude oil prices.

Many field experiments have failed or have shown performances far below expectations because of inadequate well patterns, poor knowledge of reservoir characteristics, or degradation of chemicals, for instance, polymer biodegradation, leading to the loss of mobility control. However, some pilot tests, implemented with properly designed chemicals and with accurate reservoir characterization, have been reported to be technically successful, with recoveries in the order of 50% of the oil at the start of the flood. Nevertheless, the economics remained largely unfavorable.

Two field operations must be specially mentioned because they were very well documented and involved quite different reservoir and design conditions. In both cases, the micellar flood recovered around two-thirds of the residual oil.

The Chateaurenard Field test [50]. The process was applied in a high-permeability sand containing a very low-salinity brine and a crude oil with a viscosity of 40 mPa s. A microemulsion made of brine, crude oil, petroleum sulfonates, and alcohols was injected, followed by a polyacrylamide polymer slug.

The Loudon Field test [51]. This test worked in a low-permeability sandstone reservoir containing a high-salinity brine and a 5 mPa s crude oil. In that case, a dilute aqueous surfactant solution containing xanthan gum was used, followed by a polymer drive.

Many other micellar/polymer pilot tests have also been described in detail, such as those of Salem Field [52], Ranger Field [53], and Glenn Pool Field [54].

As already mentioned, implementation of the process under alkaline conditions may be profitable for reducing adsorption and divalent-cation interactions with surfactants [55–58]. Another objective in conducting the process with alkaline agents is to lower the cost of chemicals by making use of surface-active agents generated in the reservoir if the crude has a high enough acid number.

In the mid-1990s, the research and development activity for the microemulsion process is extremely slowed down because of the low crude oil price and because of the feeling that this economic situation is going to last for several decades. Only a very few countries, such as Norway, are maintaining some investigations in this field.

VII. CONCLUSION AND FUTURE OUTLOOK

Microemulsion flooding is a technically proven process. It appears to be highly efficient for recovering oil. When conducted according to the state of the art, it can mobilize more residual oil than any other EOR method. According to the most recent investigations, this efficiency results from an appropriate combination of surfactants, polymers, and alkaline agents.

However, considerable technical advances still remain to be achieved, especially in the field of products (surfactants and polymers), phase behavior and displacement modeling, reservoir characterization, and process implementation.

The economic situation may postpone the development of the chemical process, but it can be considered as an ongoing technology open to great improvements. New and more cost-effective surfactants and polymers can be anticipated, as well as improved reservoir simulators and better reservoir characterization.

During the first half of the 21st century, oil will remain a major and difficult-to-replace source of energy. Consequently, the development of such a technology looks possible insofar as significant technical progress improves its efficiency and economics.

REFERENCES

1. See Bibliography (p. 353).
2. J. J. Taber, Soc. Petrol. Eng. J. 9:3–12 (1969).

3. G. L. Stegemeier, in *Improved Oil Recovery by Surfactant and Polymer Flooding* (D. O. Shah and R. S. Schechter, eds.), Academic Press, New York, 1977, pp. 55–91.

4. P. A. Winsor, Trans. Faraday Soc. *44*:376–398, 451–471 (1948); *46*:762–772 (1950); Chem. Rev. *68*(1):1–40 (1968).

5. J. L. Salager, in *Encyclopedia of Emulsion Technology* (P. Becher, ed.), Vol. 3, Marcel Dekker, New York, 1988, pp. 79–134.

6. A. M. Bellocq, D. Bourbon, B. Lemanceau, and G. Fourche, J. Colloid Interface Sci. *89*(2):427–440 (1982).

7. A. Graciaa, L. N. Fortney, R. S. Schechter, W. H. Wade, and S. Yiv, Soc. Pet. Eng. J. *22*(10):743–749 (1982).

8. M. Kahlweit et al., J. Colloid Interface Sci. *118*(2):436–453 (1987).

9. M. Bourrel and R. S. Schechter, *Microemulsions and Related Systems*, Marcel Dekker, New York, 1988.

10. L. E. Scriven, Nature *263*:123–125 (1976).

11. C. J. Glover, M. C. Puerto, J. M. Maerker, and E. L. Sanvik, Soc. Pet. Eng. J. *19*(3):183–193 (1979).

12. W. C. Tosch, S. C. Jones, and A. W. Adamson, J. Colloid Interface Sci. *31*(3): 297–306 (1969).

13. G. J. Hirasaki, Soc. Pet. Eng. J. *22*:471–482 (1982).

14. M. Bavière, B. Bazin, and C. Noïk, SPE Reservoir Eng. *3*(2):597–603 (1988).

15. R. L. Reed and R. N. Healy, in *Improved Oil Recovery by Surfactant and Polymer Flooding* (D. O. Shah and R. S. Schechter, eds.), Academic Press, New York, 1977, pp. 383–437.

16. H. Kunieda and K. Shinoda, J. Colloid Interface Sci. *75*(2):601–606 (1980).

17. H. Kunieda and S. Friberg, Bull. Chem. Soc. Jpn. *54*(4):1010–1014 (1981).

18. D. Langevin, D. Guest, and J. Meunier, Colloids Surfaces *19*:159–170 (1986).

19. C. Huh, J. Colloid Interface Sci. *71*(2):408–425 (1979).

20. P. A. Winsor, *Solvent Properties of Amphiphilic Compounds*, Butterworth, London, 1954.

21. A. Capelle, Tenside Detergents *20*(3):124–127 (1983).

22. B. R. Voca, J. P. Canselier, C. Noïk, and M. Bavière, Progr. Colloid Polym. Sci. *76*:144–152 (1988).

23. S. J. Lewis, L. A. Verkruyse, and S. J. Salter, SPE/DOE 14910, Fifth Symposium on EOR, Tulsa, Oklahoma, April 20–23, 1986.

24. A. Bhardwaj and S. Hartland, J. Dispers. Sci. Technol. *14*(1):87–116 (1993).

25. S. J. Salter, Paper SPE 6843, 52nd Annual Fall Techn. Conference and Exhibition, Denver, October 9–12, 1977.

26. M. Bavière, W. H. Wade, and R. S. Schechter, J. Colloid Interface Sci. *81*(1): 266–279 (1981).

27. J. Biais, P. Bothorel, B. Clin, P. Lalanne, and J. L. Trouilly, Mol. Cryst. Liq. Cryst. *152*:9–36 (1987).

28. M. C. Leverett, Petrol. Trans. AIME *132*:381–403 (1938).

29. F. G. Helfferich, Soc. Pet. Eng. J. *21*(1):51–62 (1981).

30. G. J. Hirasaki, Soc. Pet. Eng. J. *21*:191–204 (1981).

31. W. B. Gogarty, H. P. Meabon, and H. W. Milton, J. Pet. Technol. February: 141–147 (1970).
32. G. Chauveteau and K. S. Sorbie, in *Basic Concepts in Oil Recovery Processes* (M. Bavière, ed.), Elsevier Applied Science, London, 1991, pp. 43–87.
33. G. A. Pope, L. W. Lake, and F. G. Helfferich, Soc. Pet. Eng. J. *18*(12): 418–434 (1978).
34. G. J. Hirasaki, Soc. Pet. Eng. J. 22:181–192 (1982).
35. C. Noïk, M. Bavière and D. Defives, J. Colloid Interface Sci. *115*(1):36–45 (1987).
36. M. Delshad, M. Delshad, D. Bhuyan, G. A. Pope, and L. W. Lake, SPE/DOE 14911, Fifth Symposium on Enhanced Oil Recovery, Tulsa, Oklahoma, April 20–23, 1986.
37. D. Defives and B. Bazin, Second European Symposium on Enhanced Oil Recovery, Paris, November 8–10, 1982, Technip, Paris, pp. 113–119.
38. P. H. Krumrine, J. S. Falcone, and T. C. Campbell, Soc. Pet. Eng. J. 22: 503–513 (1982).
39. P. J. Shuler, D. L. Kuehne, and R. M. Lerner, J. Pet. Technol. January: 80–88 (1989).
40. R. C. Nelson, Chem. Eng. Prog. March: 50–57 (1989).
41. G. J. Hirasaki, H. R. van Domselaar, and R. C. Nelson, Soc. Pet. Eng. J. *23*(3):486–500 (1983).
42. M. Bavière, J. C. Moulu, T. Paal, G. Tiszai, G. Gaal, L. Schmidt, and G. Gesztesi, U.S. Patent 4,648,451 (1987).
43. S. Thomas, S. M. Farouq Ali, and S. B. Supon, presented at the ACS Symposium on Advances in Oilfield Chemistry, at the Third Chemical Congress of the North American Continent, Toronto, June 5–11, 1988.
44. S. Thomas and S. M. Farouq Ali, J. Can. Pet. Technol. *31*(8):53–60 (1992).
45. N. Van Quy and J. Labrid, Soc. Pet. Eng. J. *23*(3):461–474 (1983).
46. D. Camilleri, A. Fil, G. A. Pope, B. A. Rouse, and K. Sepehrnoori, SPE Reservoir Eng. 2(4):441–451 (1987).
47. D. Bhuyan, L. W. Lake, and G. A. Pope, SPE Reservoir Eng. 5(5):213–220 (1990).
48. M. R. Islam and A. Chakma, J. Pet. Sci. Eng. *5*:105–126 (1991).
49. G. Moritis, Oil Gas J. April: 51–79 (1992).
50. D. Chapotin, J. F. Lomer, and A. Putz, presented at SPE/DOE 14955, Fifth Symposium on EOR, Tulsa, Oklahoma, April 20–23, 1986.
51. T. R. Reppert, J. R. Bragg, J. R. Wilkinson, T. M. Snow, N. K. Maer Jr., and W. W. Gale, SPE/DOE 20219, presented at the Seventh Symposium on Enhanced Oil Recovery, Tulsa, Oklahoma, April 22–25, 1990.
52. J. W. Ware, SPE 11985, presented at the 58th Annual Technical Conference and Exhibition, San Francisco, October 5–8, 1983.
53. S. M. Holley and J. L. Cayias, SPE Reservoir Eng. 7(1):9–14 (1992).
54. J. H. Bae, SPE/DOE 27818, presented at the Ninth Symposium on Improved Oil Recovery, Tulsa, Oklahoma, April 17–20, 1994.

55. F. F. J. Lin, G. J. Besserer, and M. J. Pitts, J. Can. Pet. Technol. *26*(6):54–65 (1987).
56. S. R. Clark, M. J. Pitts, and S. M. Smith, SPE 17538, presented at the Rocky Mountain Regional Meeting, Casper, Wyoming, May 11–13, 1988.
57. B. Bazin, C. Z. Yang, D. C. Wang, and X. Y. Sue, SPE 22363, presented at the International Meeting, Beijing, China, March 24–27, 1992.
58. M. Bavière, P. Glénat, V. Plazanet, and J. Labrid, SPE/DOE 27821, SPE Reservoir Eng., *10*(3):187–193 (1995).

BIBLIOGRAPHY

D. O. Shah and R. S. Schechter, eds. *Improved Oil Recovery by Surfactant and Polymer Flooding*, Academic Press, New York, 1977.
P. Neogi, in *Microemulsions: Structure and Dynamics* (S. E. Friberg and P. Bothorel, eds.), CRC Press, Boca Raton, FL, 1987, pp. 197–212.
C. A. Miller and S. Qutubuddin, in *Interfacial Phenomena in Apolar Media* (H. F. Eicke and G. D. Parfitt, eds.), Marcel Dekker, New York, 1987, pp. 117–185.
E. C. Donaldson, G. V. Chilingarian, and T. F. Yen, eds., *Enhanced Oil Recovery*, Elsevier, New York, 1988.
L. W. Lake, *Enhanced Oil Recovery*, Prentice Hall, Englewood Cliffs, NJ, 1989.
M. Bavière, ed., *Basic Concepts in Enhanced Oil Recovery Processes*, Elsevier Applied Science, London, 1991.

17

Use of Microemulsions for the Extraction of Contaminated Solids

KARIN BONKHOFF, MILAN J. SCHWUGER, and GÜNTER SUBKLEW Institut für Angewandte Physikalische Chemie, Forschungszentrum Jülich GmbH, Jülich, Germany

I. INTRODUCTION

During the past three decades all over the world, protection of the environmental compartments of water, soil, and air has become of growing importance to many people. However, there are some important differences in the advances in obtaining clean soil and water in comparison with the goal of unpolluted air. Reducing the content of harmful substances in the atmosphere by low-emission techniques and stack gas cleaning can only be directed toward the present time and the future; a retrospective reduction of atmospheric pollution is not possible. With respect to water and soil, the conditions are different in some cases. Avoiding input of pollutants into these compartments is one part of the strategy. The other part is the destruction and removal of pollutants resulting from past activities. Mainly soil and sediments must be listed here.

In Germany, for example, the number of sites thought to be contaminated by residues from waste disposal or industrial activities is increasing continuously. At the end of 1993, the provisional result of a nationwide listing was 139,000 sites suspected of contamination, 86,000 caused by waste disposal and 53,000 by industry. In addition, there are about 4000 sites thought to be polluted by armament factories and a further 10,000 military sites that are being tested for chemotoxic substances. In the future, the number of sites with an urgent need for decontamination is estimated to be about 25,000, deduced from experience in the past [1].

At sites contaminated by uncontrolled land filling, construction engineering measures to make the site secure are mainly applied. Remediation of soil charged with organic and/or inorganic pollutants is done mostly at sites of former industrial activities. Methods of biotechnology, physical chemistry, or thermal treatment are put into practice for on-site or off-site soil decontamination, individually or in combination. Sometimes selected techniques are also applied to clean wastewater, ground water, and exhaust air. The objective of these steps is to avoid a negative environmental impact of the soil during decontamination and after reinstallation. The definitive elimination and destruction of the separated contaminants is carried out by thermal treatment under oxidizing or reducing conditions (incineration, gasification, or pyrolysis) or by microbiological degradation.

The Institute of Applied Physical Chemistry at the Research Centre Jülich implements intensive research activities in the extraction of organic contaminants from solid industrial residues by the use of microemulsions. The application of this method is shown by way of example in the isolation of organic contaminants from fine soil particles during soil remediation.

II. SOIL-WASHING TECHNIQUES

Soil washing has emerged as an efficient treatment technology for the remediation of contaminated soils. The process is based on two basic assumptions. First, a significant part of the pollutant is associated with the fine fraction of the soil, and second, the other part of the contamination, which is assigned to the coarse fraction, is adsorbed mainly on the surface of these particles. By mixing the washing solution and the soil, three process steps can be distinguished. Intensive stirring breaks up larger soil agglomerates, setting free highly contaminated fine particles. It also scrubs off the pollutant from the surface of the coarse fraction [2]. In addition, the contaminants can be dissolved in the washing solution. Separation gives a "clean" sand fraction besides the washing solution charged with the highly contaminated fine particles. In connection with biological treatment, the fine fraction can be removed by flocculation and sometimes treated in a slurry bioreactor, while the water is also cleaned in a bioreactor before flowing back into the washing process. Alternatively, the separated fine fraction can be deposited or treated in an incineration plant.

A. Without Surfactants

This method can easily be applied to soils consisting mostly of relatively large grains like sand, but remediation becomes difficult for soils with a high amount of silt. Only polar pollutants, which are easily soluble in water, can be remediated by this process.

B. Surfactant Enhanced

Another technique is washing with water with the addition of surfactants. Here, a solubilization process is involved in the washing mechanism. In general, this method is more effective than washing with pure water. The solubilization capacity of such a micellar solution is determined by the volume of the micellar core and the micellar concentration.

Previous investigations of surfactants applicable for soil remediation have generally been focused on the physical mechanism of contaminant removal, release from the solid surface by altering the wettability or reducing the interfacial tension in particular. Surfactants have often been investigated and shown good results for the purpose of enhanced oil recovery. However, the properties of such surfactants, like resistance to biodegradability and tolerance to large temperature jumps, make then unsuitable for soil remediation [3]. Rickabaugh et al. [4] performed batch and bench-scale test for removing chlorinated hydrocarbons with 14 different surfactants. Blends of anionic and nonionic surfactants showed better removal properties than each individual nonionic or anionic surfactant. Cationic surfactants were not appropriate for this purpose. Peters et al. [5] screened 21 surfactants (anionic, cationic, and nonionic) for their use in mobilizing organic pollutants from soil prior to bioremediation. They found the anionic surfactants to be best suited for removing diesel. Abdul et al. [6] selected 10 nonionic and anionic surfactants for the removal of petroleum products from sandy aquifers, only 6 of which were biodegradable. Besides the properties of solubility and surface tension, they performed tests of soil dispersion, which is an important parameter for the in situ cleanup of soils.

As most of the investigations were performed with surfactant concentrations of about 2% (w/w), this technique of surfactant flushing can be economical only if a process of surfactant recovery is included [7]. Gannon et al. [8] tested the extraction for removing *p*-dichlorobenzene, biphenyl, and naphthalene from sodium dodecyl sulfate (SDS) solutions. Removal was successful, but the extraction times were not practicable. They used the anionic surfactant SDS because anionic surfactants are less soluble in organic extraction solvents than nonionic surfactants. Underwood et al. [9] improved the extraction technique for the same contaminants from SDS solution into hexane with a continuous countercurrent flow column.

C. With Microemulsions

A microemulsion intensifies the advantages of surfactant solution, such as decreasing interfacial tension and increasing wettability of soil. Microemulsions are known to be excellent solvents for polar and nonpolar organic substances. As they are formed by water, oil, and surfactant, they exhibit sites for the solubilization of ions or dipoles as well as for hydrocarbons or amphiphiles. The important difference between soil washing with surfactant solutions and the microemulsion treatment is the additional solubility of the pollutant in the oil component of the microemulsion, which further enhances uptake [10]. Therefore, this extraction technique is applicable to soils containing high amounts of silt and clay in particular. Due to their low interfacial tension, microemulsions show excellent wetting behavior for the silt fraction of a soil. As it is inevitable that a small amount of microemulsion remains on the soil particles, microemulsions applied in soil remediation must be composed of biodegradable oils and surfactants.

III. CONCEPT OF SOIL REMEDIATION
WITH MICROEMULSIONS

Starting with the washing techniques and taking into account the properties of a microemulsion, a concept of soil remediation with microemulsions was developed. Soil contaminated with, e.g., polycyclic aromatic hydrocarbons (PAHs) is split up into a low-polluted fraction (large grain size) and a highly polluted fraction (small grain size), which may be the clay compound of the soil. The material to be cleaned and the microemulsion are intensively mixed. During the extraction process, the pollutants pass from the soil particles into the microemulsion. After the extraction step, the contaminant-rich microemulsion and the solid soil particles are separated. Washing with water and inoculation with microorganisms lead to biologically active soil. Therefore the small quantities of oil and surfactant adsorbed at the soil compartments must be biodegradable in a reasonable time by natural processes.

The microemulsion carrying the contaminant can be separated into a surfactant-rich phase (low content of contaminants) and an oil-rich phase (high content of contaminants). The surfactant-rich phase is recycled to form a new microemulsion. The oil-rich phase and the water from the soil-rinsing step are treated either separately or together. It is planned to degrade the contaminants together with the residual surfactant and oil by selected microorganisms. Alternatively, a thermal treatment of the highly contaminated oil phase may be undertaken. As result of the biodegradation process, PAH-free water and a surplus of adapted microorganisms are produced [11–15].

A. Basic Considerations

The research activities on remediation with microemulsions presented here are limited to on-site or off-site treatment, e.g., to the discharge from a washing-plant hydrocyclone. This material is highly polluted with PAHs, and the extraction results are focused mainly on pyrene as one represen-

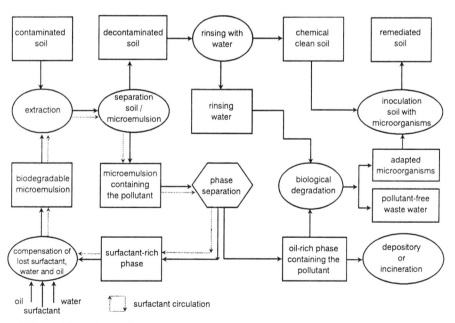

FIG. 1 Flow sheet of the remediation process with microemulsions.

tative of the PAHs. The pyrene was analyzed by ultraviolet (UV) spectroscopy, high-performance liquid chromatography (HPLC), or gas chromatography. Appropriate microemulsions for the extraction were evaluated and investigated, keeping in mind the necessity of biological degradation of all the components and the right phase behavior for the concept. In Table 1 the surfactants and oils used in this work are listed.

B. Contaminant Isolation

The efficiency of microemulsions for soil remediation was evaluated by extraction experiments performed on a laboratory scale with 15 mL of extraction solvent and 2.5 g of soil. The samples were shaken for 2 h and then centrifuged. The temperature was always 25°C. The oil components were chosen through solubility experiments, and different surfactants with various hydrophile-lipophile balance (HLB) values were applied as listed in Table 1.

1. Solubilization Capacity

The determination of the solubility of PAHs, especially pyrene, in different extraction media, such as water, surfactant solutions, hydrocarbons, native oils, and microemulsions, should underline the advantages of microemulsions for pollutant isolation. In Table 2 the maximum amounts of pyrene solubilized in several solutions are given at a temperature of 25°C.

The very low solubility of pyrene in water [16] can be raised by adding a nonionic surfactant [17]. The hydrophobic solute can be dissolved in the nonpolar phase in the interior of the surfactant micelles, if the surfactant

TABLE 1 Selected Nonionic Surfactants

Trade name and supplier	Chemical type (EO = number of ethoxy groups per molecule)	HLB	CMC (ppm)
Igepal Ca-520, Aldrich	Branched alkyl-phenyl-ethoylate $i\text{-}C_8\text{-}C_6H_4\text{-}EO_5$		
Marlipal 24/X0, Hüls	Alkyl-polyethoxylate $C_{12/14}EO_{2 \text{ to } 10}$	6.2–13.9	11–19
APG, Hüls	Alkyl-polyglucoside $C_{10/12}\text{-}APG$	13	
	$C_{12/14}\text{-}APG$	20	
Brij 52, ICI	Alkyl-polyethoxylate $C_{16}EO_2$	5.3	

TABLE 2 Solubilization Capacity of Pyrene

Solution [% (w/w)]	g/L
Water	0.0001
10% $C_{12/14}EO_7$	0.7
10% $C_{12/14}EO_7$ + 3% rape oil	2.7
12% i-C_8-C_6H_4-EO_5 + 44% isooctane	3.8
n-Hexane	17.2
n-Dodecane	27.9
Rape oil	50.5
Sunflower oil	65.5
Rape oil–methyl ester	69.5

concentration exceeds the critical micelle concentration (CMC). The solubility of a single component, like pyrene, p-dichlorobenzene, or biphenyl, in micellar surfactant systems should rise linearly with increasing amount of surfactant [18–21]. Investigations with mixtures of hydrophobic substances show only small differences in the individual solubilities of the single components with increasing surfactant concentrations as function of the mole fractions in the mixture [22]. A 10% (w/w) solution of $C_{12/14}EO_7$ was chosen as a pure surfactant system. Adding 3% (w/w) rape oil to this micellar surfactant solution enhanced the solubility overproportionally.

The microemulsion with 42% (w/w) water + 42% (w/w) isooctane + 16% (w/w) i-C_8-C_6H_4-EO_5 is not biodegradable, but as the phase behavior is well known from the literature it was used as a "reference microemulsion" [23]. Although this microemulsion contains a lot of surfactant and oil [58% (w/w) together], pyrene solubility is only a little better than in the system mentioned above with 13% (w/w) organics. The solubility capacities of the pure native oils and their derivatives, which are larger than those of the linear hydrocarbons, make them suitable for an extraction component in a microemulsion.

2. Influence of Added Oil

Systematic extraction experiments were performed beginning with surfactant solutions [5% (w/w)] without additives and then adding 0.5% (w/w) rape oil (RO) and at least 0.5% (w/w) rape oil–methyl ester (RME). The equilibrium concentrations of pyrene after extraction are represented in a bar diagram in Fig. 2.

The general trend revealed after applying the pure surfactant solutions is a maximum concentration of pyrene for the surfactant $C_{12/14}EO_7$ and decreasing values for the more hydrophilic and more hydrophobic surfactants ($C_{12/14}EO_4$ forming liquid crystalline phases). With the more hydro-

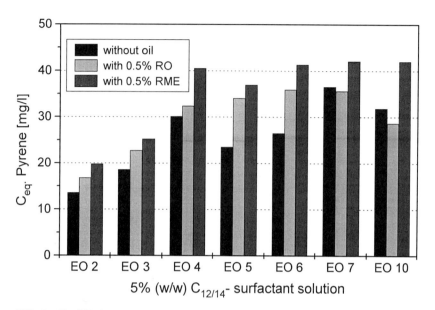

FIG. 2 Equilibrium concentrations of pyrene after extraction for different systems.

phobic surfactants $C_{12/14}EO_2$ to $C_{12/14}EO_6$, the addition of rape oil increases the equilibrium concentration of pyrene, whereas with the more hydrophilic surfactants such as $C_{12/14}EO_7$ and $C_{12/14}EO_{10}$, the rape oil has no pronounced effect. Adding rape oil–methyl ester in all extraction experiments leads to a significantly higher pyrene concentration being detected in comparison with the pure surfactant solutions.

3. Influence on the Extraction Results of the Soil to Extraction Solution (s/e) Ratio and the Number of Extraction Steps

Another parameter for extraction is the ratio of soil to extraction solution (s/e). The equilibrium concentration of the contaminants depends strongly on this ratio if the composition of the extraction solution is kept constant. To investigate this influence in more detail, experiments with four systems at four different s/e ratios were performed. The results are given in Fig. 3.

The pyrene concentration does not rise linearly with increasing amount of surfactant. The system with 10% (w/w) surfactant leads to higher equilibrium concentration of pyrene than the system with 5% (w/w) surfactant, but with respect to the amount of detergent the 5% (w/w) system is more effective. The variation of the s/e ratio from 1/2 to 1/10 causes a strong

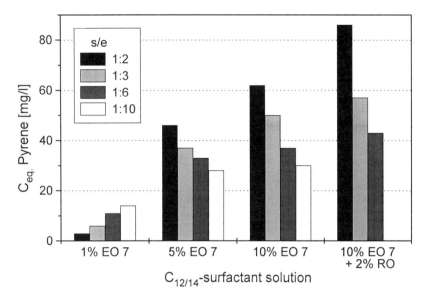

FIG. 3 Equilibrium concentrations of pyrene after extraction for different s/e ratios.

decrease in the equilibrium concentration of pyrene for systems with more than 5% (w/w) surfactant, which is in accordance with measurements by Liu and Roy [24]. For systems with lower initial surfactant concentration [1% (w/w)] the adsorption of surfactant onto soil leads to the opposite effect.

Calculating the extraction results by keeping the total extraction volume constant shows that an extraction performed with an s/e ratio of 1/2 is much more effective in removing pyrene from soil than an s/e ratio of 1:10. Figure 4 shows the advantages of a small s/e ratio and a multiple extraction process for a 10% (w/w) $C_{12/14}EO_7$ + 3% (w/w) rape oil–methyl ester solution. The extraction results for the different s/e ratios were taken from a one-step extraction (see Fig. 4) and kept constant for each subsequent step. Although the process with an s/e ratio of 1:2 supplies the lowest absolute extraction result of 32% (w/w) for one step, under the conditions of a fixed total volume of extraction medium its loading capacity is the best.

C. Extraction Kinetics

Aside from equilibrium solubility, the kinetics of PAH transport from the soil surface into the microemulsion is important. The low interfacial tension

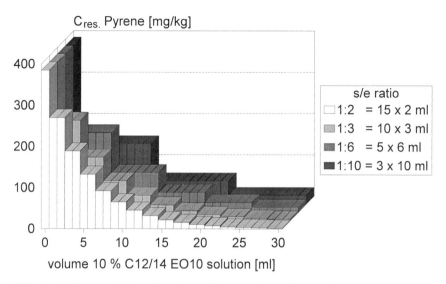

FIG. 4 Influence of the s/e ratio and the number of extraction steps for a 10% (w/w) $C_{12/14}EO_7$ extraction solution.

of a microemulsion should cause good wetting behavior for the soil, so the removal of the contaminants and the following solubilization should occur within a short time. To check this assumption, several kinetic measurements were performed. Real contaminated soil was extracted with different surfactant solutions ($C_{12/14}EO_7$ and $C_{12/14}EO_{10}$) and microemulsions. The s/e ratio was set at 1:10. Figure 5 shows the extraction results for pyrene for several solutions with $C_{12/14}EO_2$ as a function of extraction time.

Based on the assumption that the pyrene adsorption is described by the initial part of a Langmuir isotherm, the extraction kinetic can conveniently be fit by the exponential expression

$$C_{pyr} = a \times [1 - \exp(-b \times t)]$$

Here C_{pyr} is the concentration of pyrene in the extraction solution at time t. The parameter a shows the maximum equilibrium pyrene concentration, and b is a measure of the extraction kinetics, including the adsorption and desorption rates (Table 3).

For the systems with $C_{12/14}EO_7$, the addition of rape oil and especially of rape oil–methyl ester leads to a pronounced acceleration of the extraction kinetic. The equilibrium pyrene concentration is increased only in the system with rape oil–methyl ester. The kinetics of the systems with $C_{12/14}EO_{10}$ are not influenced significantly by added oil. As already shown

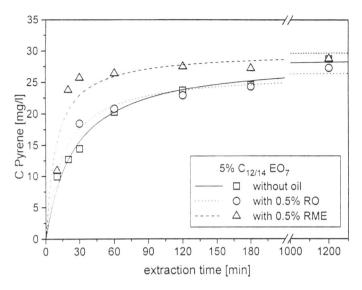

FIG. 5 Extraction kinetics for pyrene with different extraction solutions. Extraction solutions with $C_{12/14}EO_7$.

in Fig. 2, the equilibrium concentration of pyrene in the extraction systems with $C_{12/14}EO_1$ is lowered by rape oil and increased by rape oil–methyl ester.

D. Adsorption Phenomena

1. Sorption Properties of PAHs

Besides the pure solubility of PAHs in different extraction solvents, the effect of added soil on the equilibrium concentration of PAHs in solution

TABLE 3 Parameters of the Extraction Kinetics

Extraction system	a [mg/l]	b [1/min]
5% (w/w) 24/70	23.9 ± 1.2	0.037 ± 0.005
5% (w/w) 24/70 ± 0.5% RO	23.3 ± 0.8	0.049 ± 0.009
5% (w/w) 24/70 ± 0.5% RME	27.7 ± 1.3	0.072 ± 0.013
5% (w/w) 24/100	19.6 ± 1.5	0.065 ± 0.019
5% (w/w) 24/100 ± 0.5% RO	16.4 ± 0.8	0.067 ± 0.012
5% (w/w) 24/100 ± 0.5% RME	20.9 ± 1.1	0.067 ± 0.013

has to be investigated. The sorption of PAHs and other organic substances depends largely on the aqueous solubility and on the organic carbon content of the soil. Increasing solubility causes, in general, a decrease in adsorption. A high content of organic carbon in the solid means a large amount of PAHs is adsorbed [25].

In order to quantify these effects, various amounts of pyrene were added to pure water, to a surfactant solution, and to a surfactant solution with rape oil–methyl ester. Equal amounts of kaolin were added as a model clay. After shaking the samples for several hours and subsequent centrifugation, the concentrations of pyrene in the solutions were detected either by UV spectroscopy or by HPLC. The amounts of pyrene adsorbed on kaolin were calculated from the different concentrations in the solutions in the absence and presence of soil.

The adsorption of pyrene onto soil from pure water is lowered by adding surfactant, as would be expected from the solubilization measurements [17]. The mass of adsorbed PAHs on soil is shifted to lower values with increasing surfactant concentration [26,27]. Additional oil raises the solubility of pyrene in the solution and further reduces adsorption on kaolin.

2. Adsorption of Surfactant

An important point in the soil remediation concept with microemulsions is knowledge of the adsorption properties of the surfactants. On the one hand, the loss of surfactant by adsorption onto soil can change the composition of the microemulsion in such a way that it falls apart during extraction. This will lead to a surfactant-rich phase, which is less effective for removal of the contaminant, and to an oily phase from which oil will also preferably adsorb onto soil. The adsorption properties are therefore important in determining the composition of the utilized microemulsion. On the other hand, the residual surfactant, which cannot be removed even in subsequent rinsing steps, means additional costs for the soil remediation process. Particularly for soil with a high amount of carbon (humus), for which the adsorption of nonionic surfactants is known to be very strong, this contribution cannot be neglected.

To clarify these facts, experiments with the nonionic $C_{12/14}EO_7$ in the presence of two different real contaminated soils [fine soil fraction, 14% (w/w) C; real soil, 3.3% (w/w) C] were conducted. The initial concentration of the surfactant varied from 10 to 100 g/L, which is a typical range of application in soil-washing plants. Figure 6 shows the two adsorption isotherms corresponding to the different soil types. The equilibrium concentrations of the surfactant were analyzed by HPLC and by measuring the total amount of carbon (TC).

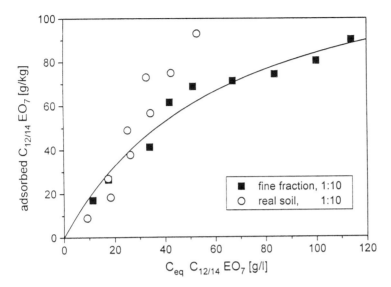

FIG. 6 Adsorption isotherms for $C_{12/14}EO_7$ on two different soils.

The adsorption isotherm of $C_{12/14}EO_7$ on fine soil show a Langmuir-type increase with rising bulk solution surfactant concentration. The corresponding isotherm on real soil gives slightly higher values up to 110 mg/g at an equilibrium surfactant concentration of 60 g/L. The lack of homogeneity of the real soil causes the scattering of the measurements. This adsoption behavior is consistent with neither reported plateau values of the sorbed surfactant if the equilibrium concentration exceeds the CMC (for surfactants forming micelles, which should apply to $C_{12/14}EO_7$) nor the adsorption isotherm obtained from surfactants forming lamella-like $C_{12}E_4$ [28]. Abdul and Gibson [29] investigated the sorption of a related commercial surfactant ($C_{10-12}EO_7$) onto sandy aquifer and found an increasingly curvilinear isotherm. It must be mentioned again that the cited results from the literature were observed for surfactant solutions in a concentration range about the same as the CMC, whereas in the case of the $C_{12/14}EO_7$ system, the surfactant concentration (up to 120 g/L) is some orders of magnitude higher than the CMC (0.015 g/L). For the adsorption of $C_{12/14}EO_7$ onto sand in the lower concentration range, we also found a Langmuir-type adsorption isotherm with a plateau value of about 0.55 g surfactant/kg sand at surfactant concentrations exceeding 0.3 g/E.

3. Uptake of PAHs as a Function of Surfactant Concentration

The linear dependence of the solubility of PAHs with increasing surfactant concentration changes into a nonlinear relationship if soil is present in the system. Besides the partition of the PAHs (defined by K_p for a fixed surfactant concentration) between soil and bulk solution, the adsorption of the surfactant (defined by K_s) also takes place. On the one hand, additional sorbed surfactant on the soil surface can reduce further PAH adsorption by lowering the available surface area. On the other hand, it can cause synergetic effects for PAH partition by hydrophobization. These two contributions result in a nonlinear increase in PAH uptake from the soil into the solution with increasing surfactant concentration. Measurements with different contaminated soil types and various amounts of surfactants were made to determine the equilibrium pyrene concentration as a function of surfactant concentration. The results for the fine soil fraction are shown in Fig. 7.

The equilibrium surfactant concentration rises linearly for both systems. However, the pyrene concentration in the bulk phase terminates in a plateau value. Edwards et al. [21] obtained similar results for the uptake of different PAHs into a bulk solution of a model surfactant. They estimated a math-

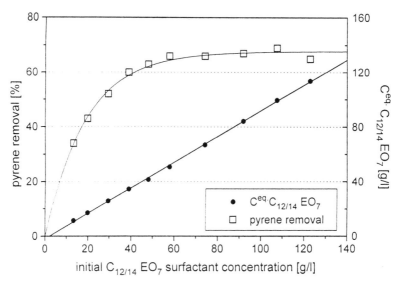

FIG. 7 Pyrene uptake from a fine soil fraction at various $C_{12/14}EO_7$ concentrations.

ematical model, including two different partition coefficients and some other parameters from independent measurements, which is in good agreement with their experimental data. This relationship is important for technical applications, because it shows the failure of an extraction solution to double the removal of pollutants by increasing the surfactant concentration by a factor of 2.

E. Phase Behavior

An important point of the remediation concept is the phase separation of the microemulsion after extraction by lowering the temperature. This behavior of a microemulsion is well known from the literature for model systems [23]. The phase behavior of the reference microemulsion was investigated to test this separation step under real conditions. The best way to do this is to note the coexisting phases of samples of the ternary system (water, isooctane, i-C_8-C_6H_4-EO_7) with the different overall surfactant concentration γ as a function of temperature at constant α in a so-called "fish cut" [30–32], with:

$$\alpha = \frac{\text{mass oil}}{\text{mass (oil + water)}} \quad \text{and} \quad \gamma = \frac{\text{mass surfactant}}{\text{mass (oil + water + surfactant)}}$$

From this phase diagram the corresponding concentrations of the three components forming a microemulsion in a certain temperature region can be calculated. For the reference microemulsion a symmetrical fish shape (Fig. 8a), as known from the literature for pure surfactant-alkane-water systems, did not result (Fig. 8c, model phase diagram). This is caused by free alcohol, included in most commercial surfactants as a by-product. The free alcohol raises the polarity of the oil, and therefore the repulsive forces between the hydrophilic parts of the surfactant and the oil become smaller. The consequence is a decreasing mean temperature of the three-phase temperature region as a function of γ. As this effect depends on the overall surfactant concentration, only the tail of the fish is shifted to lower temperatures at higher surfactant concentrations [33].

For real applications, commercial $C_{12/14}EO_2$ with an HLB value of 6.2 as the surfactant and rape oil as the oil component were tested. Here a three-phase region is observed for $\alpha = 0.3$ at a temperature of about 50°C (Fig. 8b). But this pseudo-three-component system is complicated, not only because of the effect of free alcohol but also because of the wide distribution of the homologues of the surfactant, which also causes a largely skewed fish [34]. The oil component itself is a mixture with various fatty acids. The detected three-phase region thus lies close to the water-surfactant side of the phase triangle at low temperatures and is shifted toward

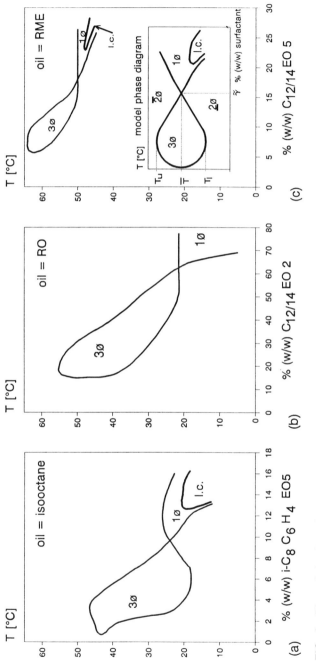

FIG. 8 Phase behavior of the ternary systems: (a) water + isooctane + i-C_8-C_6H_4-EO_7, $\alpha = 0.5$; (b) water + rape oil + $C_{12/14}EO_2$, $\alpha = 0.3$; (c) water + rape oil–methyl ester + $C_{12/14}EO_5$, $\alpha = 0.5$, and a schematic model phase diagram.

the oil-rich side with increasing temperature. The region of existence of the microemulsion is limited within a small concentration range due to the extensive region of liquid crystalline phases.

Similar phase behavior can be detected in a system with rape oil–methyl ester and $C_{12/14}EO_5$ as the surfactant. As the HLB value of rape oil–methyl ester is higher than that of rape oil, a surfactant with a higher HLB value (10.6) was likewise chosen for this system. Figure 8c shows the fish cut for this system with $\alpha = 0.5$. A ternary system with more hydrophilic oil and surfactant causes the appearance of liquid crystalline phases to shift to lower temperatures.

F. Additional Investigations

Besides the given results, a wide range of further experiments was performed. One point of interest was the extraction dynamics for the reference system. But since a different soil sample with a higher amount of contamination and a different s/e ratio was used, the aqueous concentrations of PAHs are not directly comparable to the results shown in Fig. 2. For this reference system the analysis of the pollutants was expanded to include other four- to six-ring PAHs. Because the initial concentration of the various PAHs on the soil was different, the extraction results (total amount of a single PAH in the aqueous phase/total initial amount of the single PAH on the soil) were compared. All four-ring PAHs show extraction results of about 70% at the end of the experiment (70 h), but for the five- and six-ring PAHs only 60% was removed from the soil. This trend is consistent with the individual solubility of the PAHs in the oil component, which has a direct influence on the solubility of the single PAH in the microemulsion and therefore on the extraction results [12,35].

Alkyl polyglucosides, which are known to be easily biodegradable [36,37], were also tested in investigating the phase behavior of nonionic surfactants with rape oil and rape oil–methyl ester. But the technical surfactants used, $APG_{10/12}$ and $APG_{12/14}$, both with a degree of polymerization of 1.3, did not show any three-phase region in the investigated temperature region, neither with rape oil nor with rape oil–methyl ester. Balzer [38] and Clemens [39] found high sensitivity of this surfactant group to electrolyte, which is unusual for nonionic surfactants, and a strong dependence of the cloud temperature on the alkyl chain length of the surfactants. As the HLB values of 14.5 and 13.0 for these two surfactants seem to be too high, a surfactant mixture of $APG_{10/12}$ and a technical alkyl polyethoxylate (Brij 52) with an HLB value of 5.3 was tested. A three-phase region and the corresponding microemulsion can be detected in a temperature range of about 50°–80°C with a mixture ratio of 3:7 (APG/Brij 52). By changing the surfactant ratio to 4 : 6 (APG/Brij 52), the fish is raised to a temperature of 80°C and higher. In the first case, the whole region of the three-phase body could not be measured, because below 50°C a solid surfactant-rich phase precipitates. Varying the HLB value of the surfactant, depending on the oil used, is thus the right way to evaluate a system forming a microemulsion, but the group of alkyl polyglucosides is unsuitable as a surfactant for application in soil remediation.

IV. OUTLOOK

The extraction experiments discussed show the enhanced effect of a microemulsion as an extraction solvent for organic hydrophobic pollutants, enriched in the fine grain size material of soil even after a washing process. This method of extracting pollutants from solid particles is not limited to PAHs but is also applicable to polychlorinated biphenyls (PCBs) and other organic substances. On the other hand, each type of contaminant and each mixture of chemotoxic components to be extracted require a well-adapted microemulsion; in particular, the surfactant and oil must be selected very carefully. This affects not only the physicochemical properties of the different components, like adsorption behavior on the soil particles and solubilization capacity of the oil, but also the microemulsion itself. The microemulsion stability must be guaranteed within a defined temperature and concentration range, and phase separation should be achieved completely by lowering the temperature by about 10°C.

Another important property is the biodegradability of oil, surfactant, and the mixture of both microemulsion components in the presence or absence of contaminants. Besides the degradability of these substances in a solution (bioreactor), the degradation of the residual amounts of surfactant and oil adsorbed at the soil must be guaranteed.

The application of microemulsions as extraction agents combined with the microbiological degradation of harmful contaminants has a huge potential for future developments in careful soil remediation methods.

REFERENCES

1. V. Franzius, Umwelt Technologie Akutell 6:463 (1993).
2. D. D. Chilcote and T. J. Chresand, in *Gas, Oil, Coal and Environmental Biotechnology*, (C. Akin, ed.), Institute of Gas Technology, Chicago, 1990, pp. 263–277.
3. J. K. Currie, A. L. Bunge, D. M. Updegraff, and W. H. Batal, Hydrocarbon Contam. Soils 2:641 (1992).

4. J. Rickabaugh, S. Clement, and R. F. Lewis, in *Surfactant Scrubbing of Hazardous Chemicals from Soil*, Proc. 41st Purdue Industrial Waste Conference, *41*:377 (1986).

5. R. W. Peters, L. Shem, C. D. Montemagno, and B. A. Lewis, in *Gas, Oil, Coal and Environmental Biotechnology*, (C. Akin, ed.), Institute of Gas Technology, Chicago, 1990, pp. 121–147.

6. A. S. Abdul, T. L. Gibson, and D. N. Rai, Ground Water *28*(6):920 (1990).

7. W. E. Ellis, J. R. Payne, and D. G. McNabb, Treatment of Contaminated Soils with Aqueous Surfactants, U.S. EPA Report No. EPA/600/2-85/129, PB 86-122561, 1985.

8. O. K. Gannon, P. Bibring, K. Raney, J. A. Ward, D. J. Wilson, J. L. Underwood, and K. A. Debelak, Sep. Sci. Technol. *24*(14):1073 (1989).

9. J. L. Underwood, K. A. Debelak, and D. J. Wilson, Sep. Sci. Technol. *28*(9): 1647 (1993).

10. K. Stickdorn and M. J. Schwuger, Tenside Surfactants Detergents *31*(4):218 (1994).

11. W. D. Clemens, F.-H. Haegel, M. J. Schwuger, K. Stickdorn, G. Subklew, and L. Webb, in *Contaminated Soils '93*, Vol. II, Fourth International KfK/TNO—Conference on Contaminated Soil, Berlin, May 3–7 (F. Arendt, G. J. Annokkée, R. Bosman, and W. J. van der Brink, eds.), Kluwer Academic Publishers, Dordrecht, 1993, pp. 1315–1323.

12. W. D. Clemens, F.-H. Haegel, K. Stickdorn, G. Subklew, L. Webb, Commun. Jorn. Com. Esp. Deterg. *24*:35–44 (1993).

13. K. Bonkhoff, W. D. Clemens, F.-H. Haegel, and G. Subklew, in *Proceedings of the International Conference and Course on Cost Efficient Acquisition and Utilization of Data in the Management of Hazardous Waste Sites*, Waste Policy Institute, Herdon, VA, May 23–25, 1994, 34–36.

14. W. D. Clemens, F.-H. Haegel, M. J. Schwuger, C. Soeder, K. Stickdorn, and L. Webb, WO 94/04289 3. 3. 1994 (priority 22. 8. 1992), patent, "Process and plant for decontaminating soil materials contaminated with organic pollutants."

15. W. D. Clemens, F.-H. Haegel, P. Nolte, K. Stickdorn, and L. Webb, in *Contaminated Soils '93*, Vol. II, Fourth International KfK/TNO Conference on Contaminated Soil, Berlin, May 3–7 (F. Arendt, G. J. Annokkée, R. Bosman, and W. J. van der Brink, eds.), Kluwer Academic Publishers, Dordrecht, 1993, pp. 1375–1376.

16. W. Karcher, R. J. Fordham, J. J. Dubois, P. G. J. M. Glaude, and J. A. M. Lightart, in *Spectral Atlas of Polycyclic Aromatic Compounds* (W. Karcher, ed.), Reidel, Dordrecht, 1985, pp. 92–95.

17. W. D. Clemens, F. H. Haegel, and M. J. Schwuger, Langmuir *10*:1366 (1994).

18. M. J. Rosen, in *Surfactants and Interfacial Phenomena*, (M. J. Rosen, ed.), Wiley, New York, 1989, pp. 170–205.

19. D. A. Edwards, R. G. Luthy, and Z. Liu, Environ. Sci. Technol. *25*:127 (1991).

20. K. D. Pennell, L. M. Abriola, and W. J. Weber, Jr., Environ. Sci. Technol. *27*(12):2332 (1993).

21. D. A. Edwards, Z. Liu, and R. G. Luthy, Water Sci. Technol. *26*(9–11):2341 (1992).
22. J. L. Underwood, K. A. Debelak, D. J. Wilson, and J. M. Means, Sep. Sci. Technol. *28*(8):1527 (1993).
23. R. Schomäcker, K. Stickdorn, and W. Knoche, J. Chem. Soc., Faraday Trans. *87*(6):847 (1991).
24. M. Liu and D. Roy, Miner. Metall. Process. *9*(4):206 (1992).
25. R. T. Podoll, K. C. Irwin, and S. Brendinger, Environ. Sci. Technol. *21*:562 (1987).
26. S. Laha, Z. Liu, et al., in *Gas, Oil, Coal and Environmental Biotechnology*, (C. Akin, ed.), Institute of Gas Technology, Chicago, 1990, pp. 279–295.
27. Z. Liu, S. Laha, and R. G. Luthy, Water Sci. Res. *23*(23):475 (1991).
28. Z. Liu, D. A. Edwards, and R. G. Luthy, Water Res. *26*(10):1337 (1992).
29. S. A. Abdul and T. L. Gibson, Envir. Sci. Technol. *25*:665 (1991).
30. M. Kahlweit, E. Lessner, and R. Strey, J. Phys. Chem. *87*:5032 (1983).
31. M. Kahlweit, R. Strey, and D. Haase, J. Phys. Chem. *89*:163 (1985).
32. M. Kahlweit, R. Strey, and G. Busse, J. Phys. Chem. *94*:3881 (1990).
33. M. Kahlweit, Tenside Surf. Det. *30*(2):83 (1993).
34. H. Kunieda and M. Yamagato, Langmuir *9*:3345 (1993).
35. L. Peschke, F. H. Haegel, and J. Fresenius, Anal. Chem., *351*:622–624 (1995).
36. P. Busch, H. Hensen, and H. Tesman, Tenside Surf. Det. *30*:116 (1993).
37. K. Fukuda, O. Söderman, B. Lindman, and K. Shinoda, Langmuir *9*:2921 (1993).
38. D. Balzer, Langmuir *9*:3375 (1993).
39. W. D. Clemens, thesis, University of Düsseldorf, 1995.

18

The Role of Microemulsions in Detergency Processes

NURIA AZEMAR   Departamento Tecnología de Tensioactivos, Centro de Investigación y Desarrollo, Consejo Superior de Investigaciones Científicas, Barcelona, Spain

I. INTRODUCTION

There has been considerable research work directed toward the development of improved surfactant systems for soil removal at low temperatures (approximately 15° to 35°C) in the past years. This trend is a result of the greater interest in energy savings, environmental concern, and the increasing use of temperature-sensitive fabrics [1–3]. In order to meet the new requirements, renewed attention has been directed toward the mechanisms of detergency.

Detergency can be defined as the removal of unwanted substances (soil) from a solid surface brought into contact with a liquid [4]. It is a complex process that depends on several factors, such as the nature and concentration of the washing solution, additives (builders, enzymes, antiredeposition agents), nature of the solid surface, hydrodynamic conditions, mechanical action during washing, water hardness, temperature, and electrolyte level [5]. The substrate may vary from a smooth, hard surface like that of a glass plate to a porous, soft surface like that of a fabric. The soil may be liquid or solid and is usually a combination of both. Liquid (oily) soil may contain skin fats (sebum), fatty acids, mineral and vegetable oils, fatty alcohols, etc. [6]. The cleaning effect for both oily and particulate soil is caused primarily by the presence in the washing solution of surfactants, which act by altering interfacial tensions at the various phase boundaries within the surfactants [5]. Due to the great variability of substrates and soils, there is no one single mechanism of detergency but a number of different mechanisms depending on the nature of the substrate and soil [6].

Restricting the attention to oily soils, it has generally been accepted that the predominant mechanism in their removal is the so-called roll-up mechanism. It was first described by Adam [7] and consists of the displacement of soil from a solid substrate through a mechanism of preferential wetting by the liquid. In the roll-up mechanism, the balance of surface energies at the oil/fabric/water contact line of an oil film or drop on the fabric is altered by addition of surfactant, causing the contact angle in the oil phase to increase, and the oil drops break off by agitation in the washing bath [8]. If the contact angle is less than 180° but more than 90°, the soil will not be displaced spontaneously but can be removed by hydraulic currents in the bath. When the contact angle is less than 90°, part of the oily soil remains attached to the substrate, even when it is subjected to the hydraulic currents of the washing solution, and mechanical work or some other mechanism is required to remove the residual soil from the substrate [6].

Although roll-up has been accepted as the predominant mechanism in oily soil removal, other mechanisms such as solubilization, emulsification, or intermediate phase formation have also been proposed [4–6]. However, under certain conditions (e.g., at low temperatures, with surfactants less hydrophilic than the typical anionic surfactants, with synthetic fabrics), those mechanisms might be the predominant ones [9].

Solubilization with surfactant micelles has been considered the most important secondary mechanism for the removal of small amounts of oily soil. The extent of solubilization of this type of soil depends mainly on the chemical structure of the surfactant, its concentration in the bath, and the temperature. It has been reported that the removal of soil by solubilization from hard and textile surfaces becomes significant only above the critical micelle concentration (CMC) and reaches its maximum only at several times the CMC [6]. However, solubilization is almost always insufficient to remove all the oily soil, and the remaining oily soil may be suspended in the bath by emulsification. In this mechanism it is very important that the interfacial tension between oily soil droplets and bath be low, so that emulsification is achieved with a minimum of mechanical work. The ability of the bath to emulsify the oily soil is insufficient to keep all the soil from redepositing on the substrate. In contrast, solubilization can result in complete removal of the oily soil from the substrate [6].

The formation of intermediate phases during the detergency process at the washing solution–oil interface was also considered to play a secondary role in the mechanisms of oily soil removal. Lawrence [10] applied the formation of mesomorphic phases in ternary systems to detergency, proposing a soil removal mechanism based on the penetration of aqueous surfactant solutions into oily soil to produce liquid crystalline phases with the polar soil constituents. Owing to the adsorption of surfactant at the water–oil interface forming a close-packed monolayer, the local surfactant concentration is high enough for extensive penetration into the oily soil and formation of liquid crystalline phases, despite the very low overall surfactant concentration in the washing solution. Subsequently, the mesomorphic phases are swollen and broken by agitation and emulsified into the aqueous solution, allowing contact of the remaining soil with the washing solution.

Under the present washing conditions, liquid crystalline phases as well as other surfactant aggregates could play a predominant role in the mechanisms of soil removal. Microemulsions, due to their characteristic properties, were promising systems to investigate for detergency purposes. The most significant properties of microemulsions for their interest in detergency are the solubilization capacity for both polar and nonpolar soil compounds, the extremely low values of interfacial tension achieved between aqueous and oil phases, and spontaneous formation when the components are brought into contact.

The main aim of this chapter is to review studies concerning the role of microemulsions in detergency. The first part deals with the role of microemulsions as intermediate phases in the mechanisms of soil removal. This subject was treated by Miller and Raney [1] in a comprehensive review covering different types of polar and nonpolar soils and by Solans and Azemar [11], who discussed nonpolar soils. The second part of this chapter deals with the role of microemulsions as washing solutions. The results of detergency experiments are discussed on the basis of phase behavior of the corresponding surfactant systems.

II. MICROEMULSIONS AS INTERMEDIATE PHASES

The investigations that showed the role microemulsions may play as intermediate phases in detergency processes were based on studies of model systems [1,9,11,12]. The washing solutions consisted of aqueous pure nonionic surfactant (i.e., polyethylenglycol alkyl ether) solutions. Hydrocarbons such as hexadecane and squalene were chosen as models for nonpolar hydrocarbon-based soil, and the solid substrate to which model soil was directed applied was a textile fabric.

The washing solution/soil systems in these experiments were ternary water/nonionic surfactant/hydrocarbon systems. It is well known that nonionic surfactants change from hydrophilic to lipophilic with increasing temperature [13]. Accordingly, the phase behavior of these ternary systems is greatly dependent on temperature [14]. At lower temperatures, a surfactant-rich phase [an aqueous micellar solution or oil-in-water (O/W) microemulsion] coexists with an excess oil (O) phase. With a rise in temperature, solubilization of oil in the micellar solution increases and this phase splits into two phases at a critical end point; three liquid phases, consisting of water (W), surfactant phase or middle-phase microemulsion (D), and oil phase (O) are in equilibrium. With a further increase in temperature, the D phase merges with the oil phase at another critical end point and two liquid phases, water (W) and reverse micellar solution or W/O microemulsion, are in equilibrium. The water-in-oil interfacial tension is at a minimum, and the type of emulsion changes from W/O to O/W and vice versa in the three-phase temperature interval. The hydrophile-lipophile balance (HLB) temperature of phase inversion temperature (PIT) is the temperature at which hydrophile-lipophile properties of the surfactant just balanced, that is, the midtemperature of two critical end points [14].

The first step in the investigations of textile detergency was to study the effect of nonionic surfactant and temperature on detergent efficiency [9,12]. Miller et al. [9] carried out radiotracer detergency studies of hexadecane and squalane removal from a polyester/cotton fabric, using washing solutions that contained aqueous 0.05 wt% tetra- and pentaethylenglycol dodecyl ether surfactant ($C_{12}E_4$ and $C_{12}E_5$, respectively) solutions. Their experiments were carried out under dynamic conditions, with a Terg-o-Tometer apparatus. Solans et al. [12] also studied nonpolar soil removal (hexadecane, squalane, mineral oil) from polyester/cotton fabrics. Their experiments were under static conditions (without agitation) and they used as washing solutions, aqueous 1 wt% tetra-, penta-, and hexaethyleneglycol dodecyl ether ($C_{12}E_4$, $C_{12}E_5$, and $C_{12}E_6$, respectively) solutions. Independent of nonionic surfactant, soil, and the experimental technique used, it was found that detergent efficiency increased with temperature, reaching an optimum, after which it experienced a pronounced decrease, as shown in the schematic representation of Fig. 1. Keeping the other variables unchanged, the optimum in detergent efficiency was displaced toward higher temperatures as the degree of ethoxylation of the surfactant increased. The most salient feature of these investigations was the corre-

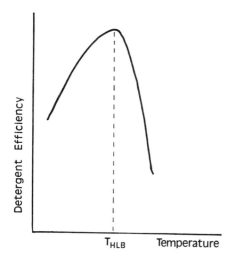

FIG. 1 Schematic representation of the influence of temperature on detergent efficiency of monodispersed polyoxyethylene n-alkyl ethers.

lation observed between the highest levels of soil removal and HLB temperature (PIT) of the corresponding washing solution/soil system (Table 1).

The correlation between detergency and phase inversion temperature found with pure nonionic surfactants was verified using technical-grade surfactants. The detergency tests were carried out using surfactants with an average degree of ethoxylation of 4 and hydrocarbon chain lengths of 10, 12, and 14. The removal of hexadecane as a function of temperature

TABLE 1 HLB Temperature of Ternary Water/ Nonionic Surfactant/Oil Systems and Temperature of Optimum Detergent Efficiency

System	T_{HLB} (°C)	T_{OPT} (°C)
$H_2O/C_{12}E_4$/hexadecane	32	32
$H_2O/C_{12}E_5$/hexadecane	53	55
$H_2O/C_{12}E_6$/hexadecane	71	75
$H_2O/C_{12}E_4$/squalane	47	50
$H_2O/C_{12}E_4$/mineral oil	43	45

followed the same trends: the optimum in detergent efficiency was achieved at the phase inversion temperature of each washing solution/soil system [12].

The high levels of soil removed near the HLB temperature were attributed to the phenomena occurring at this temperature, namely ultralow interfacial tensions achieved when aqueous, surfactant, and oil phases coexist and maximum solubilization of oil in water and vice versa leading to microemulsion formation [14]. Initially, the optimum in detergent efficiency was also attributed to the presence of a liquid crystalline phase in the washing solution, a binary water/nonionic surfactant system. This interpretation seemed quite reasonable, since the contribution of mesomorphic-phase formation in detergency has long been taken into account. Several studies [10,15,16] based on model oily soils have shown that the formation of a lamellar crystalline phase during the detergency process or their presence in the washing solution [9,17,18] play a decisive role in improving soil removal. Considering the phase behavior of binary water/$C_{12}E_4$ and water/$C_{12}E_5$ systems as a function of temperature [19], it was readily seen that at the temperature at which optimum detergent efficiency was obtained [12], the washing solution consists of two phases, an aqueous phase and a lamellar liquid crystalline phase. However, when the $C_{12}E_6$ system was considered, no liquid crystalline phase was present initially in the washing solution between 0 and 100°C and yet an optimum in detergent efficiency was also observed at the HLB temperature of the corresponding system (Table 2). Therefore it was concluded that the optimum in detergent efficiency observed at the HLB temperature might be due to the properties of the ternary washing solution/soil system rather than to those of the binary washing solution system.

TABLE 2 Temperature of Optimum Hexadecane Removal and Types of Phases Present in Washing Solutions Consisting of Aqueous 1 wt% Nonionic Surfactant

Washing solution (1% aqueous nonionic surfactant)		Temperature of optimum hexadecane removal (°C)
Surfactant	Type of phases	
$C_{12}E_4$	W-L$_\alpha$	35
$C_{12}E_5$	W-L$_\alpha$	55
$C_{12}E_6$	W-L$_1$	75

As stated above, at the HLB temperature ultralow interfacial tensions are attained and maximum solubilization of oil-soluble compounds in water is achieved. These are conditions in which microemulsion formation is favored. Although the concentration of surfactant needed to form a microemulsion is higher than the concentrations generally used in detergency, local surfactant concentrations at the soil-substrate interface produced during a washing process could promote the formation of a D phase microemulsion as the intermediate phase. In this context, the contacting experiments carried out by Miller et al. [20] are very illustrative and supported these assumptions. In those experiments pure straight-chain hydrocarbons were in contact with 1% aqueous solutions of pure nonionic surfactant and the water-oil interface was observed as a function of temperature by video-enhanced microscopy. It was reported that at the HLB temperature of the corresponding ternary system a high concentration of a surfactant-rich phase (a microemulsion) was present near the surface of contact between oil and aqueous phases. The experiments of Solans et al. [11,12], using anionic instead of nonionic surfactants as the washing solution, confirmed the relationship between microemulsion formation and optimum oily soil removal. They predicted that the optimum detergency should be obtained at optimum salinity (the salinity at which a balance between hydrophilic and lipophilic properties of the ionic surfactant and cosurfactant is reached). The phase transitions that are observed in anionic surfactant systems with salinity are similar to those of nonionic surfactants with temperature; that is, the phenomena occurring at the phase inversion temperature (maximum solubilization, low values of interfacial tension, microemulsion formation) also occur at optimum salinity [21,22]. In these investigations 1 wt% sodium dodecyl sulfate/n-pentanol and brine were used as washing solutions and hexadecane as a model soil [12]. Previously, a phase behavior study of the washing solution/soil system was carried out as a function of salinity in order to find the optimum salinity, that is, the concentration of brine at which maximum instability of three liquid phase emulsions is produced. In Fig. 2 the detergent efficiency as a function of brine (NaCl) concentration at three temperatures is shown. Independent of the temperature, an optimum detergent efficiency was obtained at 10 wt% NaCl, the optimum salinity of this system at which the conditions for microemulsion formation are favored.

The most immediate and practical conclusion from these investigations was that for a given soil, optimum detergent efficiency could be achieved at a chosen temperature by selecting an appropriate surfactant solution. Therefore, in view of current trends in laundering habits toward lower temperatures, the factors that lower the phase inversion temperature of

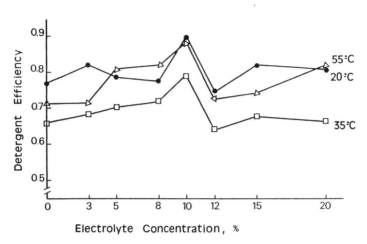

FIG. 2 Detergent efficiency of 1 wt% sodium dodecyl sulfate/n-pentanol (weight ratio 1/2) in brine as a function of electrolyte concentration at three temperatures. Model soil: hexadecane.

nonionic surfactant systems were investigated in order to decrease the optimum detergent temperatures. Raney and Miller [17] studied the detergency properties of mixtures of $C_{12}E_5$ with hydrophobic additives triethyleneglycol dodecyl ether ($C_{12}E_3$) and n-dodecanol ($C_{12}E_0$) for the removal of n-hexadecane from polyester/cotton fabrics. They found optimum detergent efficiency at temperatures lower than that for $C_{12}E_5$ alone. The optimum temperatures corresponded to the phase inversion temperatures of the pseudoternary washing solution/soil system. It should be noted that, in contrast to that of the ternary water/surfactant/oil system, the HLB temperature of the pseudoternary systems is not independent of the oil/surfactant weight ratio [23]. Therefore the HLB temperature values given for the systems water/surfactant/oil/additive were extrapolated to the low soil-to-surfactant ratios used in these investigations. Analogous results were obtained with technical-grade surfactants [24]. In this context, Azemar et al. [25] studied the effect of addition of several lyotropic electrolytes in depressing the optimum detergent temperature. It is well known that electrolytes may produce two effects in a binary water/nonionic surfactant system: (1) an increase in the mutual solubility of surfactant in water, that is, the "salting-in" effect, and (2) a decrease in the solubility between them, "salting-out" effect. Salting-out electrolytes lower the cloud point of nonionic surfactants by dehydrating the ethylene oxide group [26]. Consequently, the phase inversion temperature of ternary water/nonionic surfactant/oil systems is similarly affected by addition of this type of salts [27]. Figure 3 shows the effect of sodium citrate in depressing the HLB temperature of three water/nonionic surfactant/hexadecane systems. The effect of electrolyte on the temperature of optimum detergent efficiency was determined using aqueous surfactant solutions containing electrolyte as washing solutions. Optimum detergency temperatures as a function of surfactant degree of ethoxylation without and with sodium citrate (0.05 and 0.02 M) are shown in Fig. 4. By comparing the results of Figs. 3 and 4, one can observe the correlation between optimum detergent efficiency and phase inversion temperature of the corresponding washing solution/ soil system. As expected, the optimum detergent efficiency of a given non-ionic surfactant aqueous solution was shifted toward lower temperatures by the addition of sodium citrate.

Fabric detergent studies were also performed using pure triolein [28] as a model for triglyceride soils, basic components in oily laundry soil (se-bum). In these studies, the washing solution/soil systems consisted of ter-nary water/nonionic surfactant/triglyceride systems.

The phase behavior of systems with amphiphilic oils, although rather more complex than that of the corresponding systems with nonpolar oils,

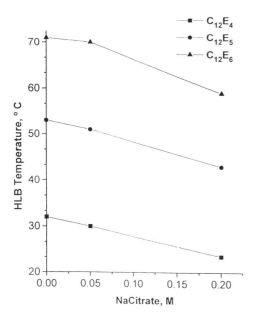

FIG. 3 HLB temperature as a function of sodium citrate concentration (M) of ternary aqueous electrolyte solution/nonionic surfactant/hexadecane systems.

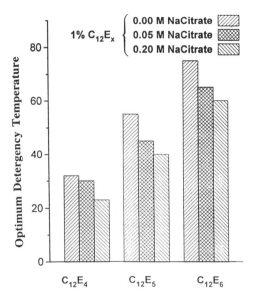

FIG. 4 Optimum detergency temperature for the removal of hexadecane of aqueous sodium citrate/nonionic surfactant systems.

is at present well understood. Kunieda and Haishima [29] showed that in water/nonionic surfactant/triglyceride systems, two types of three-liquid-phase regions exists. In addition to the W+D+O region, another surfactant phase, D', coexists with water and oil phases at lower temperatures. However, this oil phase is not a simple excess oil phase; it contains considerable amounts of water and surfactant. The D' phase is related to the L_3 phase of the corresponding binary water/nonionic surfactant system and can solubilize only small amounts of oil. The W+D'+O region turns to the W+D+O region via the four-phase region W+D'+D+O with increasing temperature. The finding of four-phase behavior and coexistence of two microemulsion phases in three-component systems is remarkable. Similar phase behavior has been reported for ternary water/nonionic surfactant/alkanol systems [30]. The ability to form the D' phase in systems with polar or amphiphilic oils has been attributed to the increased flexibility of the surfactant-oil bilayers, which makes it possible to assume the configuration of the spongelike microstructure of this phase. When the bilayers are more rigid, i.e., with nonpolar oils, a lamellar structure forms instead of a D' phase. In contrast to the D' phase, the D phase (middle-phase microemulsion) is able to solubilize higher amounts of oil and therefore is more favorable for detergency.

Radiotracer detergency studies of triolein removal from polyester/cotton fabrics showed [28] that triolein removal was very low for $C_{12}E_3$ and $C_{12}E_4$. A D phase is not formed in the $C_{12}E_3$ system, and it forms only in a very narrow temperature range (0.2° near 55°C) in the $C_{12}E_4$ system. The same study for the $C_{12}E_5$ system showed an optimum in detergent efficiency in a small range of temperatures at 65°C, which corresponded to the temperature of intermediate D phase formation, the phase inversion temperature of the system. As with the hydrocarbon soils, it was concluded that optimum removal of long-chain liquid triglycerides could be obtained under conditions at which an intermediate D phase forms during the detergency process. Formation of this D phase was promoted by using mixtures of surfactants with varying hydrocarbon chain lengths and also if sufficient hydrocarbon was present with the triglyceride in the initial oily soil [1,28].

III. MICROEMULSIONS AS WASHING SOLUTIONS

Because of the characteristic properties of microemulsions, namely thermodynamic stability and ability to solubilize both water- and oil-soluble compounds, it was predicted that they might be efficient systems as washing solutions. Traditionally, organic solvents have been used in the so-called dry-cleaning processes for textile cleaning. However, in addition to the difficulties in the removal of water-soluble soils and the redeposition of particulate soil, other major disadvantages involved in these processes are the health risk of chlorinated solvents and the flammability of the low-molecular-weight hydrocarbons, the most common solvents used in dry cleaning [31]. One important advantage of microemulsions with respect to other cleaning agents is that aqueous and nonaqueous processes could be combined in only one process.

In this context, investigations of the use of O/W and W/O microemulsions in a new process for industrial cleaning was reported about 15 years ago [31]. Studies of microemulsion efficiency in soil removal from textile fabrics [32], in raw wool scouring [33], and in skin degreasing processes [34] have been reported. The current raw wool scouring methods are based on the use of surfactant solutions containing electrolytes as washing solutions in a process carried out at high temperatures (about 50–60°C). Similarly, the conventional skin degreasing processes involve prolonged treatment of the substrate in white spirit media, followed by treatment in aqueous salt solutions of nonionic surfactants at high temperature with vigorous mechanical agitation. In view of the high energy consumption and high volume of contaminant effluent of these soil removal processes, there is a strong need to optimize their efficiency.

In order to elucidate the role of microemulsions as washing solutions, a preliminary step of these studies [32–34] was to determine the solubility regions corresponding to microemulsions with low surfactant concentration in different water/nonionic surfactant/oil systems as a function of temperature, electrolyte concentration (sodium tripolyphosphate, sodium citrate, sodium chloride) and other additives. Several microemulsions were selected from the corresponding phase diagrams to evaluate their washing efficiency between 20 and 35°C. Microemulsion efficiency was compared to that using aqueous nonionic surfactants ($C_{12}E_4$, NPE_8) or commercially liquid detergent solutions at the same temperatures. Independent of the substrate, microemulsions showed higher efficiency in removing contaminants than conventional detergent solutions. Figure 5 is an illustration of microemulsion efficiency. Photomicrographs of raw wool fibers (a) and scoured wool fibers in water (b), surfactant solution (c), and microemulsion (d) media are shown. Wool scoured in microemulsion media was essentially free from surface contaminants, contrary to wool scoured in other media [33]. It should be noted that the experiments on soil removal efficiency of microemulsions [32–34] were performed under conditions of null

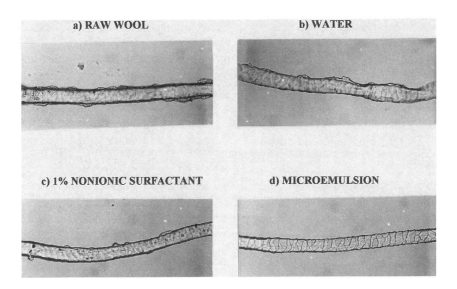

FIG. 5 Photomicrographs (×550) of raw wool fibers (a) and scoured wool fibers at 20°C in water (b), 1 wt% nonionic surfactant (c), and a microemulsion system (d).

or very low mechanical agitation. Therefore, processes using microemulsions would have the additional advantage of low energy requirements.

Although optimization of microemulsion formulation parameters is required for the industrial use of microemulsions as cleaning systems, the results of these investigations have clearly shown that by conveniently selecting a microemulsion system, soil can be removed efficiently under conditions of minimum mechanical energy and at low temperature.

ACKNOWLEDGMENT

Financial support from DGICYT (grant PB92-0102) is gratefully acknowledged.

REFERENCES

1. C. A. Miller and K. H. Raney, Colloids Surfaces A Physicochem. Eng. Aspects 74:169–216 (1993).
2. T. P. Matson and M. F. Cox, J. Am. Oil Chem. Soc. 61:1270–1272 (1984).
3. J. E. Zweig, H. L. Benson, T. K. Brunk, and K. R. Cox, Soap Cosmet. Chem. Spec. 3:35–47 (1985).
4. E. Kissa, in Detergency, Theory and Technology, (W. G. Cutler and E. Kissa, eds.), Marcel Dekker, New York, 1987, pp. 2–81.
5. Kirk-Othmer Encyclopedia of Chemical Technology, Vol. 22, Wiley, New York, 1980, pp. 387–431.
6. M. J. Rosen, in Surfactant and Interfacial Phenomena, Vol. 10, Wiley, New York, 1989, pp. 363–391.
7. N. K. Adam, J. Soc. Dyers Color 53:121 (1937).
8. L. Thompson, J. Colloid Interface Sci. 163:61–73 (1994).
9. K. H. Raney, W. J. Benton, and C. A. Miller, J. Colloid Interface Sci. 117:282–290 (1987).
10. A. S. C. Lawrence, Nature 183:1491 (1959).
11. C. Solans and N. Azemar, In Organized Solutions (B. Lindman and S. E. Friberg, eds.), Vol. 19, Marcel Dekker, New York, 1992, pp. 273–288.
12. C. Solans, N. Azemar, J. L. Parra, and J. Calbet, Proceedings of the CESIO 2nd World Surfactants Congress, Paris, 1988, Vol. 2, pp. 421–429.
13. K. Shinoda and H. Arai, J. Phys. Chem. 68:3485 (1964).
14. H. Kunieda and K. Shinoda, J. Colloid Interface Sci. 107:107 (1985).
15. D. G. Stevenson, in Surface Activity and Detergency (K. Durham, ed.), Macmillan, London, 1961, Chapter 6.
16. H. S. Kielman and P. J. F. van Steen, in Surface Activity Agents, The Society of Chemical Industry, London, 1979, p. 191.
17. K. H. Raney and C. A. Miller, J. Colloid Interface Sci. 119:539–549 (1987).
18. F. Shambil and J. Schwuger, Colloid Polym. Sci. 265:1009–1017 (1987).

19. D. J. Mitchell, G. J. T. Tiddy, L. Waring, T. Bostock, and M. P. McDonald, J. Chem. Soc. Faraday Trans. I 79:975–1000 (1983).
20. W. J. Benton, K. H. Raneym, and C. A. Miller, J. Colloid Interface Sci. *110* :363 (1987).
21. M. L. Robbins, in *Micellization, Solubilization and Microemulsions* (K. L. Mittal, ed.), Vol. 2, Plenum, New York, 1976, p. 713.
22. C. A. Miller and P. Neogi, in *Interfacial Phenomena, Equilibrium and Dynamic Effects*, Vol. 1, Marcel Dekker, New York, 1985, pp. 140–179.
23. H. Kunieda and N. Ishikawa, J. Colloid Interface Sci. *107*:122 (1985).
24. R. Bercovici and H. Krüßmann, Tensile Surf. Deterg. *27*:8 (1990).
25. N. Azemar, I. Carrera, and C. Solans, J. Dispers. Sci. Technol. *14*(6):645–660 (1993).
26. K. Meguro, M. Ueno, and K. Esumi, in *Nonionic Surfactants* (M. S. Schick, ed.), Marcel Dekker, New York, 1987, pp. 109–178.
27. K. Shinoda and H. Kunieda, in *Encyclopedia of Emulsion Technology* (P. Becher, ed.), Vol. 1, Marcel Dekker, New York, 1983, p. 337.
28. F. Mori, J. C. Lim, K. H. Raney, C. M. Elsik, and C. A. Miller, Colloids Surfaces *40*:323 (1989).
29. H. Kunieda and K. Haishima, J. Colloid Interface Sci. *140*:383 (1990).
30. H. Kunieda and A. Miyajima, J. Colloid Interface Sci. *129*:554 (1989).
31. G. Gillberg, in *Emulsions and Technology* (K. J. Lissant, ed.), Marcel Dekker, New York, 1984, pp. 1–43.
32. C. Solans, J. J. García-Dominguez, and S. E. Friberg, J. Dispers. Sci. Technol. *6*(5):523–537 (1984).
33. P. Erra, N. Azemar, M. R. Juliá, and C. Solans, J. Dispers. Sci. Technol. *13* (1):1–12 (1992).
34. N. Azemar, J. Cot, P. Erra, and C. Solans, in *Compendium of Advanced Topics on Leather Technology* (J. Cot, ed.), Vol. 1, Barcelona, 1991, pp. 287–298.

19
Emulsions and Microemulsions in Metalworking Processes

FRANCESC GUSI, A. C. AUGUET, and F. X. GAILLARD Hispano
Química, S.A., Barcelona, Spain

I. INTRODUCTION

The term "metalworking" designates those machine tool processes in which the shape of a metal is modified. In these processes, the tools are exposed to high pressures and friction; as a result, heat is generated. Consequently, the points of contact between the tool and the metal must be lubricated, to reduce the friction, and cooled, to remove the heat. Oils, namely mineral oils, are known to be good lubricants, and water is the most common cooling agent. The double target of lubricating and cooling is achieved with emulsions and microemulsions composed basically of mineral oil, water, and emulsifier. In processes in which little heat is produced, only lubrication is needed; mineral oil or the so-called synthetic lubricants (mineral oil–free formulations with water-soluble ingredients) are used as lubricants.

The purpose of this chapter is to describe some metalworking processes in which lubrication is needed and the role played by emulsions and microemulsions as lubricant formulations. Attention is focused on the following metalworking processes: metal rolling, metal cutting, and wire drawing.

II. METAL ROLLING

In metal rolling processes, a metal strip is reduced in thickness, increasing its length and keeping its width unchanged. These operations are currently carried out under cold conditions. In order to obtain plastic deformation in only one direction, it is necessary to apply a force in the rolling sense as well as a pressure between two cylinders or rolls. This process can also be carried out in hot conditions, with the metal heated at temperatures near its melting point, generally 100 or 200°C below it. In hot rolling, force is not applied in the rolling direction, but two lateral rolls are charged to avoid an increase in width.

The rolling process is performed in rolling mill machines, in which metal coils pass throughout the working rolls. The separation between rolls corresponds to the gauge obtained. The relationship between the initial and final thicknesses is called reduction and is expressed in terms of percentage. This operation is done at high speed with modern metals that make it possible to reduce the metal at high force and speeds. Rolling mills can be single (one stand) or with various roll systems working together (multistand). Single mills are referred as reversing mills because successive reductions are made by passing the strip throughout the rolls and changing the rolling sense by reversing the direction of rotation of the rolls. Multistand mills can produce five times faster the same quantity of rolled steel as single mills. The latter generally need to pass the strip five times, whereas five-stand tandem mills produce the same final reduction passing the strip once.

In metal plastic deformation processes, the metal crystal structure is changed by permanent deformation. In rolling, the crystal structure is modified in the rolling direction. The energy required in this process is very high. Moreover, the friction between crystals generates heat. This heat is transferred from the metal to the rolls, increasing its temperature and modifying its size. The energy necessary for the change in the shape of the crystalline structure is transferred via working rolls to the metal throughout the so-called roll bite or contact area. The contact area is quickly changed, because new and fresh surfaces of metal are produced. In the roll bite, there are different slipping speeds at the starting and final points, but a neutral point with no slipping is present. Slipping exists except in the neutral point. In these areas it is necessary to use a lubricant to reduce the friction as well as to remove the heat generated by the plastic deformation and friction.

In hot and cold rolling processes water is used as a coolant because of its high specific heat as well as to lubricate the roll bite. However, oil-in-water emulsions and microemulsions are considered the most appropriate formulations for refrigerating and lubricating at the same time. The oil content is transferred from the emulsion to the roll bite, leaving the aqueous phase and covering the metal surfaces, producing a strong lubricating layer. At the same time, the water phase of the emulsion refrigerates the overall system, cooling the rolls and the strip and keeping the temperature constant.

The components commonly used in rolling lubricant formulations are mineral oil, natural and/or synthetic fats, antiwear additives, and emulsifiers. The design of an appropriate emulsifier mixture is the most important aspect of the formulation. The emulsifier avoids separation of the oil phase from the overall emulsion, which would produce a loss of the oil concentration and consequently a decrease in the lubricant capacity. The emulsifier system must also allow plate-out of the oil over the metallic surface; thus, at the high temperatures produced in the roll bite, the emulsifier system allows more oil from the emulsion droplets to reach the surface.

Metal strips must be annealed after rolling, and for this reason it is very important to use emulsifiers without ashes. In fact, the heavy residue that remains over the sheet after annealing produces dirty surfaces, with carbon and inorganic deposits on the surface. The use of nonionic emulsifiers without ashes is a normal practice—i.e., ethylene oxide condensates of fatty acids, fatty alcohols, nonyl phenol, or fatty amines. Very high volumes (5×10^4 to 5×10^5 L) of rolling lubricants are used, producing wastes that have to be treated. Natural esters and fatty acids or alcohols condensed with ethylene oxide do not produce waste treatment problems. In contrast, ethylene oxide condensates of nonyl phenol are not appropriate because of toxicological problems with the wastes. Special emulsifier systems with cationic or anionic emulsifiers can also be used. When a strong positive charge on the oil particle is required, cationic emulsifiers such as fatty amines with ethylene oxide are selected to achieve good plate-out for very severe reductions or with harder materials. The HLB values of emulsifiers are around 10, ranging from 8 to 12. These values allow immediate emulsion with enough stability.

Emulsion stability is very important in rolling emulsions because lubrication capacity is decreased with stability. The most stable emulsion produces less oil plate-out; for this reason, a balance between emulsion stability and high oil consumption has to be established. Usually, mechanical agitation in the tank is sufficient to allow the use of low-stability emulsions. Specific laboratory tests have been developed to evaluate emulsion stability for rolling emulsions. In this context, specific stability is expressed as a percentage and varies from 30% for the less stable emulsions to 100% for the most stable emulsions. The normal values range from 70 to 95%.

Oil droplet diameters in rolling emulsions vary between 0.5 and 10 μm, depending on the lubrication requirements. The average diameter in a working emulsion is about 4 μm for the so-called stable emulsions, with a specific stability of 90–95%. The pH of the water used in the emulsion usually corresponds to the neutral range of 6–7.5. It is considered that the pH value of the emulsion is not changed by nonionic emulsifiers. A slightly acidic medium is generally preferred in order to achieve the appropriate lubrication.

During the rolling process, the emulsion components are severely changed by the extreme working conditions in the roll bite: hydrolysis, oxidation, and other reactions take place. For this reason, control of the emulsion is required in the process. The main variables tested are concentration, neutralization number, pH, conductivity, iron content, emulsion stability, and emulsifier concentration.

III. METAL CUTTING

Metal cutting processes, such as drilling, screwing, or grinding, are metalworking operations in which cutting tools are in contact with pieces of metal to be transformed, supporting high pressures and friction with heat generation. As in other metalworking operations, lubrication of the contact points and refrigeration are required. Microemulsions and emulsions are generally used as lubricant formulations, and they are recirculated in the metalworking system. Only the amount of formulation required is sent to working points. Two types of systems are employed: single and centralized systems. In the first system, individual tanks are used for each machine. In the centralized system, each machine is connected to a big central tank that distributes the required amount of emulsion when required. The emulsion is recovered after the process.

Emulsion and microemulsion formulations are designed according to each metalworking operation and each metal to be transformed. The components of these emulsions and microemulsions for metal cutting processes are lubricants such as mineral or natural oils, emulsifiers, rust inhibitors, bactericides, bacteriostatics, and fungicides. Thus, depending on the lubricating capacity required, the formulation will have different concentrations of components. Particle sizes of these formulations range from 80 to 600 nm. They are stable up to 60°C.

For rational use of emulsions or microemulsions in metal cutting, the lubricant and coolant capacity of the formulation as well as the emulsion tank system characteristics have to be considered. The most important aspects to take into account in selecting the appropriate formulations are (1) tool design, (2) material design, (3) lubricant or coolant design, and (4) emulsion tank system design, especially in centralized systems. Tool design can influence the displacement, pressure, and orientation of the coolant. This factor has to be considered not only to achieve the best results for cooling, lubrication, and chip removal but also to avoid pickup in edge tools. In the material design, the alloy is very important, because it influences hardness. The alloy employed can modify hardness and, consequently, affect cutting. In selecting the lubricant, the balance between lubrication and refrigeration capacity has to be considered. In a centralized emulsion tank system it is important to avoid chip accumulation, foam formation, release from tramp oils, etc. These aspects are less important in single-tank systems. Therefore their design offers fewer technical problems.

When designing a centralized emulsion tank system, the main factors to be considered are the amount of chip to be removed, the number of operations to be lubricated and/or cooled, the plow rate for each operation, and the average temperatures, which determine the amount of water that evaporates and therefore the water consumption. Use of demineralized water improves rates and avoids oil separation produced by the high-salinity effect. Having the liquid fall from a high level must be avoided because of foaming. It is important to be careful with the upper liquid level over the pumps, avoiding pump cavitation and air bubble pumping.

It should be noted that emulsions or microemulsions in centralized systems are in a dynamic state, exposed to high temperatures in the cutting area, to shear by recirculating pumps, to chip filtration, air contact, etc. They are important economic factors in a production plant. For economic reasons, formulations must be carefully controlled to avoid malfunctions or to correct them as soon as they appear. Table 1 shows the main parameters to be controlled and their relation to the problems to be solved.

IV. WIRE DRAWING

Drawing processes are used in metalworking industries in order to reduce the diameter of metallic wires, rods, or tubes. As in other metalworking procedures, a lubrication/cooling effect is required (1) to reduce the friction coefficient between the drawing material and the tool so that the drawing forces and energy consumption decrease and (2) to decrease the working temperature to an acceptable level by a cooling effect.

Two lubrication methods are used in wire drawing procedures: dry and wet. In the dry process, solid or semisolid products are applied to the metal surface, which has been previously phosphated, at the entrance of the tool.

TABLE 1 Control Parameters in Centralized
Emulsion and Microemulsion Tank Systems

Control parameter	Effects
pH	Corrosion
	Bacterial growth
Bacteria	Corrosion
	Bad smell
	Poor lubrication
Fungi	Pipe blocking
Chloride	Corrosion
Conductivity	Corrosion
	Oil separation
Tramp oils	High paper consumption
	Bacterial contamination
Concentration	Lubrication
	Bacteria
	Corrosion
Water hardness	Foam formation
	Soap formation

These products are composed mainly of metal soaps, such as sodium and/or calcium stearates, anticorrosives, extreme pressure additives (EP additives), and inert substances.

In the wet process, the system is submerged in the lubricant. The lubricant can be composed of mineral oil, synthetic oils, or emulsions. In this process, the cooling effect depends on the auxiliary product employed. Aqueous emulsions have a high specific heat; therefore they provide a greater cooling effect (continuous phase) than oils and also a lubricant effect (dispersed phase).

In the drawing process, refrigeration is very important because an increment in the temperature produces an increment in the friction coefficient, altering the mechanical characteristics of the treated material. It could also produce pickup in tools and as a result a worse surface finish.

The factors that influence metalworking and their interpretation will be discussed in detail by describing a particular drawing process. Attention will be focused on copper wire drawing because of the wide use of emulsions in this industrial sector and the importance of this process.

When optical fiber appeared in the market it was thought that copper wire could be replaced by optic fiber, especially in the telecommunications area. However, at present the copper wire and cable manufacturers have

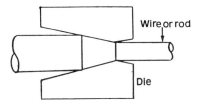

FIG. 1 Schematic representation of a stretching tool.

increased their production capacity. The consumption in the European Union, United States, and Japan in 1990 was more than 4 million tons of copper thread and cable in the following applications: wires, electromagnetic threads, electricity transportation, construction, car industry, telecommunications, etc.

Copper wire drawing is performed in tanks with a capacity that generally ranges from 0.5 to 40 m³, although it can be greater. The stretching process is done with a special tool, known as a die, shown schematically in Fig. 1. It has a smaller diameter than the wire to be treated. Once the wire has gone through the tool, it has the same diameter as the die, but it is elongated. The wire is moved by the action of a succession of disks, called cones or capstans, through the tool. They gather the wire and send it to another die to obtain a smaller diameter, and so on, until the desired thickness is achieved. The drawing process, according to the reduction and the type of wire to be obtained, can be organized in more than one step. Each one involves and cone or capstan. The stretched material undergoes a plastic deformation from the initial diameter to the diameter of the die. Table 2 indicates types of drawing according to the size of wire (typical values). Nowadays, it is possible to obtain elongations near 75 times the initial length; prototypes are being designed to reach elongations of over 150-fold.

TABLE 2 Example of Types of Wire Diameters in the Drawing Process

Type	Input diameter (mm)	Output diameter (mm)
Rod	8	5
Intermediate	4	0.4
Fine	1	0.1
Capillary	0.75	0.01–0.05

Traditionally, the material employed in dies and cones/capstans was tungsten carbide (W_2C), but now it is being replaced by polycrystalline diamond (PKD) in the dies and ceramic material in the cones/capstans. These materials are more expensive and harder than tungsten carbide, but they have contributed to increasing the life of the tool. However, they require lubricant emulsions with better qualities.

At the beginning of the conversion from dry to wet processes, mineral oil mixed with natural or synthetic fatty esters was employed to increase the lubricant power. Successively, the pure oils were replaced by soluble oils, whose main component is oil mixed with emulsifiers and other additives. The formulations are stable emulsions (microemulsions) with cooling and lubricant properties, and they have obvious advantages in safety aspects owing to their low inflammability. The emulsion properties required in the copper drawing process are lubrication, appropriate particle size, stability, unfoaming properties, detergency, etc.

The lubricant has to operate on the die, on the cone, and on the wire. There are two types of lubrication: hydrodynamic and boundary. The first one is applied in the cones. In the die the issue becomes more critical because both sorts of lubrication are involved. At this point, it is important that the lubricant forms a continuous film, which depends on the lubricant and the material affinity.

The emulsion/microemulsion particle size is another factor with a complex influence on the lubricant and distribution of the dispersion. Figure 2 shows the particle size of a formulation used in the process of drawing

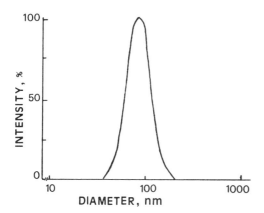

FIG. 2 Particle size distribution of a formulation used in drawing copper threads of small diameter. Mean formulation diameter 80 nm.

copper threads of small diameter (fine size). Figure 3 shows the particle size distribution of a formulation used in drawing copper threads of larger diameters as rod or wire.

Lubricant components are usually insoluble in water; therefore it is essential to formulate them with appropriate emulsifiers, which can be of a nonionic or anionic nature. The choice of these emulsifiers is somehow difficult, not only in order to get the correct stability and a determined particle size but also to attain good detergent power and unfoaming properties. As a consequence of emulsion instability, two phases appear, breaking the system homogeneity. The result is that the lubrication and cooling effect is unbalanced and wearing appears quickly in dies and/or capstans. On the other hand, the oily phase can deposit on different parts of the installation, stacking deposits of metallic copper or copper compounds. The soil increases in the tool as well as on the wire, and the possibilities of tool damage increase. The same problems can be generated by accumulation of tramp oils (i.e., hydraulic oils) that can drop into the emulsion. Therefore, the emulsifiers have to provide good chemical and mechanical stability as well as an adequate HLB depending on the insoluble substances to be emulsified. These requirements force us to use "tailor-made" emulsifiers. However, instability can also be due to an increment in the conductivity of the system or microbial contamination. If microbial pollution is not controlled, the emulsion can be destroyed.

Foam is generated from two possible sources in drawing installations. The first is a mechanical source that depends on the suitability of the installation design. The other source has a physicochemical origin in the

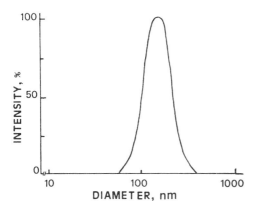

FIG. 3 Particle size distribution of a formulation used in drawing thicker copper threads. Mean formulation diameter 150 nm.

product formulation (water quality, emulsifiers, etc). Among formulation components there are fatty acids, esters, and other substances. During the process, these substances can generate copper soaps, which are water insoluble. These soaps are responsible for soil accumulation according to their dispersibility in the emulsion, which varies with the emulsion formulation. However, copper coaps are effective antifoaming agents. A good solution is to reduce their concentration by using sequestering agents or alkanolamines. The alkanolamines are usually used because they are able to block the copper ions, but they also produce foam as a secondary effect. For this reason, the amount and periodicity of incorporation of quelant agents or alkanolamines are critical. Sometimes, antifoaming agents are employed to control the foam developed, but this solution can have negative secondary effects, because these products alter the lubrication.

As a conclusion, a preliminary study of the global process is advisable and more beneficial in the long term. this study should be focused on the emulsifiers and other formulation components; their interaction, and foam generation. Also, good knowledge of the installation and the working conditions under which the emulsion will be used is required in order to obtain a formulation that has minimal foam generation and is easy to break. The presence of foam in this process has negative consequences at a mechanical level, because it produces pump cavitation and drawing speed reduction and the occluded air can produce a defect of refrigeration in the aqueous phase or a lubrication discontinuity at the points where air bubbles stick. both effects, overheating and poor lubrication, produce wear on dies and cones.

Detergency can be considered an opposite concept to lubrication because it seems that too clean a wire is not well lubricated. Again, it is necessary to apply a balance of these two concepts that, generally, is based on the experience of the formulator. the balance can be modified according to the diameter and the type of wire to be obtained. It is generally accepted that for materials with big diameters such as rods, the lubricant/detergent balance is displaced to the former factor and for smaller diameters, such as multifilar, it is displaced to the latter factor.

The soil accumulated in the installation is generated mainly by the presence of suspended metallic copper residues, insoluble copper soaps, or other degradation produces and by emulsion instability. The degradation products have a sticky nature. At key points of the process, they can affect the wire finish and can produce damage to the installation, resulting in lower productivity. Also, they can be stacked on the emulsion surface, becoming a foam support that avoids foam destruction. to reduce this soil as much as possible, the components employed in the formula have to be sufficiently thermally and chemically stable for the working conditions in order to generate a minimal amount of these residues. The emulsifiers and surface-active substances employed in the formula have to show sufficient detergent power to keep the contaminants dispersed, because their generation during the process is unavoidable. The composition of a typical formula for a drawing process is shown in Table 3. In the drawing operation, periodic and systematic control of the emulsion is required to achieve good productivity and to ensure a constant quality of the copper wire.

The main aspects of control for the water used in the drawing bath are chloride concentration and water hardness. Chloride concentration has to be lower than 100 ppm to avoid corrosion problems. Water hardness, which affects detergency power and emulsion stability, has to be lower than 10°HF; in general it is not possible to avoid the formation of calcium soaps that are water insoluble.

The bath has to be maintained below 45°C, otherwise the friction coefficient will rise. An increase in the friction coefficient negatively influences lubrication and emulsion stability. The concentration of the bath ranges from 5 to 10% according to the type of operation to be performed. It is important to assure the bath lubricant power and control the presence of tramp oils. The bath quality can be determined by the alkalinity index, the refraction index, or the emulsion acid treating method. The pH of the bath has to be maintained between 8.5 and 9.5. Conductivity values have to be lower than 300 μS; this parameter is used to control the electrolyte level in the emulsion and it is an index of the bath aging. The ionic copper content has to be maintained under 400 ppm and constitutes a measure of

TABLE 3   Example of a Typical Formulation Used in Drawing Processes

Component	Amount (%)
Mineral oil	20–60
Fatty ester	10–30
Sulfated fats, sulfited fats, sulfonates	5–15
Fatty acid soaps and alkanolamines	0–5
Nonionic emulsifiers	5–10
Corrosion inhibitors	0–1
Biocides	0.5–2
Antioxidant agents	0.5–2
Antifoam agents	0–1
Water	0–10

the soap level. Variation of the acidity index is a manifestation of fatty acid generation. Fatty acids are produced by the hydrolysis of the fatty substances employed.

Periodic microbiologic control of bacteria and fungi is essential to assure efficacy of the emulsion during the drawing process. A high microorganism level (higher than 10^6) can produce a pH decrease that could induce corrosion of the installation and even emulsion instability. The growth of microorganisms can be controlled by addition of fungicides and/or bactericides, depending on the contamination appearing in the emulsion. Global maintenance and control of the bath must be followed in close collaboration between the user and the supplier of the auxiliary produce in order to reach the best performance of the drawing process.

BIBLIOGRAPHY

L. de Chiffre, S. Lerson, K. B. Pedersen, and S. Skade, in A Reaming Test for Cutting, Technical University of Denmark, Lynby, Denmark.

K. Glossop, Lubricants for copper wire drawing. *Wire Industry*, May 1985.

G. F. Micheletti, in *Mecanizado por arranque de viruta*, Blume, Barcelona, 1980.

E. S. Nachtman and S. Kalpakjian, in *Lubricants and Lubrication in Metalworking Operation*, Marcel Dekker, 1985, pp. 17, 252.

C. Suarez, in Metal Drawing, Lab. Hispano Química, internal communication, 1990.

Ullrann's Encyclopedia of Industrial Chemistry, 5th ed. (B. Elvers, S. Hawkins, and G. Schulz, eds.), A15, VCM, New York, 1990, p. 482.

R. Zimmerman and G. H. Geiger, Alambre, 1992, Vol. 42, p. 61.

Index